Praise for *Sacred Mountains of the World*

"Hinting of another world, far above our own, mountains have worked their magic on humankind since time began. *Sacred Mountains of the World* tells their story with authority. The principal fabled peaks are described individually, and the symbolism of mountains as such is made plain. The result is a book that is insightful, often inspiring, and solidly researched."

Houston Smith, Professor of Philosophy and Religion, Emeritus, Syracuse University

"A treasure of a book that masterfully weaves together fact, history, legend, and spiritual truth."

Jacob Needleman, author of *The New Religions*

". . . Edwin Bernbaum has succeeded in unfolding the many aspects of this rich symbol, and has never forgotten to let the reader climb with him in this symbolic ascent. In our time of increasing ecological disasters—as well as awareness—such a book is a timely and important contribution. To regain the experience of mountains as living entities in symbiotic relationship with us is an important factor to save not only the Earth but ourselves as well."

Raimon Panikkar, Professor of Religious Studies, University of California, Santa Barbara

"Once in a great and precious while, a book comes across a columnist's desk that arrests the spirit and causes one to see the world in a breathtakingly new way. *Sacred Mountains of the World* is such a book."

Georgie Ann Geyer, Universal Press Syndicate

SACRED
MOUNTAINS
OF THE
WORLD

EDWIN BERNBAUM

With a New Preface

University of California Press
Berkeley   Los Angeles   London

University of California Press
Berkeley and Los Angeles, California

University of California Press, Ltd.
London, England

Library of Congress Cataloging-in-Publication Data

Bernbaum, Edwin.
    Sacred mountains of the world / Edwin Bernbaum.
        p.   cm.
    Originally published: San Francisco: Sierra Club Books,
1990.
    Includes bibliographical references and index.
    ISBN 0-520-21422-6 (pbk. : alk. paper)
    1. Mountains—Religious aspects.   I. Title.
[BL447.B47   1998]
291.3'5'09143—dc21   97-35987
                                                  CIP

Editor: Linda Gunnarson
Book design: Desne Border
Map: Earth Surface Graphics
Composition: Classic Typography

Printed and bound in Hong Kong by Toppan Printing Company (HK) Ltd.

9 8 7 6 5 4 3 2 1

Acknowledgment is gratefully made for permission to reprint
the following material:

Plate 12, *Holy Ones Standing on Top of Holy Mountains* (painting of four sacred mountains of the Navajo in Chapter 9 of this
book), is reprinted from *Where the Two Came to Their Father:
A Navaho War Ceremonial Given by Jeff King*, edited by Maud
Oaks, with commentary by Joseph Campbell, Bollingen Series
1. Copyright © 1943, © renewed by Princeton University Press;
plate reprinted with permission of Princeton University Press.

Quotations from *Annapurna*, by Maurice Herzog, translated
by Nea Morin and Janet Adam Smith. Copyright © 1952 by
E. P. Dutton; reprinted by permission of Maurice Herzog.

"Climbing a Mountain," by Tao-yün, translated by Arthur
Waley in *A Hundred and Seventy Chinese Poems*. Copyright ©
1919 by Alfred A. Knopf; reprinted by permission of Alfred A.
Knopf and Constable Publishers.

"Meru," by W. B. Yeats, reprinted with permission of Macmillan Publishing Company and A P Watt Limited on behalf of
Michael B. Yeats and Macmillan London Ltd from *The Poems
of W. B. Yeats: A New Edition*, edited by Richard J. Finneran.
Copyright © 1934 by Macmillan Publishing Company, renewed
1962 by Bertha Georgie Yeats.

Poem # 69, translated by Burton Watson in *Cold Mountain: 100
Poems by the T'ang Poet Han-shan*. Copyright © 1970 by
Columbia University Press; used by permission.

Verses from "Mount Fuji," translated in *The Manyoshu: The
Nippon Gakujutsu Shinkokai Translation of One Thousand
Poems*. Copyright © 1969 by Columbia University Press; used
by permission.

"My Retreat at Mount Chung-nan," by Wang Wei, translated
by Witter Bynner in *The Jade Mountain: A Chinese Anthology*.
Copyright © 1929 by Alfred A. Knopf; reprinted by permission
of Alfred A. Knopf.

Verse by Saigyo, translated by Royall Tyler in "A Glimpse of
Mt. Fuji in Legend and Cult," *Journal of the Association of
Teachers of Japanese*; reprinted by permission of Royall Tyler.

Frontis: Stars circling in a time exposure over Mount Kailas,
Tibet. (Galen Rowell/Mountain Light)

Page vi: Buddhist shrine beneath Mount Taweche, Nepal. (©
Robert Holmes)

*For my parents, Betty and Maurie,
my wife, Diane, and sons, David and Jonathan,
and for all who share a love of the mountains*

CONTENTS

CHAPTER 10

CHAPTER 11

PART II

THE POWER AND MYSTERY OF MOUNTAINS

CHAPTER 12

CHAPTER 13

CHAPTER 14

CHAPTER 15

Preface to the 1997 Edition

JUST AS VARIOUS ROUTES OF ASCENT converge at the top of a mountain, the diverse interests that have shaped the course of my life come together in this book. Some of my earliest and most influential memories are of magical views of snowcapped peaks in the Andes, where I lived as a young child and where I began mountain climbing as a teenager. Later, the Peace Corps offered me the opportunity to live in Nepal, within sight of the highest mountains on earth. As a result of my climbing and trekking experiences in the Himalayas, I wrote a book on Tibetan myths of hidden valleys and their symbolism and completed a doctorate in Asian studies in the area of comparative religion and mythology. The idea of *Sacred Mountains of the World* emerged as a natural extension of my interests, while the material for the book grew out of a course I taught at the University of Colorado and seminars I gave at the University of California Berkeley Extension and the Smithsonian Institution.

Because of the broad appeal and fascination of the subject, I wrote this book for a general audience as well as for specialists. Readers intrigued by myth and religion, art and literature, mountaineering and travel, wilderness and the environment, cross-cultural studies and personal development will find something in this book that touches on their interests. It is also my hope that the book will prove useful to scholars in diverse fields—from art history, comparative literature, and anthropology, to geography, environmental studies, and the history of religions—by providing an overview of sacred mountains that helps situate their research in a broader context. In these pages both the general reader and the specialist will encounter an interesting and fruitful approach to understanding how symbols transform our experience of reality.

The Introduction prepares the reader for the journey that follows. Drawing on experiences of people in both traditional and modern societies, it examines the physical and spiritual qualities that give mountains their extraordinary power to awaken a sense of the sacred. Part I explores the rich, diverse significance of sacred mountains in cultures throughout the world. Each chapter introduces the mountains in a particular region and their meaning to those who revere them, then focuses on a few representative peaks. Some of these peaks—Olympus, Fuji, Sinai, and Kailas—are well known. Others such as Kaata in Bolivia and Muztagh Ata in western China are less famous and have been chosen for a variety of reasons: religious and historical importance, geographic balance, illustration of themes, mountaineering significance, bearing on environmental issues, and idiosyncratic personal interest. I was unable to discuss in depth a number of major regions and mountains in this book. Korea, for example, deserves a chapter of its own, but I did not have the space to do it justice. Because of such limitations, I could only mention sacred mountains such as Chomolhari in Bhutan, Omine in Japan, and Mount Diablo in California, near where I live. Five of the eleven chapters in Part I are devoted to Asia, the continent with the largest population, the greatest land mass, and the highest peaks on earth.

Part II, The Power and Mystery of Mountains, begins with a chapter identifying the major themes found in traditional views of sacred mountains—themes such as mountain as center of the cosmos, abode of the gods, and place of revelation—and establishes an approach to understanding how their symbolism awakens a sense of the sacred. The next chapter draws on this approach to take a fresh look at well-known works of literature and art to see how they use mountain imagery to transform our perceptions of reality. The following chapter explores the spiritual dimensions of mountaineering. The final chapter examines the ways in which the contemplation of sacred mountains can help us to appreciate the value of wilderness, treat the environment with care and respect, and live deeper, more meaningful lives.

Since the first publication of *Sacred Mountains of the World* in 1990, there has been a growing interest in the significance of sacred places and a deepening recognition that they provide a sustainable basis for the conservation of biological and cultural diversity. For people who revere a mountain or other natural site, that place assumes a special value that makes it worth protecting at all costs—an ultimate value that may, in fact, transcend all cost. At the very least people leave such sites alone out of fear of provoking supernatural punishment. This kind of protection has made sacred places important centers for preserving traditional knowledge about the environment and maintaining valuable gene pools for replacing plant and animal species lost elsewhere. In addition, it strengthens cultures whose core values and beliefs center on sacred places.

Of all the different kinds of natural sacred sites, mountains form the most diverse and complete environments and ecosystems. They include shrub lands, forests, meadows, deserts, tundra, glaciers, rivers, and lakes, and they range in climate from the tropical to the arctic. Furthermore, mountains are the largest features of the natural landscape that can be seen in their entirety. A single peak, viewed from base to summit, can function, in fact, as a microcosm of the environment as a whole. Sacred mountains therefore give us some of the most detailed, comprehensive pictures of what traditional societies value and seek to preserve in nature. They highlight spiritual and cultural factors that deeply influence how people view and treat the environment.

The sense of wonder and awe awakened by mountains has played a key role in inspiring the modern-day environmental movement. John Muir, for example, founded the Sierra Club primarily to preserve pristine valleys and peaks in the Sierra Nevada as places of spiritual inspiration and renewal. Many parks and wilderness areas in mountain regions have been set aside for protection because of the cultural and spiritual values placed on them by modern societies—values that often outweigh economic, political, and scientific considerations. People see them as places where they can go to leave behind the materialistic, competitive concerns of the urban world and immerse themselves in a primordial reality that renews them spiritually and physically.

To be effective in the long run, environmental programs need to be based on the kind of respect for the land reflected in traditional views of sacred mountains. Without it, conservation efforts lack the rock-solid commitment needed to sustain them. To this end *Sacred Mountains of the World* provides a wealth of useful information and background material for scientists, environmentalists, managers of protected areas, policy makers, and others who recognize the need to take the spiritual and cultural significance of mountains into account in doing research and protecting the environment.

A few matters regarding terminology, style, and content bear mention. First, this book recounts a great number of myths about sacred mountains. I use the word *myth* to refer to stories, ideas, and beliefs that express what people in a particular culture or tradition take to be ultimately real. On a superficial level myths may be factually true or false. On a deeper level they embody strongly held views of reality—the basic, often unconscious, but necessary assumptions that guide people in their lives and interactions with the world around them. This holds true for modern technological societies as well as traditional religious cultures. Second, for simplicity and ease in reading, I have avoided the use of unusual diacritical marks such as the ones used to transliterate Sanskrit and Japanese terms. Those who know them will recognize the words and be able to supply such marks for themselves.

Writing *Sacred Mountains of the World* has enriched my life and broadened my horizons in ways I never could have imagined. I hope that the reader will have a similar experience in reading this book.

Edwin Bernbaum
September 3, 1997
Berkeley, California

ACKNOWLEDGMENTS

I AM PROFOUNDLY GRATEFUL to the many people who have helped me with the research and writing of *Sacred Mountains of the World.* To my surprise and delight, some of the greatest rewards of working on this project have been the friendships that have come out of it. Discovering the interest and excitement that others share with me has confirmed my belief that the subject of sacred mountains has as much to do with people and their relationship to mountains as with the mountains themselves.

Maurice Herzog's book *Annapurna* inspired my childhood interest in mountain climbing that ultimately led me to the subject of sacred mountains. Years later, in Nepal, I had the good fortune to spend time with Annulu, a Sherpa who embodied in his person many of the values associated with sacred peaks. Raimon Panikkar encouraged me to write *Sacred Mountains of the World.* Lynne Kaufman and Edmund Worthy arranged seminars I gave at the University of California Berkeley Extension and the Smithsonian Institution that led to the development of this book. I wish to thank them and Noel Miner, as well as the people with whom I have climbed and trekked through the years. And for the insight and joy they have given me, I would like to offer my gratitude to the mountains themselves.

Kalsang Namgyal, Kaldhen Sherpa, the Tengboche Rimpoche, and Gautam Vajracharya helped me with research on sacred mountains in Nepal and Tibet. I benefited from discussions on various Himalayan topics with Harold Arnold, H. Adams Carter, Matthew Kapstein, Frances Klatzel, Frits Staal, Stan Stevens, and Hugh Swift. Anna Seidel and Raoul Birnbaum shared their expertise on the religions and mountains of China. Judith Boltz, William Doub, Lewis Lancaster, William Powell, and Marilyn Rhie also assisted me in my research on the subject. Allan Grapard and Henry Smith generously discussed their work on Japanese topics related to sacred mountains. Minoru Harashida, Kiyoshige Ida, Shiro Tanabe, and Fumiko Umezawa went out of their way to help me in Japan. I received additional assistance from H. Byron Earhart, Christine Guth, Laurel Kendall, and Richard Payne. Eric Oey guided me through the intricacies of Southeast Asian cultures. I am indebted to Ronald Bernier, Robert del Bontá, Imade Budi, Eric Crystal, Patrick Harrigan, Robert Reed, David Sanford, Wayne Surdam, and Barend van Nooten for their thoughtful insights and suggestions on sacred mountains of South and Southeast Asia.

Pinchas Giller, John Lindow, David Robertson, Martin Schwartz, and Zeph Stewart offered valuable suggestions on the role of mountains in the cultures of the Middle East and Europe. I am also indebted to David Enelow, Patricia Fletcher, China Galland, Maria Kotzamanidou, and Hector McDonnell. Ian Allain, Alan Bechky, and Willy Makundi shared their intimate knowledge and experience of African mountains. My sister, Marcy Bernbaum, and brother-in-law, Eric Zallman, helped me track down material in Nairobi and Washington, D.C.

For a deeper understanding and appreciation of Navajo and Hopi views of mountains I am indebted to Wilson Aronilth, Jr., Theresa Boone, Marilyn Harris, Phyllis Hogan, C. Benson Hufford, Wilmer Kavena, Hartman Lomawaima, and John Wood. Will Channing, Leanne Hinton, and José Lucero helped me with research in the Southwest. Peter Nabokov shared his extensive knowledge of Native American cultures and sacred geography. I benefited from a sweat ceremony and conversations with Charles Thom. Johan Reinhard was more than generous with the results of his research on high-altitude ruins and mountain worship in South America. Winston Crausaz directed me to important sources on the mountains of Mexico. Margaret McKinzie-Hooson and Margaret Orbell provided valuable information on New Zealand. Noa Emmett Aluli and Ralph Palikapu Dedman described their efforts to preserve the integrity of sacred sites in Hawai'i. I am indebted to Jeffrey Falt, Kal Muller, and Rudy Wenk for material on Oceania.

For help with matters of art I thank James Cahill, Roger Keyes, Roger Lipsey, Mimi Lobell, and Martin Amt. A number of people helped me locate art, photographers, and images to supplement my own photographs, among them Stephen Bezruchka, Ina Cooper, Robert Holmes, Fredric Lehrman, Ken Scott, Jean Sulzberger, Hugh Swift, and Michael Warburton. I am particularly indebted to Richard Anderson and Galen Rowell for their advice and help with photographic research. I want to thank the photographers whose work enhances the pages that follow and whose names appear in the captions, as well as others who took the trouble to send me their photographs for consideration.

I also would like to express my gratitude to the staffs at Mountain Travel in El Cerrito and at InnerAsia in San Francisco for sharing their knowledge and facilities and for arranging trips I led that gave me the opportunity to visit certain sacred mountains. As a research associate, I made extensive use of libraries and resources at the University of California, Berkeley.

I want to thank the following people for reviewing various parts of the manuscript and offering valuable criticism and advice: Alan Bechky, Raoul Birnbaum, Jeffrey Falt, Pinchas Giller, Allan Grapard, Stuart Kelman, Roger Keyes, John Lindow, Roger Lipsey, Margaret McKinzie-Hooson, Peter Nabokov, H. Arlo Nimmo, Margaret Orbell, Richard Payne, Johan Reinhard, David Robertson, Anna Seidel, William Shumate, Henry Smith, Zeph Stewart, and Hugh Swift.

My editor, Linda Gunnarson, played a crucial role in helping me craft this book. I am grateful for her astute suggestions, critical eye, and unflagging perseverance and attention to detail, as well as her feel for the subject of sacred mountains. Like an apparition on a mountain, my agent, Joe Spieler, appeared at just the right moment to make this book a reality. I am indebted to him and Jon Beckmann, the publisher, for prodding me into making it a better work than it otherwise would have been. I would like to thank Mary Anne Stewart for her copy editing, Katherine Wright for her proofreading, Susan Ristow and Eileen Max for their work in production, Desne Border for the book's handsome design, Deborah Gwynn-MacDougall for her editorial assistance, and Peter Beren and Sam Petersen for their interest and encouragement.

Finally, I especially want to thank my family for putting up with me during those periods of frustration when I moaned and groaned my way up the mountain. My wife Diane's unfailing love and encouragement helped me lift my eyes and recover the vision needed to complete the journey involved in writing *Sacred Mountains of the World*. She listened with infinite patience as I read and reread passages out loud and offered essential criticism. In the liveliness of their enthusiasm and the love they gave me, my sons David and Jonathan played a role of paramount importance by constantly reminding me of the sense of wonder and delight that sacred mountains and this book are all about.

INTRODUCTION

As the highest and most dramatic features of the natural landscape, mountains have an extraordinary power to evoke the sacred. The ethereal rise of a ridge in mist, the glint of moonlight on an icy face, a flare of gold on a distant peak — such glimpses of transcendent beauty can reveal our world as a place of unimaginable mystery and splendor. In the fierce play of natural elements that swirl about their summits — thunder, lightning, wind, and clouds — mountains also embody powerful forces beyond our control, physical expressions of an awesome reality that can overwhelm us with feelings of wonder and fear.

People have traditionally revered mountains as places of sacred power and spiritual attainment. Sinai and Zion in the Middle East, Olympus in Greece, Kailas in Tibet, T'ai Shan in China, Fuji in Japan, the San Francisco Peaks in Arizona — all have acquired a special stature as natural objects of religious devotion. Speaking of the spiritual character of these mountains, of their ability to arouse spontaneous feelings of reverence and awe, Lama Anagarika Govinda, a Western practitioner of Tibetan Buddhism, wrote:

The power of such a mountain is so great and yet so subtle that, without compulsion, people are drawn to it from near and far, as if by the force of some invisible magnet; and they will undergo untold hardships and privations in their inexplicable urge to approach and to worship the center of this sacred power. Nobody has conferred the title of sacredness on such a mountain, and yet everybody recognizes it; nobody has to defend its claim, because nobody doubts it; nobody has to organize its worship, because people are overwhelmed by the mere presence of such a mountain and cannot express their feelings other than by worship.[1]

Throughout the world people of traditional religious cultures have looked up to mountains as symbols of their highest spiritual goals. Reflecting such a view of the heights, a ninth-century Japanese account describes the quest of a Buddhist monk named Shodo to climb Nantaizan, a sacred peak formerly known as Fudaraku:

In this very same province is a mountain called Fudaraku, whose peaks soar into the Milky Way, whose snow-covered summit touches the emerald walls of the sky. Bearing in its bosom the roaring thunder which marks the passing hours, it is the abode of the Phoenix, twisted like the horn of a sheep. Rare is the presence of demons, and none the traces of human steps. . . . The Master of the Law [Shodo] . . . urged his will onward. . . . "If I do not reach the top of this mountain, I shall never be able to achieve Awakening!" After having uttered this vow, he moved across the flashing snows and walked over the young leaves shining like jewels; when he had gone half the way up, his body was exhausted, his strength left him. He rested for two days and finally came to see the summit: his ecstasy was like that in a dream, he felt a vertigo like that of Awakening.[2]

For Shodo and others who followed him, the summit of the sacred mountain was a place to attain an inspiring glimpse of enlightenment, the ultimate goal of the Buddhist path.

Even today, in the secular, modern world, mountains are regarded as embodiments of humanity's highest ideals and aspirations. Expeditions to Mount Everest and other high peaks stand out as symbols of supreme efforts, of attempts by men and women to overcome their limitations and attain transcendent goals. Whether they realize it or not, many who hike and climb for sport and recreation are seeking an experience of spiritual awakening akin to that sought by people of traditional cultures. Maurice Herzog, the leader of the 1950 French expedition that made the first ascent of Annapurna — the first of the highest

Himalayan peaks to be climbed—describes a dramatic example of such an experience as he approached the unclimbed summit:

I felt as though I were plunging into something new and quite abnormal. I had the strangest and most vivid impressions, such as I had never before known in the mountains. There was something unnatural in the way I saw Lachenal [Herzog's companion] and everything around us. I smiled to myself at the paltriness of our efforts, for I could stand apart and watch myself making these efforts. But all sense of exertion was gone, as though there were no longer any gravity. This diaphanous landscape, this quintessence of purity—these were not the mountains I knew: they were the mountains of my dreams. . . .

An astonishing happiness welled up in me, but I could not define it. Everything was so new, so utterly unprecedented. It was not in the least like anything I had known in the Alps, where one feels buoyed up by the presence of others—by people of whom one is vaguely aware, or even by the dwellings one can see in the far distance.

This was quite different. An enormous gulf was between me and the world. This was a different universe—withered, desert, lifeless; a fantastic universe where the presence of man was not foreseen, perhaps not desired. We were braving an interdict, overstepping a boundary, and yet we had no fear as we continued upward. I thought of the famous ladder of St. Theresa of Avila. Something clutched at my heart.[3]

These two accounts—traditional Eastern and modern Western—illustrate how people experience the sacred in mountains and in cultures around the world. In many of these experiences we find descriptions of an encounter with something totally apart from the world we know—what the German scholar of religions Rudolf Otto termed the "wholly other"—an inscrutable mystery that attracts and repels us with intense feelings of wonder and awe. This source of fascination and fear may appear divine or demonic, assuming the form of a god or the shape of a demon. Whether it reveals a vision of heaven or hell, the encounter with the sacred moves us to the depths of our being to disclose a realm of existence beyond the power of words to describe.[4]

Floating above the clouds, materializing out of the mist, mountains appear to belong to a world utterly different from the one we know, inspiring in us the experience of the sacred as the wholly other. Their dark forests, jagged cliffs, and twisted glaciers evoke impressions of a strange and alien universe. Their summits, barren and lifeless, often brilliant with snow, are harsh and forbidding places graced with incredi-

ble beauty, where only those with extraordinary powers or skills can survive. A description of the Alps at twilight by the Italian climber Guido Rey reveals the awesome impression that mountains can make as manifestations of the wholly other:

. . . the peaks seemed to shine alone in the colorless vault, and to hang as if they did not touch the earth; they were like unreal shapes, created from nothing, like phantoms that live by night in the terrible heights of the sky and only appear now and then to sleepers in their dreams.

I did not recognize the beautiful forms I had seen by day; they had increased beyond measure, they had changed their appearance, they no longer belonged to our world; they were shadows of other unknown mountains cast by an unexplained phenomenon onto our sky by some distant star.

An irresistible shudder assailed us . . .[5]

Of all the features of the landscape, mountains most dramatically inspire a sense of awe in the presence of forces capable of annihilating us in an instant. Like the ark of the covenant and other sanctuaries of divinity, they must be approached with caution and respect. Those who are careless in the heights do not live long: a slight mistake, a disregard of the weather, and one can fall or freeze to death. Accordingly, people of traditional cultures have commonly regarded mountains as the dangerous haunts of gods and demons. A Chinese poem composed in the fourth century B.C. conveys a vivid sense of the holy terror that the experience of mountain climbing can provoke:

Climb higher and gaze into the distance,
Your heart will be gripped with fear.
Cirques of chasms surrounded by peaks,
Frowning cliffs all around;
Loose rocks that lean over the abyss,
Escarpments that overhang each other

. . . .

Clinging like a climbing bear,
you remain frozen in place,
Perspiration dripping down to your feet.
You feel yourself lost, reeling,
Transfixed with anguish, out of yourself;
And your spirit, shaken loose,
plunges into terrors without cause.[6]

Mountains may so overwhelm us with their size and grandeur that we feel like insects crawling upon them. Gaston Rébuffat, one of Herzog's companions on Annapurna, wrote:

And up on the mountain we began our ant-like labours.

A Buddhist monk at a mountain monastery in Tibet fingers prayer beads, expressing religious devotion. (Edwin Bernbaum)

Star streaks at twilight over the peaks of Shivling and Meru evoke a sense of mystery. Indian Himalayas. (Robert Mackinlay)

What is a man on an ice-world up in the sky? At that altitude he is no more than a will straining in a spent machine.[7]

More casual travelers often remark on how insignificant they feel in the presence of high and impressive peaks. Mountains rise over the surrounding countryside in undisputed splendor, sovereigns of the valleys, plains, and lesser hills beneath them; they are commonly described as *majestic* and *mighty*. Unlike the reign of human kings, theirs seems eternal and incorruptible, like that of the highest gods, who sit enthroned upon their lofty summits.

The majestic power of the sacred is reflected in the forces that form the mountains themselves. Volcanoes, in particular, erupt with a fiery wrath that consumes everything in its path. We can do nothing to stop this energy—we can only get out of its way. The same holds true for the icy fury of an avalanche or a raging blizzard of snow. Such events are often interpreted by traditional peoples as expressions of divine displeasure. Even a person who does not believe in supernatural entities will be moved by the natural power of mountains—and pause a moment to wonder if he or she fully understands it.

The power of the sacred can take the form of all-pervading love as well as all-consuming wrath. In fact, mystics often speak of their experience of divine love as a scorching heat they can hardly bear. A famous passage by the French philosopher Blaise Pascal describes such an experience:

From about half past ten in the evening until about half past midnight—FIRE. God of Abraham, God of Isaac, God of Jacob, not the God of philosophers and scholars. Certitude. Certainty. Feeling. Joy. Peace.[8]

The sacred does not simply present itself to our gaze: it reaches out to seize us in its searing grasp.

Like the sight of a mountain peak breaking free from the earth to leap toward the open sky, the experience of the sacred can send our spirits soaring to sublime heights of bliss and rapture, uplifting us with visions of beauty and goodness beyond our wildest dreams. The sacred can also give us a sense of reassuring serenity and fulfillment. The fascination that it inspires leads to feelings of love and devotion so intense that we would give anything, even our lives, to remain in its presence. Mountains, in particular, have the power to arouse such feelings of overwhelming devotion.

Two Sherpas walk above the clouds, transcending the world of ordinary life. Tesi Lapcha Pass, Nepal. (Edwin Bernbaum)

Despite the hardship and fear encountered on the heights, people return again and again, seeking something they cannot put into words. Religious pilgrims are drawn to a power or presence they sense in a peak; tourists come to gaze on splendid views; trekkers return to wander in a realm set apart from the everyday world. The fascination of mountains casts a particular spell on mountaineers, who knowingly risk their lives for the sense of exultation they experience in ascending a high and dangerous mountain—or just being there on its heights. One well-known British mountaineer, Frank Smythe, reportedly languished and died of a broken heart when he was denied permission to climb Kangchenjunga, a Himalayan peak that he had set his mind on climbing.[9] Reflecting the sentiments of many of her fellow mountaineers, the Australian climber Freda Du Faur expressed the religious nature of the fascination mountains had for her:

From the moment that my eyes rested on the snow-clad Alps [of New Zealand] I worshipped their beauty and was filled with a passionate yearning to touch these shining snows, to climb to their heights of silence and solitude, and feel myself one with the mighty forces around me.[10]

Immersed in a landscape of infinite grandeur, we may find it easy to let go of our feelings of separateness and merge with the mountains around us or feel at home in their awesome presence. One evening at sunset, while climbing high in the Himalayas, I lingered outside my tent to watch the light fade off the surrounding peaks. Across a pool of dark clouds, the highest summits burned with a red glow that seemed to warm my body, as if I were standing before a fire. The light blazing on their snows gradually cooled to pink, then suddenly went out, and the peaks appeared to turn into gray mounds of ash. At that moment, just as I expected them to take on a cold, hostile cast, a lavender glow, shading to green near the horizon, rose in the sky to the north, over Tibet, and I felt a friendly presence envelop me, as if the mountains themselves were extending me their welcome.

Not everyone experiences the sacred as the wholly other. Many find the sense of mystery it evokes right here in the midst of the world we think we know, even in what lies closest to us—ourselves. We live with ourselves every moment of our lives, but we scarcely know ourselves as we truly are. To know oneself, according to Hindu philosophy, is to attain the supreme realization of one's identity with the mysterious essence of the universe itself. To reduce the mystery of the sacred to the wholly other would condemn each person to remain forever alienated from his or her true self. Realizing their intimate relationship with God, Jewish mystics strive to sanctify each moment of their lives in order to become aware of a divine spark in all things and living creatures. The Navajo and the Hopi of the American Southwest revere every rock and feature of the landscape in which humans live.

The sacred is profoundly mysterious, not just as the wholly other but as an embodiment of the unknown itself. It is the aura of mystery, of something beyond our ken, that attracts us. We are drawn to the sacred precisely because it is unknowable—something that remains mysterious even when we are in its presence. We find this fascination reflected in the twinge of disappointment we feel when an unclimbed peak has been climbed. Something about the peak that gave it a special quality vanishes, and it becomes in some sense ordinary, like the rest of the world, no matter how distant or exotic it may be. There is a profound attraction in the very fact that a peak is unexplored or unclimbed. In a similar way, the sacred by its very nature eludes all our attempts to define and grasp it. Without some inner core of inscrutable mystery, it ceases to be sacred.

Mountains have a special power to evoke the sacred as the unknown. Their deep valleys and high peaks conceal what lies hidden within and beyond them, luring us to venture ever deeper into a realm of enticing mystery. Mountains seem to beckon to us, holding out the promise of something on the ineffable edge of awareness. There, just out of sight, over the next ridge, behind a summit, lies the secret, half-forgotten essence of our childhood dreams. Rudyard Kipling captured this aspect of the mystery of mountains in these well-known lines:

Something hidden. Go and find it.
Go and look behind the Ranges—
Something lost behind the Ranges.
Lost and waiting for you. Go![11]

That something lost behind the ranges may be a material treasure, some part of ourselves, or the highest spiritual truth—whatever holds the answer to our deepest longings.

The unknown also possesses a darker, more dangerous side; instead of our salvation, it may hold our damnation. The person who ventures into an unexplored range or tries to climb an unclimbed peak always harbors some fear that instead of what he seeks, he will find disaster and death. Even the hiker who goes into mountains that others know well may feel a trace of apprehension. What he finds there in the unknown may shatter his illusions and the comfortable world they uphold. Driven beyond the limits of physical endurance, he may discover things about himself—weaknesses and fears—that he would rather not know.

Although it may threaten everything we hold dear, the experience of the sacred opens up a new vision of reality that can free us from the stifling confines of the world to which we cling. We find this aspect of the sacred epitomized in the views that open around us as we climb to the top of a mountain. New and previously unknown vistas unfold around us as we emerge into the sun from the dark recesses of a narrow valley. The horizon recedes into the distance, revealing ridge after ridge of mountain ranges without end. As we gain a high promontory or a lofty summit, we may take a deep breath and feel as if we could take off and soar.

At a deeper level, the experience of the sacred as the unknown opens us to the ultimate mystery of reality itself. Just when we think we have grasped the nature of things as they are, some new aspect appears to confound our knowledge and understanding. As we climb a mountain, no matter how far our view may expand, something always lies hidden beyond the next horizon. In the encounter with the sacred, we suddenly intuit a reality that extends beyond the limits of what we know or can ever fathom.

The sacred is not merely the unknown, but the unknown that we regard as ultimately real. As the historian of religions Mircea Eliade has written, "the *sacred* is equivalent to a *power*, and, in the last analysis, to *reality*."[12] What we truly revere has a quality that sets it apart from everything else, making it seem as solid and real as a mountain made of granite. This quality of mountains contributes to the aura of sanctity that tends to settle upon their summits. The Bible describes them as the "eternal hills," implying that they have a more enduring reality than the plains around them. As the 125th Psalm declares:

They that trust in the Lord
Are as mount Zion, which cannot be moved,
* but abideth for ever.*[13]

As a manifestation of ultimate reality, the sacred is often distinguished from the profane, which appears, in comparison, fleeting and unreal. We may live most of our lives in the world of ordinary experience, but at moments, particularly when we are in the mountains, our usual preoccupations may seem in-

Taktsang, the "Tiger's Nest" Monastery, a place of pilgrimage and meditation in the Himalayas of Bhutan. (Edwin Bernbaum)

consequential. The warm, rough touch of rock on fingers when climbing, the quiet of twilight beneath high peaks, the cold and misery of being caught in a thunderstorm may strike us as more concretely real than the concerns and occupations of our everyday lives. Most of us at some point have shared the sentiments of the author of Ecclesiastes, who wrote, "I have seen all the works that are done under the sun; and, behold, all is vanity and a striving after wind."[14] The Hindu Upanishads express a similar dissatisfaction with the world of profane existence:

From the unreal lead me to the real.
From darkness lead me to light.
From death lead me to immortality.[15]

Mountains can offer a vision of something pure and eternal, beyond the corruptions of time.

Just as traditional cultures frequently separate the sacred from the profane, so they cordon off certain mountains with rules and rituals. Only those of spiritual power and purity may venture up these mountains without fear of provoking the anger of the gods or the spite of demons. In North America, for example, many Native Americans believed that only people with the proper ritual preparation might climb to the summits of sacred peaks without being struck down by spirits. In the Bible only Moses is regarded as holy enough to step onto the hallowed ground of Mount Sinai and ascend the mountain to converse with God.

Nevertheless, the sacred may erupt without warning into the profane world of material existence. We may be walking along a trail when something—a stone or a tree—catches our attention. As we glance toward it, a translucent beauty beyond anything we have ever imagined shines through it, giving it a heightened reality, and we feel ourselves in the presence of the sacred. The paradigm of such an experience occurs in the Biblical episode of the burning bush, which takes place, significantly, on a mountain. There Moses beholds a bush that burns yet is not consumed, thereby revealing the presence of God. Likewise, our stone beside the trail takes on an unearthly glow yet remains for all that a stone, a part of this world. The place where someone has such an experience often becomes a pilgrimage shrine sacred to others. Because of their awe-inspiring power, mountains are prime places for this kind of encounter with the sacred. A beautiful example appears in Herzog's description of

Mount Kailas, Tibet, most sacred mountain in the world for more than half a billion people in India, the Himalayas, and Tibet. (Hugh Swift)

The golden summit of Tanzania's Mount Meru floats like a vision of heaven above the green hills of Africa. (Clive Ward)

the striking change of perception he experienced near the summit of Annapurna:

The snow, sprinkled over every rock and gleaming in the sun, was of a radiant beauty that touched me to the heart. I had never seen such complete transparency, and I was living in a world of crystal. Sounds were indistinct, the atmosphere like cotton wool.[16]

As an expression of ultimate power and reality, what we regard as sacred possesses ultimate value and meaning. It embodies whatever we cherish above everything else, whatever stirs our deepest feelings and awakens our highest aspirations. This will differ from person to person, from culture to culture: for a Christian it may be selfless love, for a Buddhist ultimate enlightenment, for a Taoist harmony with the underlying way of nature. Many regard the sacred as the sheer embodiment of the beautiful and the sublime. Whatever it may be, the ultimate value expressed in the sacred provides meaning, direction, and purpose in life. It gives us a sense of place and inspires our greatest efforts. Here lies one of the great attractions of mountain climbing: the ascent to the summit offers an inspiring model of a path leading to a lofty goal, a path such as we would wish to follow through the confusing maze of everyday life.

Because of their power to awaken an overwhelming sense of the sacred, mountains embody and reflect the highest and most central values of religions and cultures throughout the world. Mount Sinai occupies a special place as the awesome site where God appeared in cloud and thunder to give Moses the Torah, the law and teachings that form the core of the Jewish religion. The graceful cone of Mount Fuji represents for many a sublime symbol of the beauty and spirit of the Japanese nation. The remote peak of Mount Kailas, rising aloof above the Tibetan Plateau, directs the hearts and minds of millions in India and Tibet toward the realm of the highest gods and the utmost attainments of spiritual meditation. The Hopi and Navajo view the San Francisco Peaks of Arizona as a divine source of water on which their lives and communities depend. For many in the modern world, Mount Everest symbolizes the highest goal they may strive to attain, whether their pursuit be material or spiritual.

Like the sacred values they express, the mountains revered by cultures around the world appear infinite in number and kind. They range from the highest peaks on earth to hills that barely rise above the surrounding landscape. They are regarded traditionally as places of revelation, centers of the universe, sources of life, pathways to heaven, abodes of the dead, temples of the gods, expressions of ultimate reality in its myriad manifestations. The following chapters explore the diverse ways in which people have experienced the sacred on the high and lonely reaches of our planet.

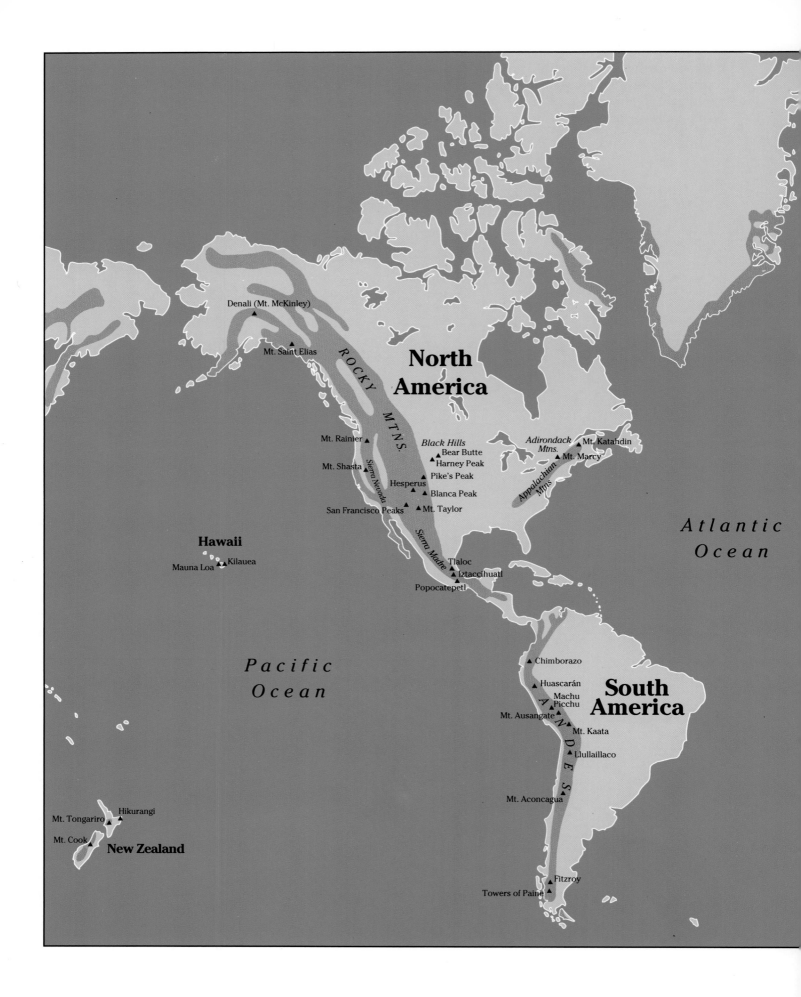

North
America

South
America

Pacific
Ocean

Atlantic
Ocean

Hawaii

New Zealand

Denali (Mt. McKinley) ▲

Mt. Saint Elias ▲

ROCKY MTNS.

Mt. Rainier ▲

Mt. Shasta ▲

Sierra Nevada

Hesperus ▲

San Francisco Peaks ▲▲

Black Hills
▲ Bear Butte
▲ Harney Peak
▲ Pike's Peak
▲ Blanca Peak
▲ Mt. Taylor

*Adirondack
Mtns.*
▲ Mt. Katahdin
▲ Mt. Marcy

*Appalachian
Mtns*

Sierra Madre

▲ Tlaloc
▲ Iztaccihuatl
Popocatepetl ▲

Mauna Loa ▲▲ Kilauea

▲ Chimborazo

▲ Huascarán
Machu
▲ Picchu
Mt. Ausangate ▲
▲ Mt. Kaata

*A
N
D
E
S*

▲ Llullaillaco

Mt. Aconcagua ▲

Hikurangi
Mt. Tongariro ▲ ▲

Mt. Cook ▲

▲ Fitzroy
Towers of Paine ▲

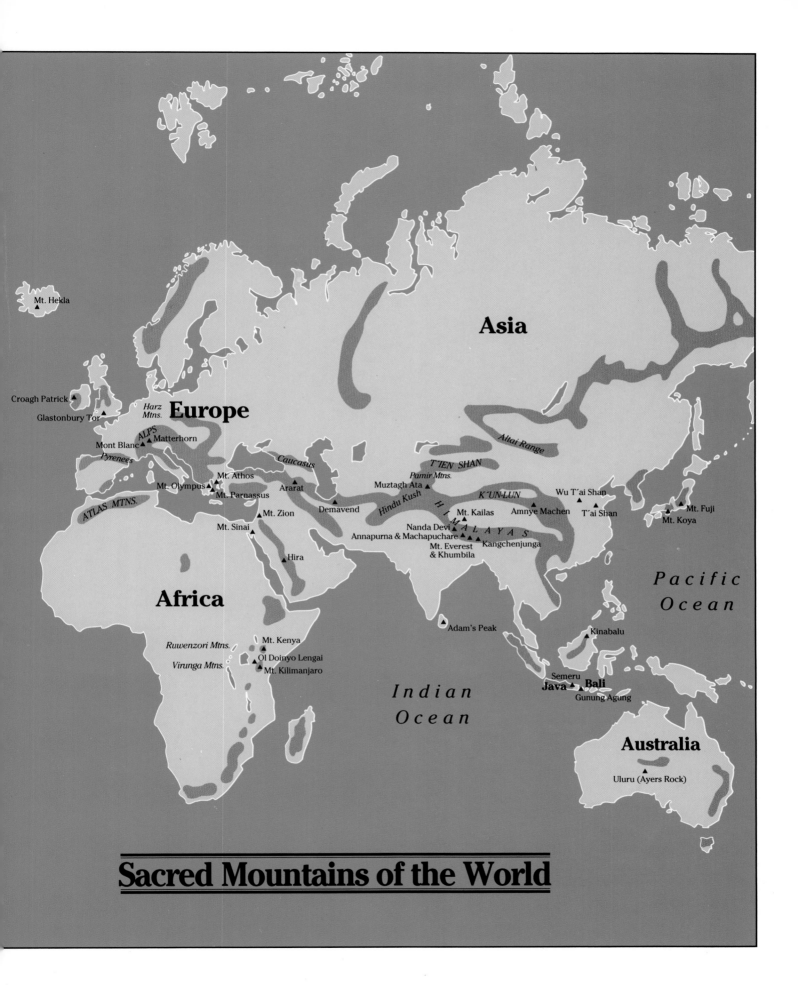

Sacred Mountains of the World

· Part I ·

Sacred Mountains Around the World

THE HIMALAYAS
ABODE OF THE SACRED

A N ENORMOUS RANGE 1500 miles long, the Himalayas rise in the monsoon-drenched jungles north of Burma to sweep in a great arc of snow and ice northwest along the borders of India and Tibet through Bhutan, Sikkim, and Nepal up to the dusty glaciers of the Karakoram on the remote desert frontier between Pakistan and China. From the plains of India, the mountains appear as luminous tracings on the far blue sky, wisps of light hinting at another world far above ours. On closer approach, their silhouettes dwindle behind intervening hills to reappear in more substantial form in flashes of white, glimpsed now and then through the opening of a dark green valley. From the vantage point of a high ridge gained by an arduous climb, the mountains emerge sharp and solid against the horizon, their glaciers glistening in the sun, too brilliant for eyes to bear. At twilight, after the colorful displays of sunset, the peaks' jagged snows soften to a lavender glow as they fade into the depths of night. No wonder that millions of devout Hindus and Buddhists regard the Himalayas as the dwelling place of the gods and the pathway to heaven.

As the loftiest mountains on earth, the Himalayas have come to embody the highest ideals and aspirations. The sight of their sublime peaks, soaring high and clean above the dusty, congested plains of India, has for centuries inspired visions of transcendent splendor and spiritual liberation. Invoking such visions, the *Puranas*, ancient works of Hindu mythology, have this to say of Himachal, or the Himalayas:

In the space of a hundred ages of the gods, I could not describe to you the glories of Himachal . . . that Himachal where Shiva dwells and where the Ganges falls like the tendril of a lotus from the foot of Vishnu. . . . There are
no other mountains like Himachal for there are found Mount Kailas and Lake Manasarovar. . . . As the dew is dried up by the morning sun, so are the sins of mankind by the sight of Himachal.[1]

In the *Mahabharata*, the great epic of Indian literature composed between the fourth century B.C. and the fourth century A.D., Prince Arjuna goes up to the Himalayas to seek the highest powers from the great god Shiva. Addressing a mythical mountain on whose slopes he has practiced spiritual austerities, he expresses views of the range that hold true for many people today:

Mountain, thou art always the refuge of the good who practice the law of righteousness, the hermits of holy deeds, who seek out the road that leads to heaven. It is by thy grace, Mountain, that priests, warriors, and commoners attain to heaven and devoid of pain walk with the gods. King of mountains, great peak, refuge of hermits, treasury of sacred places, I must go, farewell.[2]

The physical history of the Himalayas has a dramatic quality matching their spiritual grandeur. Millions of years ago the summit of Mount Everest, today the world's highest mountain, lay beneath the Tethys, an ancient sea separating Asia from the Indian subcontinent. Through the gradual movement of the earth's tectonic plates—still ongoing today—the two great land masses collided to fold and thrust up the peaks of the Himalayas and the Tibetan Plateau. Fossils found in sedimentary rocks near the tops of the highest mountains attest to the submarine origins of the range. The fracturing of the earth's crust also led to injections of magma, forming here and there, as a result of glacial action, magnificent walls and peaks

Shivling, an embodiment of the divine phallus of Shiva, God of the Mountains and Lord of the Universe. (Robert Mackinlay)

of granite—but no volcanoes. The youngest mountains on earth, the Himalayas have risen so recently, within the last few million years, that the watershed lies a hundred miles north of their crest. As a consequence, preexisting rivers have cut through the range, creating the deepest valleys in the world, such as the Kali Gandaki Valley between Annapurna and Dhaulagiri in central Nepal, nearly four miles deep.

The Himalayas form a line marking the collision of cultures as well as land masses. From the south and west have come peoples of Indian and Persian backgrounds; from the north and east, groups of Tibetan and Chinese origins. During the third millennium B.C., the indigenous Dravidians of India built the great cities of the mysterious Indus Valley civilization. Figurines found in the ruins of two of these cities, Mohenjodaro and Harappa in Pakistan, suggest that the inhabitants worshipped a cross-legged god who evolved into Shiva, the present-day Hindu deity most closely associated with the Himalayas. Around 1500 B.C., coming from the direction of Persia, a nomadic Indo-European people called the Aryans invaded India through the passes of the Hindu Kush and conquered the Dravidians. Out of the interaction of their two cultures rose the major ancient religions of India— Hinduism, Buddhism, and Jainism. The *Rig Veda*, a collection of hymns composed by the Aryans, contains what may be the earliest mention of the Himalayas, referred to as Himavat, the "Snow Mountains," one of the names used for the range in later texts.[3]

Around 500 B.C., drawing on pre-Hindu traditions and his own experience, Siddhartha Gautama, prince of a kingdom at the foot of the Nepalese Himalayas, founded the religion of Buddhism. Known as the Buddha or "Awakened One," he taught the path to nirvana or enlightenment, the ultimate goal of freedom from the suffering caused by attachments to illusory notions of the self. In the centuries that followed, Hinduism as we know it arose with the emergence of sects devoted to the worship of two major gods— Vishnu the Preserver and Shiva the Destroyer. These two, along with Brahma the Creator, represented the three principal manifestations of the supreme deity who creates, sustains, and destroys the universe. Although Buddhists and Hindus shared similar goals of liberation or release from suffering, Hindu philosophy emphasized the realization of the *atman*, or true self, as the means of attaining that end. Followers of Buddhism and Hinduism, as well as the related religion of Jainism, looked to the heights of the Himalayas as a favored abode of the gods and the ideal place to pursue their spiritual practices.

North of the Himalayas, at some indeterminate time, people with a language and culture related to Chinese moved into the valleys of southern Tibet.

Many of their religious practices centered around the worship of mountains as warrior gods imbued with the power of kings. According to ancient Tibetan myth, the first ruler of Tibet was a god who descended from the sky on a magic rope woven of light. He landed on a peak of Yarlha Shampo, a sacred mountain at the head of the Yarlung Valley south of the Tsangpo River. According to later accounts, he was actually a prince who wandered over the Himalayas from India. When the people asked him where he had come from, he pointed back toward Yarlha Shampo, from which they concluded that he was a heavenly deity who had descended onto the mountain. Whatever their origins, the first historical kings of Tibet emerged from the mists of legend around the seventh century A.D. to establish the Yarlung Dynasty near this sacred peak. When Padma Sambhava, an Indian sage and magician, introduced Buddhism into Tibet in the eighth century A.D., he subjugated a number of mountain deities who opposed his efforts, including Yarlha Shampo, and converted them into protectors of the new religion, roles they have retained to this day.[4]

Beginning around the twelfth century A.D., in response to Muslim invasions of India, great numbers of Hindus sought refuge in the lower ranges of the Himalayas. There they encountered rustic tribes with shamanistic rituals focused on invoking spirits of the mountains and other features of the landscape. Many of these tribes they converted to Hinduism, introducing the caste system of India. Despite their efforts, a number of indigenous beliefs and practices continued, with local deities taking on Hindu names. In succeeding centuries emigrants from Tibet moved into the higher ranges of the Himalayas and spread Tibetan Buddhism throughout the area. As a result, the Himalayan region today is characterized by complex overlays of different peoples, cultures, and religious traditions.[5]

The Himalayas are sacred for followers of five Asian religions—Hinduism, Buddhism, Jainism, Sikhism, and the indigenous Bon tradition of Tibet. These religions revere the mountains as places of power where many of their most important sages and teachers have attained the heights of spiritual realization. According to Jain mythology, Rishabhanatha, the first of twenty-four saviors of this age, achieved enlightenment on Mount Kailas, the most sacred peak in the Himalayan region. There, in the vicinity of the same mountain, Shenrab, the legendary founder of Bon, is said to have taught and meditated. Sikhs, followers of an Indian religion that developed from the interaction of Islam and Hinduism in the fifteenth century A.D., revere Hemkund, a mountain lake near the source of the Ganges, as the place where Guru Gobind Singh, the last of their ten principal teachers, prac-

A wandering Hindu holy man transfixes the viewer with his penetrating eyes. (William Frej)

ticed meditation in a previous life. The Himalayas abound with caves and shrines where Buddhist sages, such as the Tibetan yogi Milarepa, have meditated and attained enlightenment.

Hindus—by far the largest group in India, with more than 600 million adherents—regard the entire range as the god Himalaya, father of Parvati, the wife of Shiva. King of the mountains, Himalaya lives high on a peak with his queen, the goddess Mena, in a palace ablaze with gold, attended by divine guardians, maidens, scent-eating creatures, and other magical beings. His name, composed of the words *hima* and *alaya*, means in the Sanskrit language of ancient India the "Abode of Snow." As a reservoir of frozen water, the body and home of the god Himalaya is the divine source of sacred rivers, such as the Ganges and Indus, that sustain life on the hot and dusty plains of northern India. The ancient poets and sages regarded the range as more than a realm of snow; they saw it as an earthly paradise sparkling with streams and forests set beneath beautiful peaks:

After he had crossed through the impassable jungle at the foot of the great mountain, Arjuna dwelled on the peak of the Himalaya in all his splendor. He saw blossoming trees there, which resounded with the sweet songs of birds, and streams full of whirlpools, the color of blue beryl, resonant with geese and ducks, ringing with the cries of cranes, echoing with the calls of the cuckoo, and loud with curlews and peacocks. When that great warrior Arjuna saw those streams with their lovely woods, filled with sacred, cool, and pure water, he became joyous of spirit.[6]

Above and beyond the earthly paradise of the Himalayas lie the heights of heaven. At the end of the *Mahabharata*, the heroes of the epic, five brothers and their common wife, climb up through the range to ascend the mythical mountain of Meru. One by one they die, victims of their failings, such as gluttony and pride. Only the most virtuous, Yudhishthira, the embodiment of righteousness, reaches heaven in his physical body. After passing a series of tests, in which he refuses to abandon his faithful dog or to leave his brothers in hell, he is reunited with his family and friends. Receiving a divine form, free of imperfection, he enters the celestial abode of the gods, to dwell there in joy and bliss beyond the reach of sorrow and suffering.[7]

The north face of Mount Everest viewed from the ruins of Rongbuk Monastery in Tibet. (Edwin Bernbaum)

Mount Everest and Khumbila

The Himalayas hold hundreds of sacred mountains, a number of which stand out above the rest. The highest peak, and the one that for the West represents the range as a whole, is Mount Everest. As the loftiest point on earth, rising to 29,028 feet above sea level, it has acquired a semisacred status in the eyes of the modern world. At over 17,000 feet, the foot of Mount Everest stands higher than the highest summit of the Alps. The mountain itself lies in the eastern Himalayas, nearly hidden behind a screen of lesser peaks lining the border between Nepal and Tibet. From a few spots in the valleys of Khumbu just to the south, the top few thousand feet of Everest emerge over intervening ridges as a brownish black pyramid of rock flecked with snow and streaming with clouds. From the plains of India and southern Nepal, Everest can scarcely be distinguished from lower peaks in front of it that either conceal it completely or appear to rise to greater heights. Only from the north, from the valley of Rongbuk in Tibet, does Everest reveal itself in solitary splendor as an unmatched pyramid of rock

and ice. Great ridges of snow sweep smoothly up to converge in a graceful tip that in the last glow of twilight seems to hover among the stars. From the windswept ruins of Rongbuk Monastery, perhaps the highest monastery in the world, Everest appears framed between rocky walls that guide the eye straight to the mountain itself. No other snow peaks soar up to distract the mind from its contemplation of the highest point on earth. In the words of George Leigh-Mallory, a well-known British climber who disappeared near the summit of Everest in 1924:

At the end of the valley and above the glacier Everest rises not so much a peak as a prodigious mountain-mass. There is no complication for the eye. The highest of the world's greatest mountains, it seems, has to make but a single gesture of magnificence to be lord of all, vast in unchallenged and isolated supremacy.[8]

Until 1852 the British knew Mount Everest only as Peak XV on the Survey of India. Then a clerk computed its height from trigonometric readings and realized, to his amazement, that he had discovered — with a piece of paper and a pen — the highest mountain in the world. The British Surveyor General of

India named the peak for his predecessor, Sir George Everest. Since then, Everest has come to be associated in the West with the pinnacle of achievement, and those who have climbed it, beginning with Tenzing Norgay and Sir Edmund Hillary in 1953, are set apart in the eyes of the world.

Tibetans and Sherpas, a people of Tibetan origin and culture who live just south of Everest, know the mountains by another, more significant, name—Chomolungma or Jomolangma. Reflecting a Western tendency to assume that the local people must revere the highest peak on earth as the most sacred, modern writers and mountaineers commonly mistranslate this name as "Goddess Mother of the World." It actually has a number of other possible meanings in Tibetan, two of which seem the most likely. *Chomo* or *jomo* means "lady" and refers to the goddess who dwells on Everest, while *lung* in *lungma* can mean "wind" or "place." Chomolungma would stand for "Lady of the Wind" or "Goddess of the Place"—and indeed, a characteristic feature of the mountain is the great plume of cloud and snow that blows almost continually off its summit. A more likely meaning, however, comes from the name of the goddess of Everest, Miyolangsangma. The Tibetan language has a tendency to condense long words and phrases into one or two syllables that can be difficult to decipher. *Langma* in Jomolangma is probably short for Miyolangsangma, making the Tibetan name of Mount Everest a cryptic reference to the goddess of the mountain as the "Lady Langma."[9]

Miyolangsangma belongs to a group of relatively minor Tibetan goddesses known as the Five Sisters of Long Life. According to Buddhist texts, they dwell on mountain peaks above glacial lakes along the border between Nepal and Tibet. The leader of the group, Tashi Tseringma, resides on Gauri Shankar, a prominent peak west of Everest, and on Chomolhari, the major sacred mountain of western Bhutan. Murals at the Buddhist monasteries of Rongbuk and Tengboche on the north and south sides of Mount Everest depict Miyolangsangma as a golden goddess riding a tiger and holding a bowl of roasted barley flour and a mongoose spitting jewels. The items that she carries indicate that she bestows wealth and nourishment on those who venerate her, and the tiger symbolizes her supernatural power. Every year at the beginning of the religious festival of Mani Rimdu at Tengboche Monastery, the monks offer prayers to the goddess and dedicate a yak to her, which they release to wander the mountains in freedom as her sacred charge. Her name, in fact, means literally the "Unmovable Goddess Who Is a Benefactress of Bulls"—or, in this case, yaks.[10]

Except for the Sherpas who lived near Everest's foot, the people of Nepal took little notice of the mountain until they learned from the British of its status as the highest peak on earth. In response to outside interest, for the peak had no sacred significance for most Nepalis, government officials formulated a modern name, Sagarmatha, that designates today not only Everest but the national park created around it and the zone or province in which it is found. In Nepali, Sagarmatha means literally the "Forehead of the Sky" or the "One Whose Forehead Reaches Up to the Sky," indicating its unmatched height. The word, unusual in Nepali, brings to mind a term in the sacred language of Sanskrit—*sagaramanthana* or *sagaramatha*, meaning the "churning of the ocean." This term refers to a well-known Hindu myth in which Vishnu, the supreme deity as god of preservation, took the form of a turtle so that the gods and demons could churn the ocean to extract the nectar of immortality. According to this account, they used a mythical mountain named Mandara as a churning stick, which they placed on top of the turtle's back. In the course of spinning the peak to churn up the nectar, they also created butter, the horse of the sun, and a wish-fulfilling tree. The secondary Sanskrit associations of the name Sagarmatha would equate Mount Everest with this cosmic mountain of creation.[11]

The Sherpas themselves have traditionally paid much more attention to Khumbila, a jagged rock peak 19,294 feet high—nearly 10,000 feet lower than Mount Everest. Unlike its higher neighbor, which rises on the periphery of the Sherpa homeland, Khumbila stands right at the center of the Sherpa valleys of Khumbu. The three principal villages of Khumjung, Kunde, and Namche Bazaar actually rest on Khumbila's flanks. Because of its central location, the mountain is regarded by the people who dwell there as the abode of the local deity who watches over their land and protects them from the forces of evil. Paintings by local artists show Sherpas praying and making offerings to this deity, whose name, Khumbu'i Yulha, means the "Country God of Khumbu." Although many Sherpas make their living by accompanying mountaineering expeditions to Everest—Tenzing Norgay, who first climbed it with Sir Edmund Hillary, was a Sherpa—they would never dream of attempting Khumbila, which they regard as too sacred to be desecrated by climbers. Such an attempt would incur the wrath of the god and put their country in jeopardy. Some Sherpas believe that if they transgress or offend the deity, even in minor ways, he will send down yetis (abominable snowmen) to punish them.

Some of the stories that Sherpas tell of the yeti reveal its role as a divine or demonic messenger of the gods who watch over the practice of Buddhism. The Incarnate Lama of Tengboche, abbot of the main

Tibetan Buddhist monastery on the south side of Everest, once told me a tale with a religious moral about a lama, or Buddhist priest, who went on a house call one snowy night. On the way back he came to a mani wall of stones carved with special prayers or sacred spells called mantras. In accordance with Buddhist practice, he walked by it in a clockwise fashion with his right shoulder toward it. At the far end of the wall, which rose higher than a man, the lama found in the snow the fresh tracks of a yeti who had just passed going in the opposite direction on the other side. Since the lama had walked in the correct Buddhist manner, the abominable snowman had not seen him and had therefore not attacked and killed him. "The moral of the story," the Tengboche lama concluded, "is obvious: walk around prayer walls the right way."[12]

Khumbila is typical of hundreds of Buddhist peaks in the Himalayas and Tibet that are well known and sacred only to the people who live near them. We find them scattered throughout the range as seats of local deities who watch over the land and villages around them. As protectors and providers of food and water, these local deities play roles of vital importance in the lives of the people who depend upon them for the satisfaction of their spiritual and material needs. Mount Everest, in contrast, is known throughout the world, but has little religious significance for those who farm and graze the lands beneath its slopes. Its importance as a sacred peak derives primarily from its supreme height, which has made its summit a symbol of transcendent attainment for foreigners who dwell far from the mountain itself.

Mount Kailas

One peak in the Himalayan region stands out above all others as the ultimate sacred mountain for more than half a billion people in India, Tibet, Nepal, and Bhutan. Hidden behind the main range of the Himalayas at a high point of the Tibetan Plateau northwest of Nepal, Mount Kailas rises in isolated splendor near the sources of four major rivers of the Indian subcontinent—the Indus, the Brahmaputra, the Sutlej, and the Karnali. Hindus also regard Mount Kailas as the place where the divine form of the Ganges, the holiest river of all, cascades from heaven to first touch the earth and course invisibly through the locks of Shiva's hair before spewing forth from a glacier 140 miles to the west. Not far from the foot of the peak itself, at nearly 15,000 feet above sea level, reflecting the light of its snows, repose the calm blue waters of the most sacred lake of Hindu religion and mythology—holy

Lake Manasarovar, "Lake of the Mind." The hardiest of Hindu pilgrims aspire to take the long and dangerous journey over high passes to bathe in Manasarovar's icy waters and cleanse their minds of the sins that threaten to condemn them to the suffering of rebirth. Buddhists know it as Anavatapta, the "Unheated One," a lake that the historical Buddha is said to have magically visited when he lived and taught in India during the fifth century B.C.

At only 22,028 feet, Kailas is thousands of feet lower than Everest and many other Himalayan peaks. Its extraordinary setting and appearance, however, more than make up for its modest height. Situated in a high desert beyond the reach of monsoon rains, the peak rises in a bright and flawless sky, alone above a golden plain, unchallenged by the ranges of softly rounded hills that flank it on either side. Unlike Everest, which merges with other mountains to vanish behind them, Kailas retains its grandeur when viewed from a distance. More than any other peak in the Himalayas, it opens the mind to the cosmos around it, evoking a sense of infinite space that makes one aware of a vaster universe encompassing the limited world of ordinary experience.

Placed on a pedestal of striated rock, its dome of snow shining white in the sun, Kailas looks more like a piece of marble sculpture fashioned by the hands of the gods than a mountain created by the forces of nature. Indeed, Tibetans often compare the peak to the pagoda palace of a deity or the reliquary of a saint, and they treat it as such, prostrating themselves before it. It has also served as an inspiration for numerous Hindu temples and shrines in the distant plains of India. The mere sight of the peak has a powerful effect, bringing tears to the eyes of many who behold it, leaving them convinced that they have glimpsed the abode of the gods beyond the round of life and death.

Hindus view Mount Kailas as the divine dwelling place of the great god Shiva and his wife, the beautiful goddess Parvati. There, as the supreme yogi, naked and smeared with ashes, his matted hair coiled on top of his head, he sits on a tiger skin, steeped in the indescribable bliss of meditation. From his position of aloof splendor on the summit of Kailas, his third eye blazing with supernatural power and awareness, the lord of the mountain calmly surveys the joys and sorrows, the triumphs and tragedies, the entire play of illusion that make up life in the world below. As one of the three forms of the supreme deity—Brahma and Vishnu being the other two—Shiva is the god of destruction and lord of the dance. The power of his meditation destroys the world of illusions that bind

Sherpas make offerings to Khumbila with the god of the mountain depicted on the cliff to the left. (Stan Stevens)

9

people to the painful cycle of death and rebirth. When he rises to dance, he takes on the functions of Brahma and Vishnu and creates and preserves the universe itself.

According to a well-known Hindu myth, Shiva was once meditating in the mountains near Kailas. The lesser gods learned that only a son born of Shiva's powerful semen could defeat the demons who were oppressing the earth. They therefore sent Parvati, the daughter of the mountain god Himalaya, to seduce the celibate Shiva. She spent thousands of years practicing austerities and meditating in his presence until she caught his attention. At that point Kama, the god of love, shot one of his flowered arrows at Shiva, who seeing him at the last minute burned him to ashes with a glance of his flaming eye. But it was too late: the arrow hit its mark and Shiva fell in love with Parvati. They were married in a magnificent ceremony in the mountain palace of the god Himalaya and then went to dwell on Kailas, where they made love and had a supernatural son, Karttikeya, who grew up to defeat the demons and liberate the world from evil.[13]

Indian paintings often depict idyllic family scenes of Shiva and Parvati with Karttikeya and their other son, the delightful elephant-headed god Ganesha, picnicking in beautiful meadows and forests on top of

Shiva and Parvati with their sons Ganesha and Karttikeya. Rajput Pahari painting. (Courtesy of the Freer Gallery of Art, Smithsonian Institution, Washington, D.C.)

Mount Kailas. The opening lines of the *Tantra of the Great Liberation,* a major work of Hindu mysticism, describe this paradise:

The enchanting summit of the Lord of Mountains, resplendent with all its various jewels, clad with many a tree and many a creeper, melodious with the song of many a bird, scented with the fragrance of all the season's flowers, most beautiful, fanned by soft, cool, and perfumed breezes, shadowed by the still shade of stately trees; where cool groves resound with the sweet-voiced songs of troops of celestial nymphs . . . [14]

As a sacred mountain of multifaceted significance, Kailas encompasses the extremes of spiritual asceticism and material riches. The great epics of Indian literature, the *Mahabharata* and *Ramayana,* refer to the peak as the luxurious abode of Kubera, the god of wealth who lives in the north. *The Cloud Messenger,* a poem composed in the fifth century A.D. by the Sanskrit poet and dramatist Kalidasa, the Indian equivalent of Shakespeare, relates the story of a love-lorn *yaksha,* a supernatural being of Hindu mythology, banished to southern India from his home in Alaka, the city of Kubera on the summit of Kailas. Consumed with longing for his lover back on the sacred mountain, he sends her a message with a passing cloud that he spies drifting north toward the Himalayas. In his instructions to this unusual messenger, he compares the heavenly qualities of the city to those of the cloud itself:

Its palaces resemble you in various ways: your lightning flashes in the brilliance of their beautiful women;
Colorful paintings on their walls match the hues of your rainbow;
The music of tambourines, beating in song and dance, echoes the soft rumble of your thunder;
Floors paved with jewels reflect the gleaming waters you carry within you;
While roofs reaching up to the sky equal you in your loftiness. [15]

Like the sacred mountain it so beautifully describes, *The Cloud Messenger* represents a peak of perfection in the history of Indian literature.

Kailas also has great significance for the Buddhists of Tibet and figures prominently in one of the most beautiful and widely read works of Tibetan literature — *The Hundred Thousand Songs of Milarepa.* Like the Hindus of India, the Tibetans regard Kailas as the ultimate sacred mountain, the one that they dream of beholding at least once in their lifetime, although few ever have the opportunity to do so. Situated in the high and windswept reaches of western Tibet, Kailas lies far from the centers of population in the central and eastern parts of the country. Tibetans call

Kailas by two different names: Tise and Kang Rimpoche, the "Precious One of Glacial Snow." The second name indicates the high esteem in which they hold the mountain: Rimpoche is the title reserved for the very highest lamas, or Tibetan Buddhist priests, such as the Dalai Lama, the exiled ruler of Tibet, whom Tibetans view as incarnations of enlightened spiritual beings called Bodhisattvas.

Kailas is noted in Tibetan history and literature as the scene of a famous mountain-climbing contest between Tibet's most beloved yogi, Milarepa, and Naro Bhun Chon, a priest of the indigenous non-Buddhist religion of Bon. Milarepa, who roamed the high mountains in the twelfth century A.D. clad only in a cotton shirt and the warmth of his meditation, came to pay homage to Kailas and the sacred lake near its foot. When he arrived with his disciples, Naro Bhun Chon accosted him, saying that the area was the special preserve of the Bon religion and that he would have to take up its practices if he wished to stay. Milarepa refused, and a series of contests of supernatural power ensued to determine whose teachings would prevail— those of the Buddha, which had been recently introduced from India into Tibet, or those of Shenrab, the great teacher who had founded Bon in the vicinity of Kailas thousands of years before. After a succession of embarrassing losses in which Naro refused to concede defeat, he challenged Milarepa to a race that would settle the issue once and for all: the first to reach the summit of Kailas on the fifteenth day of the month could be acknowledged the spiritual master of the mountain.

While Milarepa relaxed and enjoyed the beautiful scenery, Naro went into spiritual training, strenuously praying to his deity for the power needed to make the ascent. On the appointed day, early in the morning, the Bon priest put on a green cloak and took off flying on his shaman's drum toward the summit of Kailas. Milarepa was still asleep. His anxious disciples woke him up to tell him that Naro had reached the waist of the mountain. Quite unconcerned, Milarepa made a gesture with his hand, and the Bon priest found himself circling around the peak, unable to go higher. Then, putting on a cloak for wings, Milarepa snapped his fingers and with the first light of the sun flew in a second to the summit of Kailas—certainly the most elegant ascent in the history of mountaineering, and one of the earliest ever recorded.

When poor Naro looked up to see Milarepa sitting at ease on top of the mountain, the Bon priest fell off his drum in amazement, whereupon the drum tumbled down the south face of Kailas, leaving a series of indentations that look like a line of steps ascending the peak. Whereas Tibetan Buddhists attribute these indentations to the fall of the Bon priest, Hindus view

them as a stairway leading up to the heaven of Shiva on the summit itself.[16] Neither Hindus, Buddhists, nor any Tibetans, however, would ever contemplate trying to climb Kailas, the most sacred of mountains. Completely humbled, Naro finally acknowledged defeat, and Milarepa magnanimously allowed him to stay on a nearby peak where he might practice his religion and continue to gaze on Kailas, now indisputably Buddhist. Milarepa's feelings about the mountain are lyrically expressed in the words of one of his many songs of spiritual accomplishment:

The prophecy of Buddha says most truly,
That this snow mountain is the navel of the world,
A place where the snow leopards dance.
The mountain top, the crystal-like pagoda,
Is the white and glistening palace of Demchog . . .
This is the great place of accomplished yogis;
Here one attains transcendent accomplishments.
There is no place more wonderful than this,
There is no place more marvelous than here.[17]

Milarepa adds that the snow mountains encircling Kailas are the dwelling place of five hundred Buddhist saints who have attained nirvana, the blessed state

Demchog, Buddhist deity of Mount Kailas, embraces his female consort, Dorje Phagmo, symbolizing the ecstatic union of compassion and wisdom. (Edwin Bernbaum)

of transcendence free from suffering. According to legend, those who have gained spiritual acuity may sometimes hear the divine music of the saints' chanting in the high clear atmosphere of the Tibetan Plateau.

In his song Milarepa refers to the palace of Demchog, the principal Buddhist deity of the sacred mountain. Of awesome appearance, blue like the sky, draped with garlands of skulls and embracing his female consort, Demchog, the "One of Supreme Bliss," dances in the ecstatic realization of ultimate reality on the summit of Kailas. An important tutelary deity of Tibetan Buddhism, he guides practitioners of meditation along the short and dangerous path leading to enlightenment in this life. Lay people and lamas alike travel for weeks and even months to his sacred mountain to experience a moment of power or revelation that will show them the way to transcend the passions and illusions of this world. They view Kailas with the invisible pagoda palace of the deity on its summit as the center of a mandala or sacred circle that represents the divine space of Demchog, where they may come to know the power and wisdom that will set them free from the bondage of suffering. In the practice of meditation, Tibetan yogis will visualize and identify themselves with this deity, and their surroundings with his domain, to awaken to their own true nature and transform their experience of the world from a profane place of illusion into the sacred realm of ultimate reality.

Every peak and prominent feature surrounding Mount Kailas corresponds to the place of a particular deity dwelling in the mandala of Demchog. The Buddhas of the four directions, such as Amitabha, the Buddha of "Boundless Light," and Ratnasambhava, the "One Born of a Jewel," occupy valleys on each side of the mountain. They embody various aspects of the transcendent awareness awakened in the attainment of enlightenment, such as the wisdom of discrimination and the wisdom of equality. The highest pass on the pilgrimage route around the peak is called Drolma La, the "Pass of the Savioress," the most important female deity in Tibet. Tibetans regard three hills near Kailas as the dwellings of Manjushri, Vajrapani, and Avalokiteshvara—the three principal Bodhisattvas, or enlightenment beings, who symbolize the wisdom, power, and compassion needed to attain the ultimate goal of enlightenment for the sake of others.[18]

The mandala of Demchog on Kailas presents a vision of the world as the sacred realm of a deity. The pattern of such a mandala appears frequently in works of Tibetan art, depicting the universe as a circle of mountains, oceans, and continents arrayed around a mythical mountain at the very center, shooting up from the depths of hell to the heights of heaven—

and the great void that lies beyond. This mountain, called Meru by Hindus and Sumeru by Buddhists, plays a pivotal role in Hinduism and Buddhism as the divine axis of the cosmos. According to Hindu mythology, Brahma, the supreme deity in the form of the creator, lives on its summit, surrounded by lesser deities; in the Buddhist version, the king of the gods, Indra, the equivalent of Zeus in Greek mythology, resides in a glorious palace on its peak, thousands of miles above the earth. Our world appears in the latter version as a triangular continent shaped like India, situated in an ocean to the south of Meru, which rises far to the north, looming over seven rings of golden mountains, so high that the sun and moon must circle around its flanks.[19]

Meru and Kailas appear as separate mountains in early texts of Buddhist and Hindu mythology, but later tradition has tended to bring them together and identify them as one and the same. Today many Indians and Tibetans view Kailas as the place where the invisible form of Meru breaks through to appear in the physical plane of existence. A pilgrimage to the mountain, therefore, represents for them a journey to the very center of the universe—the cosmic point where everything begins and ends, the divine source of all that exists and has significance. In circling the peak and paying homage to a vision of Shiva or Demchog on its shining summit, they make contact with something deep within themselves that links them to the supreme reality underlying and infusing the cosmos itself.

For most Hindus and Buddhists of India and Tibet, the journey to Kailas is, in fact, the ultimate pilgrimage, both in terms of the sanctity of its goal and the difficulty of the way. Pilgrims, or *yatris*, from the plains of India must first surmount the formidable ice barrier of the Himalayas, crossing passes more than 16,000 feet high, often clad in sandals and cotton clothes. After a difficult ascent through eerie gorges twisting between towering mountains lost in mist, the *yatris* emerge from the monsoon clouds to behold a brilliant landscape of yellow, red, and purple plains stretching off toward distant peaks, shining beneath an intense blue sky. Until not long ago, they would have had to trust in the gods to protect them from bandits while crossing these plains to reach Mount Kailas on the other side. The Chinese, who took over Tibet in 1950 and only recently opened the mountain to foreign visitors, have eliminated the robbers, along with most of the shrines and monasteries in the area. Pilgrims from central and eastern Tibet do not have to contend with the steep ramparts of the Himalayas but do have to travel for weeks and even months across the bleak and inhospitable Tibetan Plateau.

The routes of the two converge at the shores of

Tibetan pilgrims perform full-length body prostrations around Mount Kailas, passing a mani wall carved with sacred mantras.
(Hugh Swift)

Lake Manasarovar, where the Indians, to the astonishment of Tibetans, strip for a ritual immersion in its clear but icy waters. The circumambulation of the holy mountain, the ultimate goal of the pilgrimage, normally takes three days, with frequent stops at shrines and temples to recite prayers and perform rituals to the gods. Some Tibetan pilgrims, to increase the religious merit accruing from their efforts, take much longer, two to three weeks, making full-length body prostrations all the way around the mountain, unfazed by the streams, boulders, and glaciers they must cross. The high point of the pilgrimage comes at the Drolma La, a pass on the northeast side of Kailas at nearly 19,000 feet, festooned with prayer flags strung between rocks and boulders marked with prayers. Just before this pass, Tibetans leave part of themselves—a lock of hair or a tooth—symbolizing their own death and rebirth to a new and more spiritual life. On the approach to the Drolma La, a narrow crevice in the rocks through which Tibetan pilgrims must squeeze their bodies separates sinners destined for hell from those who will attain heaven—or the higher goal of nirvana.

The hardy few who manage to reach Kailas and complete the circuit of the mountain—about two hundred Indians and an indeterminate number of Nepalis and Tibetans a year—come back with a sublime vision of another, sacred realm of existence and a renewed determination to strive for the highest goals of spiritual accomplishment. Lama Anagarika Govinda, a European who became a Tibetan lama and made the pilgrimage to Kailas in 1948, had this to say of his fellow pilgrims:

They return to their country with shining eyes, enriched by an experience which all through their life will be a source of strength and inspiration, because they have been face to face with the Eternal, they have seen the Land of the Gods.[20]

Lama Govinda was one of the very few Western travelers to reach Mount Kailas before the area around the mountain was officially opened to Westerners in 1984. The first Europeans to see the peak, two Jesuit missionaries named Ippolito Desideri and Manuel Freyre, who passed by it on a journey from Kashmir to Lhasa in 1715, did not think much of it:

Mount Kailas appears as a golden temple at sunset. The gully descending its face is regarded by Hindus as the stairway to the heaven of Shiva and by Buddhists as the track left by the drum of a falling Bon priest. (Edwin Bernbaum)

Close by is a mountain of excessive height and great circumference, always enveloped in cloud, covered with snow and ice, and most horrible, barren, steep and bitterly cold.[21]

The small group of explorers, military men, sportsmen, and other travelers who followed them had a very different impression of the mountain, and a number of them even fell under its spell. Captain C. G. Rawling, a member of a British military expedition that invaded Tibet in 1903, had this to say of Kailas:

It is indeed difficult to place before the mental vision a true picture of this most beautiful mountain. In shape it resembles a vast cathedral, the roof of which, rising to a ridge in the centre, is otherwise regular in outline and covered with eternal snow.[22]

During the nineteenth and twentieth centuries, a number of Westerners who did not have the advantage of traveling in official capacities had to sneak into Tibet to see the sacred peak. Most of them wore the same disguise, which they each prided themselves on so originally choosing: that of a deaf and dumb

Indian pilgrim. One of these "pilgrims" was an Austrian geologist and mountaineer named Herbert Tichy, who went to India in 1936 to do research but with the primary intention of going to Kailas. He succeeded in reaching his goal, but almost blew his disguise by taking a picture at the wrong moment. On his return to India, he received a letter from the governor of the Punjab, congratulating him as a fellow pilgrim to Kailas and regretfully informing him that he would be arrested in a few days for illegally crossing the Tibetan border. Tichy took the kind hint and immediately left for Europe.[23]

In 1988 I went to the sacred mountain. Late in the afternoon three of us climbed up a ridge for a view of the south face with its stairway leading to heaven. Polished to a smooth finish by wind and sun, the white dome of Kailas gleamed against the sky, amazingly pure in the simplicity of its form. The wind came up, ripping at our faces, and my companions decided to go down. It was autumn and very cold. I stayed alone to watch the mountain turn orange and red in the sunset. As shadows deepened behind the peak, I began to fear that I had lingered too long. But something kept me there to see the last ray of the sun flare gold on the summit. Then, no longer anxious, but strangely excited, I started down in the twilight, suspended in the sky over the darkening plain of Barkha with the waters of Lake Manasarovar and the snows of the Himalayas glimmering blue in the distance. I felt as I had twenty years before when I first went to Nepal and stayed alone, high on a ridge, to watch Mount Everest fade in the sunset — open and free. Whistling a song, I danced down the slopes of the mountain, filled with a wild feeling of laughter and joy.

Nanda Devi

In addition to the paradise she shares with Shiva on Kailas, Parvati, the mountain daughter of Himalaya, has her abode on a number of other mountains, where she appears in various forms — some beautiful, some terrifying. As Nanda Devi, the "Goddess of Bliss," she dwells in beauty on the lovely peak of that name in the Himalayas northeast of Delhi, not far from the Nepalese border. The highest mountain in India outside the principality of Sikkim, Nanda Devi soars in alluring curves of rock and ice to culminate in a delicate summit, poised at 25,645 feet, above a ring of snow peaks that form a sanctuary protecting the goddess from all but her most determined admirers. The names of these peaks reflect their relationship to the deity they serve: Nanda Ghunti, "Nanda's Veil"; Nanda Kot, "Nanda's Fortress"; Nanda Khat, "Nanda's Bed." The only break in their otherwise impregnable wall of rock and snow is the terrifying gorge of the Rishi Ganga, one of the sources of the sacred Ganges, named after seven sages of Hindu mythology who fled the oppression of demons to seek refuge with the goddess before departing this world to become enshrined as seven stars in the constellation of Ursa Major. Shepherds and porters from nearby villages who venture into the area believe that they can sometimes hear the sounds of these sages in the company of their divine protectoress — drums beating, the blare of trumpets, and the eerie barking of dogs. The few foreign mountaineers who manage to penetrate the gorge, inching their way along the sides of sheer cliffs that plunge thousands of feet into the river roaring below, find themselves in a paradise of gentle meadows filled with flowers at the foot of the sacred peak, which stands like a temple in the middle of the sanctuary itself.[24]

Nanda Devi lies in the region of Uttarakhand, the principal area of pilgrimage in the Indian Himalayas. This region of sacred peaks and rivers ranks second only to Kailas and Manasarovar in the degree of its sanctity for Hindus. Closer to the lowlands and much more accessible, it is visited by many more pilgrims, who come by the tens of thousands to bathe at Gomukh, the glacial source of the Ganges, and to worship at Kedarnath and Badrinath, lofty temples of Shiva and Vishnu sequestered in narrow valleys beneath the icy thrones of the gods themselves. One finely chiseled peak to which they pay special reverence, 21,467-foot Shivling, they regard as the divine phallus of Shiva himself, a magnificent symbol of the god's creative power. Women in sedan chairs, old men with canes, babies in baskets, Hindus from all parts of India, many of them unprepared for the cold and wind of high altitude, flock along the network of pilgrimage routes linking the holy shrines. Many are so old or weak that they never return: they die in the land of the gods, happy in the thought that by doing so they have gained merit that will bring them closer to final release from death and rebirth. The region is also the favorite haunt of holy men and wandering yogis, who come to follow the example of Shiva and meditate in the sharp clear air of the heights, within sight of the peaks that lead to heaven and the goal they seek.

As the goddess who resides on the highest mountain in the region, Nanda Devi has many shrines and temples dedicated to her. One of the better-known ones is in the hill station of Almora, which affords one of the best views of the peak itself and the mountains that surround it. Although primarily a benevolent deity, Nanda can take on the form of Durga, the wrathful goddess who absorbed into herself the power of all the gods, including Vishnu and Shiva, in order to slay a buffalo demon who was threatening the world. Villagers accordingly treat her with great respect and sacrifice goats and buffalo to her on certain festivals or whenever they feel they may have offended her. One such festival involves a strenuous pilgrimage of three weeks over a 17,000-foot pass to snowfields not far from Nanda Devi, beneath the summit of Trisul, a spectacular peak that some view as the trident of Shiva, others as the weapon of Nanda herself. Once every twelve years thousands of devotees, many of them barefoot, accompany palanquins with images of the goddess, taking her home to the mountain bearing her name. A ram with four horns, adorned with clothes and ornaments meant for Nanda Devi, leads the party to the shrine at the end of the pilgrimage and then, its task complete, wanders off to vanish into the eternal snows.

One such pilgrimage of the past ended in disaster. According to a legend prevalent in the area, a prince of a neighboring country fell in love with a beautiful princess of Garhwal, and they consummated their marriage without performing the proper ceremonies. The goddess was offended, and evil times fell on the prince's kingdom after he became king. To placate

Nanda the people went forth in pomp on a mass pilgrimage to her shrine high in the snows beneath Trisul. The king, however, did not take such practices seriously and brought along dancing girls for entertainment. This additional transgression compounded the original offense, and the goddess destroyed the party with avalanches of snow. Whether or not it happened that way, in 1954 hundreds of corpses, some mummified, some reduced to skeletons, were found in the rocky moraine of Rup Kund, a glacial lake just short of the shrine to Nanda Devi. Carbon dating showed the grisly find to be more than six hundred years old, in all likelihood the remains of a pilgrimage party that perished in either a blizzard or an enormous avalanche.[25]

The people of the region also view Nanda Devi as a benevolent source of life and renewal. According to ancient Hindu mythology, a flood once covered the entire world. As in the biblical story of Noah, a sage named Manu was warned of the impending disaster and built himself a boat in which he survived. Vishnu, the Preserver, incarnated himself as a fish and towed the craft to safety on the summit of a mountain peak. As the waters receded, Manu together with his family and the remnants of all living creatures went down the slope to repopulate the earth. The people of Uttarakhand identify the mountain of the flood as Nanda Devi itself, and one local group, the Rajis, still regard the peak as the abode of their ancestors. According to one legend, the seven sages associated with the Rishi Ganga accompanied Manu in his boat and remained behind to dwell in the company of the goddess herself.[26]

The beautiful peak of Nanda Devi has cast its spell over Westerners as well as Indians. A succession of mountaineering expeditions, beginning in 1883, tried without success to penetrate the veil of the goddess and reach her mysterious sanctuary beyond the impassable gorges of the Rishi Ganga. Finally, in 1934, the British climbers and explorers Eric Shipton and H. W. Tilman managed to force a tenuous passage through the cliffs and became the first humans ever to enter the lovely valley at the foot of the sacred peak itself. Shipton wrote of his experience:

We were now actually in the inner sanctuary of the Nanda Devi Basin, and at each step I experienced that subtle thrill which anyone of imagination must feel when treading hitherto unexplored country. Each corner held some thrilling secret to be revealed for the trouble of looking. My most blissful dream as a child was to be in some such valley, free to wander where I liked, and discover for myself some hitherto unrevealed glory of Nature. Now the reality was no less wonderful than that half-forgotten dream; and of how many childish fancies can that be said, in this age of disillusionment?[27]

Two years later, in 1936, Tilman returned with an Anglo-American expedition to climb the mountain. The day they reached the summit, an enormous storm at the end of the monsoon caused a sacred river in the area to flood and partially destroy a village, sweeping away the lives of forty people. When some members of the expedition visited the high Hindu priest of Badrinath afterward and asked him if the villagers would hold them responsible for offending the goddess and causing the deaths, he replied, "No, they will regard you as *mahatamas*, great-souled ones, for having climbed the mountain."[28]

On the way down the mountain, people at another village had asked the expedition members if they had seen the golden pagoda and pond said to be on the summit, and when told they had not, the villagers refused to believe that the expedition had actually climbed the peak. The climbers mentioned this to the priest at Badrinath, and he said, with a smile, "No, you probably wouldn't have seen those things."[29]

Some years later Nanda Devi cast its spell on another climber, giving rise to a tragic but beautiful story that brings together the cultures of East and West. Willi Unsoeld, a well-known American mountaineer who made the first ascent of the west ridge of Mount Everest, saw Nanda Devi as a young man and thought the peak so lovely that he vowed if he ever had a daughter he would name her after it. In the course of time he married and his wife gave birth to a girl, whom they named Nanda Devi Unsoeld. When Devi, as she was called, reached the age of twenty-one, she decided that she wanted to climb the peak whose name she bore. She and her father organized an expedition that included some of the best mountaineers in America, and in 1977 they set out for Nanda Devi.

After struggling through the Rishi Ganga Gorge, the first group of climbers reached the summit by a difficult new route. Devi climbed to a high camp at 24,000 feet to make the second ascent. However, after a day of being tent-bound in a storm, she felt too ill to continue. As they were preparing to descend, she suddenly sat up and said, very calmly, "I am going to die"—and she died in her father's arms. Willi tried without success to revive her until, heartbroken, he realized she was dead. His description of what followed reveals the depth of his feelings for his daughter and the mountain for which he had named her:

A camp on Nanda Devi, the abode of the Goddess of Bliss. (John Evans)

We agreed that it would be most fitting for Devi's body to be committed to the snows of the mountain for which she had come to feel such a deep attachment. Andy, Peter and I knelt in a circle in the snow and grasped hands while each chanted a broken farewell to the comrade who had so recently filled such a vivid place in our lives. My final prayer was one of thanksgiving for a world filled with the sublimity of the high places, for the sheer beauty of the mountains and for the surpassing miracle that we should be so formed as to respond with ecstasy to such beauty, and for the constant element of danger without which the mountain experience would not exercise such a grip on our sensibilities. We then laid the body to rest in its icy tomb, at rest on the breast of the Bliss-Giving Goddess Nanda.[30]

The story continues. On the way to the mountain, Devi had made a great impression on the porters and villagers along the trail. Having lived in Kathmandu with her father, who had served there as Peace Corps director, she spoke Nepali, which had enabled her to communicate with the people in their own, closely related language of Garhwali. The natural warmth of her personality and her obvious interest in them and their welfare had touched them deeply. In addition, her striking blonde hair had elicited comparisons with Gauri, the golden form of the goddess Parvati. When the local villagers heard of her death, they concluded that she had not really died. According to them, Willi's vow to name his daughter Nanda Devi had caused the deity to enter her body and become incarnate in her. Her apparent death was, in fact, the goddess's way of coming home to her mountain. And so a new myth has entered the sacred lore surrounding the beautiful peak of Nanda Devi.[31]

Annapurna and Machapuchare

The goddess Parvati dwells in yet another form on another sacred mountain that occupies a prominent place in the history of mountaineering: Annapurna One, the first of the 8000-meter peaks — the highest in the world — to be climbed. Named like Nanda Devi for the deity said to reside on its summit, this 26,545-foot peak was the scene not only of the mystical experience described by Maurice Herzog in the introduction to this book, but also of a terrible ordeal that followed in which he dropped his mittens and lost his fingers and toes to frostbite. Infused with a sense of the spiritual significance of mountain climbing, his beautifully written account of the French expedition

that climbed the mountain in 1950 remains to this day a classic of Himalayan mountaineering literature. It was appropriate that twenty-eight years later, in 1978, the abode of a female deity should have become the objective of the first American women's expedition to an 8000-meter peak. On October 15, after a long and dangerous climb, Vera Komarkova and Irene Miller succeeded in reaching the summit. Tragically, two members of the expedition, Alison Chadwick-Onyszkiewicz and Vera Watson, disappeared on the mountain two days later, and their bodies now lie enshrined with the goddess in the soft and lovely snows of Annapurna.[32]

A range of peaks that includes Gangapurna, Machapuchare, and Annapurnas One through Four, Annapurna rises in one long sweep above the lush green hills of central Nepal. Seen from the tropical valley of Pokhara in the twilight before dawn, the range's peaks appear to float like bluish-gray icebergs on a sea of liquid shadows. As the first rays of the sun touch their summits and glide down ridges of snow and rock, the mountains turn translucent and appear to glow with an inner light, as if they were wax-paper lanterns lit by candles placed within them by the goddess herself. Etched with shadowed flutings, the corrugated face of Annapurna One, the highest summit, becomes a golden backdrop to the slender pointed peak of Machapuchare, the "Fish's Tail."

The name Annapurna means in Sanskrit "She Who Is filled with Food." Unlike Nanda Devi, who can take on the wrathful form of Durga, Annapurna is regarded in Hinduism as a purely benevolent deity. A kind-hearted goddess of plenty, she is the Queen of Benares, the holy city of the Hindus on the banks of the Ganges south of Nepal. Each year, after the autumn harvest, the people of Benares celebrate a festival dedicated to her called Annakuta, the "Food Mountain," in which they fill her temple with a mountain of food — rice, lentils, and sweets of all kinds to be distributed to those who come to receive her blessings. There, installed in her place of worship, she reigns with Shiva, the presiding deity and lord of one of the oldest and most sacred of Indian cities. For more than two thousand years, Hindu pilgrims have been coming to Benares from all parts of India to bathe in the sacred waters of the Ganges and, at the end of their lives, to have their bodies cremated and their ashes cast into the river. The goddess Annapurna provides them with sustenance in life, whereas Shiva grants them liberation at death. Together the two deities satisfy all the material and spiritual needs of their devotees.[33]

Machapuchare, the "Fish's Tail," most dramatic peak in the Annapurna Range, glows with the first light of dawn. The government of Nepal has declared the sacred mountain off-limits to climbers. (Galen Rowell/Mountain Light)

Within the heart of the range, at the foot of the sheer south face of Annapurna One, lies a hidden basin of beautiful meadows and glaciers, resembling the sanctuary surrounding the peak of Nanda Devi. A great curtain of rock and ice draped between mountains soaring to over 26,000 feet completely encloses this natural amphitheater, dropping nowhere lower than 19,000 feet—except at one place. There, an incredible gorge 12,000 feet deep slices through this otherwise impregnable barrier, right beneath the overhanging cliffs and glaciers of Machapuchare, one of the loveliest peaks in the Himalayas. A difficult trail made by shepherds and hunters leads up it through a bamboo jungle across slabs of rock slick with mud, where a slip can mean a broken leg or a shattered skull.

The Gurung people who live in the villages on the way to the gorge regard the Annapurna Sanctuary as the sacred abode of various deities—Hindu, Buddhist, and animistic. Until recently the villagers would not allow meat, eggs, untouchables, or women into the sanctuary's restricted precincts. According to one local legend, the sanctuary is the repository of sacred treasures: jewels, gold, and silver guarded by serpent deities called Nagas. Once, long ago, a group of lower-caste untouchables went up there to dig for the hidden riches. They brought along meat and eggs and behaved in such an appalling way that the king of the Nagas destroyed their mine and all but one of the men. Since that time such things have been prohibited to the few shepherds who take their flocks of goats and sheep up during the summer months to graze in the flowered meadows beneath the snow peaks of the sanctuary. When I went there in 1968, before many foreigners had started trekking in Nepal, the old men of the last village were grumbling that a few Western women had recently gone into the forbidden area and offended the deities, causing rock slides to cover the trail and make it even more difficult and perilous to follow. Since then, the sanctuary has become a popular goal for climbers and trekkers, and huts with food and lodging now stand where shepherds used to wander in the mist alone with their gods.

Machapuchare, the most spectacular peak in the Annapurna Range, stands guard over the gorge leading into the sanctuary. From a cave at the entrance to the amphitheater itself, Machapuchare's delicate summit will sometimes materialize out of the mist at sunset, to appear suspended in a golden haze almost 15,000 feet directly overhead at an altitude of 22,943 feet above sea level. The mountain is so imposing that for the people living near Annapurna, it acts like a magnet, drawing to itself whatever deity they regard as the highest and most powerful. Villagers with whom I spoke referred to it variously as the abode of the Hindu gods Vishnu and Shiva, a local deity named

Pujinim Barahar, and Tara, the Savioress of Tibetan Buddhism, as well as Amitabha, the Buddha of Boundless Light.[34]

From a mountaineer's point of view, Machapuchare appears almost impossible to climb. However, a small expedition led by Wilfred Noyce, a British climber of Everest fame, nearly reached the summit in 1957. Grooves of slick blue ice stopped them only 150 feet from the top. Realizing that the deity of the mountain had defeated them, they turned back and descended without regrets, content with what they had achieved. After their attempt the government of Nepal declared Machapuchare a sacred peak, off-limits to all climbers. And it remains to this day unclimbed, one of the few places left on earth reserved for the gods.[35]

Kangchenjunga

Of the fourteen 8000-meter peaks, the most sacred, however, is not Annapurna, but Kangchenjunga, the third-highest mountain in the world. Rising in regal splendor to over 28,000 feet on the border with Nepal, it completely dominates the Himalayan principality of Sikkim and plays a central role in the religious life of the Sikkimese, who view the peak as the divine protector of their country. When a British expedition set out to climb Kangchenjunga in 1955, the Sikkimese, Indian, and Nepali governments, fearing the harm that might come of provoking the wrath of the deity, officially asked the climbers to call off their attempt. Their leader, Charles Evans, worked out a compromise whereby he promised that the mountaineers would stop just short of the summit to avoid desecrating the seat of the god himself. The climbers kept their word: when they reached a point twenty feet from the top, they declared the mountain climbed and turned back, leaving the highest snows untouched by human feet.

Seen from the hill resort of Darjeeling at sunrise, when the clouds have settled in the valleys below the mountain, Kangchenjunga rises through a series of ridges, faces, and peaks to culminate in a summit of pristine snow, serene and silent in the first light of the morning sun. Despite its incredible mass, spreading over several degrees of the northern horizon, Kangchenjunga gives an overwhelming impression of lightness and grace comparable to that of the delicately tapered peak of Machapuchare. It presents a vision not so much of a mountain as of another world floating like a cloud above ours. Tibetan texts compare the peak to a king seated on a throne draped with curtains of white silk. In its row of five summits they see the five points of a crown worn by the mountain deity who reigns over the peaceful valleys of Sikkim.

Pilgrims approach a shrine at the foot of Kangchenjunga, the most sacred of the world's 8000-meter peaks. (William Frej)

In Tibetan the name Kangchenjunga means either "Five Treasures of Great Snow" or "Five Brothers of Great Snow." The first meaning refers to five treasures said to be hidden inside the five summits of the peak: the highest summit that catches the golden light of dawn contains gold, the one that remains in gray shadow has silver, while the other three hold jewels, grains, and holy books. Appropriately enough, Vaisravana, the Buddhist god of wealth, dwells on the mountain as the guardian and dispenser of these treasures. The texts describe him as a corpulent deity with a red lance and banner, seated on a snow lion playing on the tops of mountains. As the identity of the fifth treasure, holy books, suggests, the precious items hidden within the snows of Kangchenjunga have a spiritual as well as material significance, referring ultimately to the riches of wisdom and compassion found on the way to enlightenment. According to the second meaning of the mountain's name, the treasures are guarded by five divine brothers, each perched on a different summit, riding a different mount: a lion, an elephant, a horse, a dragon, and a mythical kind of eagle.[36]

The original inhabitants of Sikkim, a people of short stature called Lepchas, have their own name for Kangchenjunga—Konglo Chu, the "Highest Veil of Ice." They trace their mythical origins back to a man and a woman fashioned from the ice of its glaciers by the creator deity of their animistic religion. According to their beliefs, behind the great wall of Kangchenjunga, spread across the northern horizon like a veil of ice, lies the mysterious kingdom of the dead to which they go when they die. Some of the shrines they build in villages high beneath its snows contain cairnlike stones, which represent the sacred mountain and the peaks around it. Priests will sacrifice ceremonial yaks at these stones to Kongchen, the deity of Kangchenjunga, imploring him to protect their people from the forces of evil that menace not only them, but the country as a whole.[37]

Sikkimese of Tibetan descent, who took over political control in the seventeenth century and ruled the principality until 1975, also regard Kangchenjunga as a protective deity in its own right—the warrior god named after the mountain itself. According to their own accounts of their history, when Lhatsun Chembo, the lama who brought Buddhism to Sikkim, was seeking a way over the mountains from Tibet, Kangchenjunga took the form of a wild goose and flew to greet him. They met on a high place just north of the present-day border, and there the deity described to him the land he was going to settle. With the information that Kangchenjunga gave him, Lhatsun Chembo was able to lead his party over the passes into Sikkim, where he appointed one of his followers to be the Chogyal or ruler of the country, which Tibetans had long regarded as a mysterious hidden sanctuary resembling the fictional Shangri-La of the Western novel *Lost Horizon*. On the way the lama used his magic powers to fly to the summit of Kabru, a snow peak near Kangchenjunga, to survey the secluded valleys that ancient prophecies had said he would open for those who would need refuge in times of trouble.

After arriving in Sikkim, Lhatsun Chembo gave offerings of thanks to the god of Kangchenjunga. His followers made the thanksgiving an annual ritual and expanded it into an elaborate ceremony of sacred dances that has continued to this day. The performance takes place on a meadow in front of the main temple in Gangtok, the capital of Sikkim, within view of the peak itself. Colorfully dressed warriors, representing the armed retinue of the mountain deity, leap and whirl with war cries and swords, clearing the area of evil influences. When they have finished their dance, the figure of Kangchenjunga emerges, portrayed by a lama in ornate robes of silk brocade, wearing a red mask with an angry grimace and a hideous mouth with four protruding fangs. Like mountain deities in Tibet, he possesses a wrathful appearance to drive away demons and other spirits that threaten the religion and well-being of the people. A third eye, symbolizing spiritual wisdom and power, blazes in the center of his forehead. He carries in one hand a lance decorated with a victory banner and in the other a jewel; five flags flutter in a war helmet set on his head. With slow and stately steps, in time to the beat of cymbals and drums reinforced by the deep rumble of twelve-foot horns, he performs the sacred dance that each year renews the power of Buddhism and protects Sikkim from the insidious forces of evil.[38]

The Himalayas are mountains of great power and beauty. That power and beauty can make itself felt in profound and mysterious ways. When I first came to the Himalayas in 1968, I went climbing in the Annapurna Sanctuary. Two of us were nervously following our companions up a broad snowslope that issued from a gully cut into the side of a ridge. Tiers of small glaciers lined the rock walls of the ravine, threatening to tumble off in avalanches.

My eye wandered up to a glacier hanging over a cliff near the crest of the ridge. As I looked at it, it shattered. The ice blurred and hung poised for a moment, as though encased in plastic film. Then, in eerie silence, the entire front of the glacier collapsed and plunged over the cliff into the gully below. Striking a foot of fresh powder snow, the ice threw up a billowing white cloud that began to rush toward us. Its sound, audible now, echoed from the peaks around us, shaking the ice axes in our hands.

"Avalanche!"

The cloud burst out of the gully and exploded across the slope, soaring a hundred feet into the air. The avalanche swept down on us, a line of ice blocks seething along its leading edge. The roar was tremendous, louder than anything I had ever heard, like the sound of bombs bursting in a hundred thunderstorms.

"There, that crevasse, jump in it!" my companion Ned Fetcher yelled and dashed toward a blue line in the slope to our right. I hesitated, staring at my pack, then grabbed it and ran after him. Despite the weight of the snow clutching at my crampons, I felt no fatigue. I was aware only of the pearly white cloud bearing down on us. For a moment, as if from a height, I saw two tiny figures running in front of it, like two soldiers in a war movie dashing across a battlefield with bombs bursting behind them—except the bomb bursts were beautifully white and the figures were carrying ice axes, not guns. We both stopped: the crevasse was too far to reach. The slope stretched wide and smooth around us. There was no place to go, no place to hide.

I suddenly realized that I was going to die.

The snow peaks around me were shining calmly in a clear, bright sky. I saw the mountains as though they were on a screen in a movie theater, which I was about to leave. A spasm of guilt shot through me, and with a terrific wrench, the whole universe seemed to flip over in the pit of my stomach, never to be the same again.

As I watched in terror, something deep within me rose up and took over. With my mind screaming *It's no use!* I carefully placed my pack across the slope as a shield and curled up behind it. All my movements had become very smooth and precise, as if I knew exactly what I was doing. As the avalanche cloud passed across the sun, the light turned gray, followed by a moment of roaring darkness. A sharp, hard blow sent me flying in a rush of snow and air, as if caught in the wild foam of a wave breaking on a beach.

I found myself swimming with the avalanche, automatically moving my arms and legs. Things were hitting me—blocks of ice, I assumed. At any moment one of them would crush me. I could not understand why I was swimming: it seemed so utterly futile. After what seemed a long time—I was carried perhaps a thousand feet—it occurred to me that I might survive. I thought of an airplane taking off into the wind and tried to turn myself into the avalanche in order to rise to its surface. But a tremendous force flipped me the other way, into a dive, and flung out my arms, away from my face.

As the avalanche ground to a stop, the snow tightened around me and set like concrete. When I tried to breathe, nothing came. The snow was jammed like a hand across my mouth and nose. I tried to dig an air space, but I could not even wriggle my fingers, which were bare—the avalanche had ripped off my gloves. Terror coursed through me: I was going to suffocate, one of the most horrible ways to die that I could imagine. As I fought for air, my lungs heaved and shook with unbearable pain.

For no apparent reason, I stopped trying to breathe, and a strange calm came over me. The pain, the anguish, and the terror all dwindled away, and I saw there was nothing, literally nothing, to fear. Death was not anything at all. I would simply become part of the snow and ice around me. The sensation, in fact, was quite pleasant. Letting myself drift into it, I began to die.

Suddenly, my hand was in front of my face, and I had air to breathe. In one motion, without any direction from me, it had sliced through the snow and cleared a space. With that little bit of air, I would have to dig myself out. As far as I knew, my companions were dead, and nobody would come looking for us for at least a week. When I tried to dig, my fingernails merely scratched the icy snow. I wrenched my body, but my arms and legs were locked in tubes of ice. I could feel my boots sticking out, flailing in the air above.

I jumped to the conclusion that I was pinned under a block of ice. I remembered reading how a French climber had chiseled his way out from under an ice block in the Alps with a pocket knife and a piton. But he had done that at half the altitude I was at in the Himalayas, without the thin air at 19,000 feet. And, because of the snow, I could not reach the knife in my pocket. I would freeze to death, locked in the ice. Panic tore through me, followed once again by the strange calm. As I ceased to struggle, my body, on its own, made an explosive wrench, and in one clean

movement I popped free.

I was rolling on my back in blue and white blurs of sky and snow. Shaking and gasping, I rose to my feet. My hands felt stiff and frozen, like leather. I looked across the sanctuary to the summit of Annapurna One, where Maurice Herzog had dropped his mittens. Would I also lose my fingers to frostbite? I stuck my hands inside my shirt and felt them warm up from the heat of my stomach.

"Ed, come up!"

I looked up the rubble left by the avalanche. Fifty feet above me, still tied on the climbing rope, was Ned, sitting upright in the snow, buried from the waist down. Feeling utterly exhausted, inclined to sit and do nothing, I struggled up to him.

"Here, dig me out," he said, pointing at the snow packed over his legs.

"I can't. My hands are frozen."

"Dig anyway."

I began to kick gingerly at the snow, afraid of gashing his legs with the spikes of my crampons. Ned scraped with his hands. The snow was so hard that we made little progress.

A booming sound brought our efforts to a halt. Another hanging glacier had collapsed, and another avalanche was roaring down on us. I tried to run as before, but the rope had become tangled around my legs, and I could not take even a step. It was useless to try to escape. This time I just stood and stared at the oncoming cloud. I was too exhausted to lie down and try to protect myself. All I could do was think, *My God, after all this, it's unfair.*

The avalanche stopped a few yards away, and snow mist blew across our faces, cold and harsh against the skin. For a moment I stood and Ned sat, both of us too shocked to move. Then, not caring whether I gashed his legs, I kicked with all my strength, and we dug him free.

We came down from the mountain and remained for three days in the Annapurna Sanctuary. From time to time I would look up to see a snow cloud sweep down the slope we had attempted to climb. In the experience of the avalanche, I had come to know something of the power and mystery of mountains that make them sacred to people of so many different cultures and religions.

CHINA

MOUNTAINS OF THE MIDDLE KINGDOM

ALTHOUGH NOT AS HIGH as the Himalayas, the mountains of traditional China possess an extraordinary beauty and richness of character that make them equally impressive — and just as evocative. Some, like the limestone pinnacles and granite peaks of the south, hang poised over rivers and plains in such incredible shapes of delicate rock that they look as if they could only exist in a Chinese landscape painting where the laws of gravity need not apply. Others, such as the imposing massif of T'ai Shan in the east, stand as monuments of ancient stone, made venerable by ages of erosion. Elsewhere ranges of grassy mountains arch up through layers of wind-deposited loess to writhe like dragons across the plains of northern China, their rough flanks riven by gorges and bristling with cliffs. In the southwestern part of the country, fault-block peaks slant up to cast long shadows over wide green basins and deep blue valleys, bearing witness to massive forces at work beneath the earth. Here and there, dotted throughout the mountainous landscape of China, the light touch of a temple on a ridge or a gnarled pine on a crag in mist lends an air of intimate beauty lacking in the stupendous views presented by the great ice ranges of the Himalayas.[1]

Unlike the Himalayas and other ranges to the west, the mountains of China do not lie on the edges of the known world, in wild and mythical regions far from the centers of civilization. Many of them rise within sight of millions of people in farms, villages, towns, and cities scattered throughout the country. Living in a landscape dominated by mountains, the Chinese have long regarded them as sacred places imbued with special power and significance. Popular belief holds that they form the body of a cosmic being — according to some, a dragon whose twists and turns create the surface of the land on which the Chinese live. Rocks are its bones, streams its blood, trees and grass its hair, while clouds and mist are the white vapor of its magic breath, the essence of life itself. An ancient definition found in the oldest Chinese dictionary explains that mountains "give birth to the ten thousand beings" — a reference to the sum total of all living creatures.[2]

For the Chinese, mountains are not just inspiring places of beauty and grandeur: they embody in concrete form the basic principle of fertility that renews and sustains the world. As the belief about the cosmic being suggests, clouds and mist that bring rain do not simply gather around the summits of peaks and hills; rather, they emanate from within them as their breath, just as streams and rivers issue from their flanks. A people who live mostly by tilling the soil, the Chinese have from prehistoric times revered mountains as divine sources of life-giving water, responsible for their well-being and survival. Until the Communist Revolution in 1949, nearly every village in China had a temple dedicated to the local mountain god who controlled the rains and protected the region from drought and flood.

The earliest Chinese references to mountains as sacred places link them to the cult of the emperor. The *Shu-ching*, a classic of traditional history compiled around the fifth century B.C., tells us that Shun, a legendary ruler of the third millennium B.C., made it a practice every five years to visit four peaks that marked the four quarters of his realm. At each peak he offered sacrifices to heaven and granted audiences that reasserted his sovereignty over the princes of the

PAVILIONS IN IMMORTAL MOUNTAINS. *Hanging scroll by Wang Hui, 1712.*
(Courtesy of the Arthur M. Sackler Gallery, Smithsonian Institution, Washington, D.C.)

local region. The title of the highest official in his court, the "[Chief of the] Four Mountains," reveals the importance of these peaks and their role as symbols for the divisions of people and territory comprising the empire. The sacrifice performed at each one also reflects the ancient Chinese view of mountains as sacred links with heaven, from which the emperor, as Son of Heaven, received his mandate to rule.[3]

Confucianism, the traditional system of ritual and ethics codified by Confucius in the fifth and sixth centuries B.C., held up Shun as the model of the perfect ruler, whose conduct all virtuous emperors should strive to emulate. Like the ideal emperor, who obtains his power and authority through them, mountains embody for Confucians the important principle of stability, on which the order of nature and society depends. The enormous mass of the mountain rests on the earth and keeps it from moving. In a similar way, the weight of divine authority invested in the emperor holds the empire in place and prevents disorder. Just as mountains have acted as a wall, protecting China from invaders to the north, west, and south, so the emperors, through the power of their virtue, protect the people from the onslaught of their enemies. As one of the most enduring features of the landscape, a mountain also symbolizes the long life and reign that a good emperor should enjoy in a properly ordered state.

As dispensers of life-giving waters, mountains exemplify the magnanimity possessed by the ideal emperor, who selflessly bestows benefits on his subjects. The *Analects* of Confucius use the image of a mountain to symbolize the peerless attribute of benevolence or *jen*, the virtue regarded by Confucians as more important than wisdom in the makeup of the superior person:

The wise take pleasure in water, but the benevolent in mountains; because wisdom moves about, but benevolence remains still. Wisdom leads to happiness; benevolence to a long life.[4]

For Confucius a steady mind and a long life have greater value than a quick intellect and a happy existence. They provide a more solid foundation for an enduring social order based on the concepts of filial piety and concern for others.

The followers of Taoism—a mystically oriented group of religious and philosophical traditions attributed to the sage Lao-tzu, a reputed contemporary of Confucius—took a special interest in mountains. Until the fourth century A.D., they regarded mountain peaks primarily as the awesome abode of various deities and supernatural beings—in particular, the divine immortals who had discovered the way of living in harmony with the Tao, the Way or spiritual

essence of reality that flows through all things. Taoists who actively sought immortality went up to the mountains to make contact with these divinities and receive the revelations and powers needed to fulfill their quests. There, in a wild landscape of primeval forests and pristine peaks, far from the worldly distractions of human life, they found the perfect place to pursue meditation and other esoteric practices. In particular, mountains provided many of them with the herbs and minerals and the setting required to concoct elixirs of immortality—alchemical potions which, they believed, would enable them to transform their bodies and attain their ultimate goal in one quick swallow.

Until the fourth century A.D., most Chinese—Taoists in particular—regarded mountains as alluring but dangerous places of power that, like the palace of the emperor, should only be approached with the greatest of caution. True, one could obtain immortality on their numinous heights, but one could just as easily—in fact, much more easily—find death instead. Mountains were, in the words of the French scholar Paul Demiéville, "a zone of sacred horror."[5] Around A.D. 320, Ko Hung, a well-known sage and compiler of Taoist lore, wrote:

Even the sages who lay claim to knowing all and go to the mountains in search of long life put themselves at risk of violent death. All the mountains, whether large or small, are haunted by supernatural beings: great ones on the great mountains, little ones on the little. And if one does not take appropriate precautionary measures, they will afflict one with sickness, injuries, vexations, terror, and anguish. Sometimes the traveler will see lights and shadows; sometimes he will hear strange noises. Great trees will crash down on him without there being any wind; rocks will fall without warning and strike him dead. Or, yet again, seized with panic, he will throw himself into the depths of a ravine, trying to avoid the attacks of tigers, wolves, and poisonous animals. One does not venture into the mountains lightly.[6]

The precautionary measures recommended by Ko Hung include equipping oneself with protective spells and amulets—most notably the talisman of the True Form of the Five Peaks, a symbolic representation of the mountains visited by the legendary emperor Shun (supplemented by a fifth peak added at a later date). For the common people such charms became important means of promoting well-being and long life, whether or not one ever went near an actual peak.[7]

At the time Ko Hung wrote his warning, a new, less daunting view of mountains was emerging, one that would dominate Chinese poetry and painting throughout the following centuries. The occupation of the imperial capital of Loyang by barbarians in A.D.

311 sent many literati, who were officials of the court, fleeing south to the vicinity of Nanjing. There, in the basin of the Yangtze River, they found a rippling landscape of well-watered hills and peaks, greener and far more appealing than the brown plains and dry mountains of the north. The political turbulence of the period made these beautiful peaks doubly attractive to intellectuals imbued with both Confucian and Taoist sensibilities. The corruption and disorder that pervaded society led many of them to feel that virtue had left the emperor to return to its source in the pure and untrammeled reaches of the mountains. There, secluded among peaks and valleys untouched by man, living in harmony with the natural order of things, they might experience the Tao and recover a tranquility and simplicity of life impossible to find in the tumult of the world below. The mountains became paradises of perfection, as the following verses from a poem composed in the fifth century so beautifully show:

In the mountains all is pure, all is calm;
All complication is cut off.
Rare are they who know to listen;
Happy they who possess wisdom.

If the cold wind stings and bothers you,
Sit in the sun: it is always warm there.
Its hot rays burn like flames,
While, opposite, in the shade, all is frost and snow.

One pauses on ledges, one climbs to the foot of high clouds;
One sits in the depths of a gorge, one passes windy grottos.
Here is the realm of harmony and joy,
Where the past and the present become eternal.[8]

Although the verses express sentiments shared by literati with Taoist and Confucian inclinations, the author of this poem was himself a Buddhist. Coming from India via the Silk Route of Central Asia, Buddhism had reached China four centuries earlier during the first century A.D. Its goal of enlightenment, resulting in a release from the painful round of birth and death, distinguished it from Taoism and Confucianism with their more worldly aims of attaining immortality through living in harmony with the Tao and creating a well-ordered society based on the virtue of benevolence. Like Taoist hermits, with whom they shared an affinity for solitary meditation, Buddhist monks found the mountains an ideal environment for spiritual practices and visionary experiences. Where Taoists saw the mysterious way of the Tao, winding in and out of the clouds, Buddhists found in the vast and empty views of peaks and sky a perfect expression of emptiness or the void—the ultimate nature of reality. Out of such affinities came new forms of

Buddhism, most notably the meditation school of Ch'an—or, in Japanese, Zen—which shows the strong influence of Taoist ideas. A poem by Han-shan, a celebrated Ch'an practitioner and mountain recluse of the T'ang Dynasty, reads:

Men these days search for a way through the clouds,
But the cloud way is dark and without sign.
The mountains are high and often steep and rocky;
In the broadest valleys the sun seldom shines.
Green crests before you and behind,
White clouds to east and west—
Do you want to know where the cloud way lies?
There it is, in the midst of the Void![9]

Han-shan, who lived sometime between the sixth and ninth centuries A.D., made his home on a mountain of the T'ien-t'ai range in Zhejiang Province of southeastern China. T'ien-t'ai lent its name to an important and influential religious sect—the T'ien-t'ai

The Talisman of the True Form of the Five Peaks,
a symbolic representation of the principal sacred mountains
of China. (Anna Seidel)

school of Buddhism. A well-known haunt of Taoist recluses and poets, the airy crags and peaks of the range attracted Chih-i, the Buddhist monk who founded this sect in the sixth century. Based on the idea that everyone possesses a Buddha nature, his teachings attempted to bring all forms of Buddhism together as a means of leading all beings to enlightenment. T'ien-t'ai attained great importance in China and spread to Japan, where, pronounced in Japanese as "Tendai," it became one of the most influential Buddhist sects in the country.

Various expressions used by Taoists and Buddhists reflect their common enchantment with mountains as the perfect environment for the practice of their respective traditions. The Taoist term for an immortal, *hsien,* is composed of two pictographs, one of which represents a man, the other a mountain. In terms of its visual implications, the word means "a man of the mountains," or a hermit. Since Buddhists tended to establish their monasteries in mountainous regions where they could practice meditation in peace, far from the distractions of the outside world, the Chinese word for mountain, *shan,* took on the meaning of a monastery. The expression "to open a mountain" means, for example, "to found a monastery or sect." "To enter the mountains" is another way of saying "to embark on religious or spiritual practice," which for Buddhists and Taoists usually took place in mountains, where monasteries and hermitages were situated.[10]

The period of turbulence that followed the fall of Loyang in A.D. 311 continued until the seventh century, with China divided into regions governed by northern and southern dynasties. With the restoration of a unified empire under the T'ang Dynasty in 618, the positive attitude toward mountains that had emerged earlier underwent a shift of emphasis that would have lasting effects. Whereas the literati had previously regarded the misty heights of peaks as a refuge from disorder and chaos, now they saw them as an escape from the stifling confines of an imperial bureaucracy that imposed too rigid an order on the people. In the poetry and paintings of this time mountains stand out as symbols of a free and natural realm of the spirit, untrammeled by the artificial constraints of society. A poem by Wang Wei, one of the foremost poets and painters of the T'ang Dynasty, describes the spiritual delight he took on leaving his official post for a mountain retreat outside the capital of Ch'ang-an, or modern-day Xi'an:

My heart in middle age found the Way,
And I came to dwell at the foot of this mountain.
When the spirit moves, I wander alone
Amid beauty that is all for me. . . .

I will walk till the water checks my path,
Then sit and watch the rising clouds—
And some day meet an old wood-cutter
And talk and laugh and never return.[11]

The view of mountains as sanctuaries of the spirit carried on as an enduring feature of the literary and artistic landscape of China, shaping the attitudes of intellectuals through the succeeding centuries, right up to the present day. We can see its influence at work even in the Communist art and literature of the twentieth century. During their long struggle to gain control of China, the Communists under Mao Tse-tung found the mountains a natural refuge from oppressive forces that sought to destroy them and their revolutionary spirit. Modern paintings and murals depicting scenes from that time commonly use mountain landscapes to inspire heroic feelings in the beholder and to express ideas of spiritual renewal needed to maintain enthusiasm for the new order. Mao himself wrote poetry in the style of T'ang Dynasty poets, such as Wang Wei, and many of Mao's poems reflect the classic view of mountains as sacred places. In one particularly evocative poem, he describes the awesome ranges of eastern Tibet that he and his followers crossed in the Long March that saved them from destruction at the hands of the Nationalist army:

Mountains!
You pierce the blue sky,
Without blunting your peaks.
But for your support,
Heaven would fall.[12]

To evoke the great height of the peaks and thereby magnify the significance of the Red Army's accomplishments, Mao drew on ancient views of sacred mountains as links with heaven and symbols of stability.

T'ai Shan
and the Five Imperial Peaks

For more than two thousand years the Chinese have singled out five peaks as the principal sacred mountains of China: T'ai Shan in the east, Heng Shan to the north, Hua Shan in the west, another Heng Shan to the south, and Sung Shan in the center. Tradition has identified four of these peaks with the four said to have been visited by the legendary ruler Shun on his tours of ritual inspection in the third millennium B.C. However, the *Shu-ching,* the fifth-century Confucian classic and the oldest surviving description of the sacrifices performed by Shun, refers to only the eastern one, T'ai Shan, by name and makes no men-

Hsüan-k'ung Ssu, the "Temple Hanging in Air," perches airily on a cliff facing Heng Shan, imperial mountain of the north.
(Edwin Bernbaum)

tion of a central peak, implying that the identification of the remaining three and the addition of a mountain in the center took place at a later date. From places of ritual importance enshrined in the imperial cult, the five peaks have become focal points for a multitude of beliefs and practices associated with Taoism, Confucianism, Buddhism, and the folk traditions of the people.[13]

Rising to a height of 6617 feet in Shanxi Province, Heng Shan, the sacred mountain of the north, stands like a sentinel on the frontier between the terraced fields of northern China and the open grasslands of Inner Mongolia. A great mass of rugged cliffs and meandering ridges, it faces the site of one of the most spectacularly situated temples in China. The name of this temple, Hsüan-k'ung Ssu, means the "Temple Hanging in Air" and gives a precise description of its appearance. It literally hangs on the side of a gorge beneath overarching cliffs. Wooden beams driven into holes drilled in the rock support a spidery-looking structure of delicate pagodas that appear to float in the air. Located on the main approach to the sacred mountain, the temple overlooks a stream

named, appropriately, the Brook of the Gods. It was built in the time of the Northern Wei Dynasty, around A.D. 400, but the original wood used to attach it to the cliff has long since rotted away and has been replaced a number of times. When I visited the temple in 1984, I could feel the platforms linking the pagodas vibrate beneath my feet. A solitary monk who greeted us with a smile was taking care of shrine rooms filled with Buddhist and Taoist images, reflecting the various traditions of the pilgrims who come to pay homage to the deities of Heng Shan.

Of the five sacred mountains, the loveliest is Hua Shan, 6552 feet high, the "Flower Mountain" of the west, located in Shaanxi Province about sixty miles from the ancient capital of Xi'an. Its slender peaks of polished rock ornamented with sprays of pines open up into the sky like the delicate petals of a flower frozen in eternal bloom. Where ridges dip between its graceful summits, little temples perched in notches overlook precipices that drop into chasms of blue space. A tenuous network of pilgrimage paths climbs through a maze of cliffs and lightly follows the cutting edges of knife-shaped blades of rock. A favorite haunt of Taoist priests and hermits before the Communist Revolution, Hua Shan offers much to the pilgrim who can overcome a fear of heights and walk the clouds. For centuries Chinese poets and painters have evoked in unnerving detail the dizzy hollow-stomach sensations inspired by its fathomless views. The landscape of the sacred mountain easily matches the wildest fantasies of the kind of airy place where sages commune with the stars in heaven.[14]

The southern mountain—called Heng Shan like the northern peak, but written with different characters—lies south of the Yangtze River in the warm and fertile province of Hunan. It rises to a modest height of 4232 feet above paddies of rice flooded with water and bordered by clumps of bamboo. The mountain of the center, Sung Shan, stands in the central province of Henan, close to the cities where the ancient emperors of China reigned, southwest of the present-day capital of Beijing. Despite the meaning of its name, "The Lofty," its summit reaches to only 4902 feet above sea level. Although we might expect the central mountain of the Middle Kingdom—as the Chinese call their country—to occupy a privileged position, Sung Shan has not been singled out for particular distinction. The one peak that has instead commanded the place of honor among all the sacred mountains of China is the mountain of the east—T'ai Shan.

Located in the province of Shandong, halfway between Beijing and Shanghai, T'ai Shan rises to a modest altitude of only 5000 feet above sea level. Although some of the others, Hua Shan in particular, surpass it in both beauty and height, the position of T'ai Shan makes it the most important of the five principal sacred mountains of China. As the eastern peak, it receives the first light of the sun, the divine source of all life. In the complex system of associations that structures much of Chinese symbolism, the direction of the east also corresponds to spring, the season of fertility and renewal. The mysterious power that refreshes and revitalizes the body and spirit of all living things flows through the sacred peak, shining forth from its summit in the first magic glow of the dawn. For thousands of years Chinese writers have eulogized T'ai Shan as the supreme mountain, surpassing all others in spiritual height and significance. During the T'ang Dynasty, Tu Fu, regarded by many as China's greatest poet, wrote:

With what can I compare the Great Peak?
Over the surrounding provinces, its blue-green hue never
*   dwindles from sight.*
Infused by the Shaper of Forms with the soaring power of
*   divinity,*
Shaded and sunlit, its slopes divide night from day.

Breast heaving as I climb toward the clouds,
Eyes straining to follow birds flying home,
Someday I shall reach its peerless summit,
And behold all mountains in a single glance.[15]

Although not very high, T'ai Shan is the highest and most impressive mountain in eastern China. Formed millions of years ago from a weathered intrusion of granitic magma, it sits solidly on the earth with a sense of immovable power and majesty, like an emperor seated on his throne. Ridges embroidered with intricate patterns of pines and cedars spread out from its flanks like the folds of a ceremonial robe. Steep slopes of gray rock, cut here and there by precipitous cliffs, emerge from its lower ramparts to stand out against the sky. Set upon them like altars beneath the blue heights of heaven, the temples on top of T'ai Shan overlook a vast panorama of hills and plains that extends to the ocean. When clouds and mist obliterate the world below, the stark landscape of the summit plateau takes on the atmosphere of a much higher mountain, and one can see why tradition has given T'ai Shan its place of honor as the Great Peak of the Middle Kingdom.

Unlike the pristine peaks of higher ranges, such as the Himalayas, T'ai Shan is a very humanized mountain, bearing the physical marks of thousands of years of religious devotion. A great staircase of seven thousand steps runs from Tai'an, the City of Peace at the foot of the peak, to the Temple of the Jade Emperor on top. The slopes and summit of T'ai Shan are covered with innumerable shrines, temples, inns, and stalls for food and religious supplies. Every notable feature of the mountain, such as a ridge, boulder, or tree, has its own name—and often a monument to mark it. A great slab of rock that spreads over a hillside has the *Diamond Sutra* inscribed upon it in large Chinese characters. One of the most important works of Buddhist philosophy, the teaching enshrined in this text reminds the passing pilgrim of the ultimate nature of reality, symbolized in the empty space of the blue sky that opens overhead as he or she climbs toward the summit of T'ai Shan.

Although the legendary ruler Shun is said to have visited four sacred mountains, emperors of the historical period focused their attention on T'ai Shan, the only one of the four mentioned specifically by name in the *Shu-ching*. Those who felt confident enough of their accomplishments came to this mountain to perform the Feng and Shan sacrifices announcing the triumph of their dynasties' aspirations. Having reached the pinnacle of power and glory as the undisputed ruler of China, a successful emperor would ascend T'ai Shan to extol the merits of his ancestors and to thank the deities of heaven and earth for their help.

The Feng sacrifice to heaven he would perform at two altars: one at the base of the mountain to announce his intention to climb the peak and the other on the very summit, where he would make burnt offerings to the vast blue sky. The Shan ritual to earth would take place, appropriately, on a low hill near the foot of T'ai Shan. As a means of ensuring the longevity of his line, the emperor would have the declaration of his dynasty's accomplishments and the expression of his gratitude engraved on jade tablets. These tablets would be sealed in a jade chest inside a box of stone and left to remain for all time on the summit of the sacred mountain.[16]

Although seventy-two legendary emperors were supposed to have come to T'ai Shan to perform sacrifices, the first historical emperor to leave a physical record of having visited the sacred mountain was Ch'in Shih-huang-ti, the self-proclaimed "First Emperor of China," who initiated the construction of the Great Wall in the third century B.C. According to the Confucians, whom he persecuted, Shih-huang-ti ordered all the existing classics of Chinese literature burned — hence the difficulty we have in determining much of the early history and religion of China. Confucian sources also claim that he had most of the scholars who opposed his harsh rule killed by being planted in the earth with only their heads exposed. In 1974 a farmer plowing his fields made the startling discovery of an underground army of life-size statues guarding the approaches to the emperor's extravagant tomb — a mountain-shaped tumulus outside the present-day city of Xi'an.

Having brought China under the rule of his dynasty, the Ch'in, Shih-huang-ti decided to commemorate the event by performing a sacrifice on the summit of T'ai Shan. When the local scholars told him that he must climb the mountain on foot and conduct the ceremony in a simple and humble way, he shocked them all by driving up the slopes in his royal chariot. The literary records tell us that the scandalized officials stood by the path and shook their fingers at him in angry disapproval. On the way down from the summit, a sudden storm lashed him with wind and rain, forcing him to seek shelter beneath a nearby pine. The scholars claimed it was an expression of divine displeasure provoked by his impudence, but in defiance of their opinions, Shih-huang-ti conferred the honorary title of an Official of the Fifth Rank on the tree. A pavilion constructed in memory of the incident stands in a grove of trees said to have descended from the honored pine, forming one of the notable features of the pilgrimage route up T'ai Shan. The stone monuments that Shih-huang-ti left as evidence of his ascent in 219 B.C. make no mention of the Feng or Shan sacrifices: in keeping with his arrogant attitude, the monuments simply glorify his visit to the sacred mountain.

The first ruler to leave a stone inscription specifically mentioning the Feng and Shan sacrifices was Wu-ti, the powerful emperor of the Han Dynasty who was responsible for opening the ancient Silk Route linking China with the West. In 110 B.C. he ascended the mountain to perform the rituals under mysterious circumstances. The single retainer who accompanied him to the top died a few days after the ascent without revealing what transpired high in the clouds. Scholars have speculated that Wu-ti, who had a passionate interest in discovering the secret of immortality, performed a ritual of exorcism designed to transfer life-shortening ills from his own person to that of his loyal attendant. Strongly drawn by something about the sacred mountain, the emperor returned at regular intervals to perform the sacrifices in 106, 102, 98, and 93 B.C. A passage inscribed on a ceremonial mirror two centuries later suggests what it was that Wu-ti was seeking on the heights of T'ai Shan:

If you climb Mount T'ai, you may see the immortal beings. They feed on the purest jade, they drink from the springs of manna. They yoke the scaly dragons to their carriage, they mount the floating clouds. The white tiger leads them . . . they ascend straight to heaven. May you receive a never ending span, long life that lasts for ten thousand years, with a fit place in office and safety for your children and grandchildren.[17]

Over the centuries a number of emperors followed Wu-ti's example and ascended T'ai Shan to conduct the Feng and Shan sacrifices announcing the success of their dynasties. The last was Chen-tsung of the Sung Dynasty, who climbed the mountain in the year A.D. 1008. None of his successors had the confidence in their achievements to perform the powerful rituals: those who considered it changed their minds for fear of provoking calamity through an act of hubris. However, many of them took a deep interest in the sacred peak and visited it for other purposes, such as the construction and renovation of temples. The emperor K'ang-hsi of the Ch'ing or Manchu Dynasty came on pilgrimages in 1684 and 1689 and confirmed to his satisfaction that T'ai Shan formed the head of a dragon whose body twisted through China, disappearing beneath the land and water to emerge here and there in various mountains, beginning with a range marking its tail in his native land of Manchuria to the north. His grandson, the great Ch'ien-lung, was the last emperor to visit the mountain. He climbed it in 1748 and 1771 and, like many of his predecessors, left monuments on the summit of the sacred peak.[18]

Emperors also made it a practice to appeal to T'ai Shan for help in times of drought, flood, and earthquakes. They either invoked the mountain from afar or climbed its slopes to make offerings, praying for rain, dry weather, or a stable earth. Their appeals often carried a plaintive tone of reproach, holding T'ai Shan personally responsible for restoring the well-being of the Chinese empire, as we can see in the following passage from an imperial petition inscribed on a tablet in the fifteenth century:

If it is by my faults that I have attracted these calamities, I most assuredly will not refuse personal responsibility; but in so far as it is yours to transform misfortune into happiness, it is truly you, O God [of T'ai Shan], who has the duty to apply yourself to it. If there should be a fault of omission and you should not accomplish a praiseworthy act, you would be as guilty as I. If, on the contrary, you transform misfortune into happiness, who will be able to equal you in merit?[19]

The emperors of China regarded T'ai Shan as the son of the Emperor of Heaven, from whom they received their mandate to rule. He functioned for them as an important deity deputized to attend to the affairs of this world and to communicate their wishes to the supreme ruler on high. T'ai Shan acquired so distinct an official identity that court officials aspiring to high positions in heaven would actually give dying relatives petitions to convey to him.

As the earthly representative of the Emperor of Heaven, T'ai Shan became the greatest of all the terrestrial gods, over whom he ruled through an immense bureaucracy, patterned on that of the human imperial court. In the popular religion of the common people—which drew on Taoism, Confucianism, Buddhism, and shamanistic beliefs and practices—he assumed the role of divine arbiter of life and death. The Chinese believed that the souls of those who died went to a hill at the foot of the peak. There T'ai Shan himself passed judgment on the good and evil a person had done in his or her life. The expression "going to T'ai Shan" became a common euphemism for dying. As the peak of the east endowed with the power of dawn, the mountain was also regarded as the source and shaper of life. Through underlings occupying the maze of offices that made up his massive bureaucracy, T'ai Shan determined everything that would happen to a person—birth, position, honors, fortune, and death. Until the Communist Revolution in 1949, every village of any importance had a temple dedicated to the divine ruler of the sacred mountain—and one of the largest and most important temples in Beijing, the Temple of the Eastern Peak, was devoted to his worship.[20]

Because of its great importance in the life of the Chinese people, T'ai Shan has a longer record of ascents than any other mountain on earth. The annals tell us that from the mythical times of the legendary origins of the Chinese empire, more than four thousand years ago, people have been coming to pay homage to the deities of T'ai Shan. Emperors and peasants, nobles and serfs, priests and hermits have ascended the sacred slopes to stand in humility and awe beneath the all-encompassing dome of heaven. Perhaps no other mountain has seen and felt the feet of so many pilgrims. Before the Communist Revolution, during the height of the pilgrimage season in March and April, ten thousand a day would climb the peak to express their piety and receive blessings from the countless temples and shrines along the way. Even today, large numbers of Chinese continue to come to T'ai Shan, most of them ostensibly to enjoy the mountain as tourists, but the brooding presence of thousands of years of tradition suggests a deeper, religious motivation, of which they may or may not be aware.

The pilgrimage route up T'ai Shan passes through a series of three heavenly gates, each marking a transition to a higher and more sacred zone of the mountain. The First Heavenly Gate, a great archway of stone covered with a pagoda roof, marks the beginning of the actual climb. Here the pilgrims leave the gentle world of the plains behind and start up the stairway leading to the rocky heights of heaven. Only above the Second Heavenly Gate, halfway up the mountain, does the ascent become truly arduous. After meandering along a pleasant stretch of undulating ground, the path plunges into the Mouth of the Dragon—the entrance to a harsh gorge filled with boulders that twists dragonlike up to the summit plateau. With shaking legs and rasping breath, the pilgrims creep up two flights of over a thousand steps that tumble down from the South Gate of Heaven, a red arch with a golden roof, set like a ruby in a notch chiseled out of the skyline above. Having hauled themselves up to this, the most famous landmark on T'ai Shan, they emerge with relief on the Heavenly Way that runs gently on to the summit itself.

Not far from the South Gate of Heaven, the path comes to a temple marking the place where more than two thousand years ago the sage Confucius stood and surveyed the world below. Literary tradition claims that he had such sharp eyesight that he was able to pick out a white horse tethered to a city gate on the far horizon. His disciple Mencius wrote more realistically of the sage's experience:

The Mouth of the Dragon. Tourists and pilgrims ascend the stairway to the South Gate of Heaven on T'ai Shan. (Susan E. Thiele)

Confucius ascended the eastern hill, and the kingdom of Lu appeared to him small. He ascended T'ai Shan, and all beneath Heaven appeared to him small.[21]

The view from the heights of the sacred mountain was so awesome that it reduced the human world and all its accomplishments to nothing.

Just beneath the highest point, nestled in a hollow sheltered in mist, rests the temple of one of the most popular deities of the sacred peak—the Princess of Azure Clouds. The daughter of the deity of T'ai Shan, she is the goddess of the dawn, embodying the delicate grace and beauty associated with the birth of each new day. Women whose daughters have been unable to conceive climb up to her temple on top of the mountain to pray for grandchildren. Two goddesses stand beside the princess as attendants in her sanctuary. One heals the eyes of those who have lost their sight, filling their ruined vision with the glory of the

The Princess of Azure Clouds, goddess of the dawn, in her temple near the summit of T'ai Shan. Women make offerings to her for the birth of grandchildren. (Anna Seidel)

morning sun. The other cures children of diseases, infusing their bodies with the healthy vigor of the mountain air.

In the thirteenth century, women in quest of the princess's blessings initiated the practice of climbing T'ai Shan as a ritual of popular pilgrimage. Before that time the upper reaches of the mountain had functioned primarily as the special preserve of the imperial cult, visited only by a restricted number of people. Many of the pilgrims who climb T'ai Shan today are old peasant women who make the temple of the Princess of Azure Clouds the focus of their ascent. They come to burn offerings of paper money before the images of the goddess and her attendants. In 1984 the local authorities tried to ban this practice by stopping elderly ladies at the base of the mountain and confiscating any notes they could find in their pockets and purses. The women got around this obstacle by stuffing extra money in the padding of their quilted jackets and proceeding up to the temple, there to continue making offerings for the benefit of their families.[22]

On the very top of T'ai Shan sits the Temple of the Jade Emperor, the heavenly ruler of this world. Outside this temple stands a blank tablet of stone, its inscription mysteriously effaced—or never written at all. Some say it commemorates the ascent of Shih-huang-ti in the third century B.C.; others say that of Wu-ti in the second century. Whichever it may be, the ancient monolith speaks eloquently of the thousands of years of religious and historical tradition imbedded in the rock of T'ai Shan. A stone railing within the courtyard of the temple encloses the rounded boulder that forms the summit of the sacred mountain. The pagoda roofs that soar above the actual peak reflect the way in which the human spirit has elevated T'ai Shan to the highest place, loftier than other mountains whose summits may lie higher above the sea, but farther from heaven.

Just to the east of the summit extends a barren promontory called the "Peak Where One Contemplates the Sun." Starkly outlined against the sky, a gray pillar set upright upon an altar of weathered stone marks the ancient site where emperors would ascend to perform the Feng sacrifice announcing to heaven the glories of their accomplishments. When a cold wind blows streaks of mist across the summit, it seems to obliterate all memory of their transitory triumphs, now vanished in the ruins of time. Not far from the altar, a sheer precipice on the eastern edge of the summit drops off into the void; it used to bear the grim name of Suicide Cliff. There people would come to perform the ultimate sacrifice by hurling themselves off its edge. Some chose this course as an expedient way to end their miseries and find instant communion with the Jade Emperor; others used it as a means of expressing filial devotion by offering their lives in exchange for the health and longevity of their parents. In the sixteenth century a local governor attempted to discourage such practices by building a retaining wall and renaming the precipice the Love of Life Cliff, the name it has today. Four stones with large Chinese characters announce that "it is forbidden to commit suicide." One wonders what possible penalty the state bureaucracy could impose on those who leap beyond the reach of human laws.[23]

Whether seeking life or finding death, the people of China have embraced T'ai Shan as the sacred mountain most intimately bound up with the hopes and fears of their everyday lives. A poem composed by the wife of a Chinese general in A.D. 400 beautifully expresses the deep feelings of mystery and devotion that this peak has inspired over the millennia:

High rises the Eastern Peak
Soaring up to the blue sky.
Among the rocks — an empty hollow,
Secret, still, mysterious!
Uncarved, and unhewn,
Screened by nature with a roof of clouds.

Time and Seasons, what things are you,
Bringing to my life ceaseless change?
I will lodge forever in this hollow
Where springs and autumns unheeded pass.[24]

Wu-t'ai Shan and the Four Buddhist Mountains

Partly in imitation of the ancient model of the five principal peaks headed by T'ai Shan, the Buddhists of China developed their own scheme of four sacred mountains situated at the four points of the compass: P'u-t'o Shan to the east, Wu-t'ai Shan in the north, O-mei Shan to the west, and Chiu-hua Shan in the south. Mountain sanctuaries that had previously attracted the attention of Taoist hermits became focal points of Buddhist meditation and pilgrimage. Great monastic complexes, some of them rivaling the palace of the emperor in the beauty and magnificence of their art, blossomed on the sites of rustic retreats. Whereas pilgrims in India and Tibet circumambulated holy peaks, such as Mount Kailas, which they treated as shrines and temples, their counterparts in China adopted the indigenous Chinese practice of climbing to the top of sacred mountains, as the emperors did to perform sacrifices on the summit of T'ai Shan.[25]

Each of the Buddhist peaks is regarded as the seat of a particular Bodhisattva — a spiritual being who has dedicated himself or herself to the task of helping all living creatures transcend suffering and attain the ultimate goal of enlightenment. Devout pilgrims seeking aid and comfort amid the miseries of life in this world may meet Kuan-yin, the female Bodhisattva of Compassion — or experience her living presence — on P'u-t'o Shan, the mountain of the east. A rocky hill jutting up 932 feet above the surrounding sea, her sacred abode lies on a small island in Zhejiang Province in southeastern China. Originally a male deity known as Avalokiteshvara in India and Tibet, the Bodhisattva changed sex after reaching China, becoming a savior figure endowed with motherly attributes.

Buddhists concerned with the fate of deceased members of their families journey to the southern mountain of Chiu-hua Shan, the "Nine Flower Mountain," which rises to a height of 4400 feet in Anhui Province. There, in a setting of spectacular peaks reminiscent of the polished pinnacles of Hua Shan, pilgrims make petitions to Kshitigarbha, the "Earth Womb" Bodhisattva, who has chosen to descend into the underworld to rescue those who have sinned from the torments of their self-inflicted damnation. The ascent of the mountain, through a labyrinth of gorges and ridges, involves the powerful symbolism of a journey through hell in quest of salvation — for oneself and others.[26]

The highest of the four Buddhist mountains is the western one, O-mei Shan. A mountain of awe-inspiring precipices, it rises to 10,167 feet above the fertile valleys of Sichuan Province, within sight of the great snow peaks that form the eastern rim of the Tibetan Plateau. There, among the mists that drift from peak to peak, Samantabhadra, the "All Good" Bodhisattva, rides on a white elephant, performing good works for the benefit of all living beings. Pilgrims who climb to the heights of O-mei Shan often obtain the treasured blessing of seeing his image, or that of the Buddha himself, projected as a glorious halo on the golden clouds of dawn.

The first of the four peaks to be established as a Buddhist pilgrimage site, and in many respects the most important, is the northern one — Wu-t'ai Shan, the "Five Terrace Mountain." Regarded as the abode of Manjushri, the Bodhisattva of Wisdom, its sacred precincts lie hidden in a welter of grassy ranges near Heng Shan in the province of Shanxi. More a mountainous region than a distinct mountain, Wu-t'ai Shan actually encompasses a number of separate peaks. From a maze of mountains that confuse the eye, Buddhists have singled out five with bare, flat-topped summits that look like terraces — hence the name Five Terrace Mountain. Arranged symbolically in the pattern of a mandala or sacred circle, the peaks bear the names of the four quarters and the center. They range in altitude from the lowest, the Western Terrace, at 8530

feet, to the highest, the Northern Terrace, at 10,033 feet. Even in summer snow occasionally falls on the higher peaks, giving Wu-t'ai Shan its reputation of being extraordinarily cold. Rising like great altars fashioned of rock and earth, the five terraces enclose a grassy plateau of high valleys about 170 miles in circumference. The main valley runs between the peaks for more than 10 miles and at one time held, along with its tributary valleys, more than three hundred monasteries and temples. Chinese texts sometimes describe the whole complex as having the shape of an enormous hand with the central region viewed as the palm surrounded by the five terraces regarded as four upright fingers and the thumb.

In fact, as a sacred site, Wu-t'ai Shan feels more like a sanctuary enclosed by mountains than a mountain itself. John Blofeld, a Western Buddhist who made the pilgrimage to Wu-t'ai Shan just before World War II, describes his first impressions on reaching it after an arduous journey:

At last, gasping and sweat-sodden, I reached the pass in the company of a few other stragglers. We found ourselves looking down on a sight which might have inspired the original conception of Shangri-La. The wide, grassy plateau lay only a few hundred feet below the pass. Wild flowers grew in such extraordinary profusion that the old cliché "carpeted with flowers" seemed the most apt description possible. Here and there, nestling against the surrounding slopes or clinging to overhanging rocks were the monasteries, some large enough to house hundreds of monks, others small temples with only three or four living rooms attached. . . . Never, even upon the flowery slopes of the Dolomites, had I seen a sight so lovely nor have I beheld its equal since, unless in some of the high Himalayan Valleys.[27]

Blofeld may have made a more apt comparison than he imagined with Shangri-La, the idyllic sanctuary hidden among mountains in the novel *Lost Horizon*. An old Chinese tradition holds that in A.D. 309 about a hundred families fled to Wu-t'ai Shan, seeking a similar kind of refuge from the turmoil of the outside world. There they established a peaceful abode far from the warfare and pillage racking most of China. The few travelers who happened to glimpse the sanctuary in the distance found it impossible to reach when they tried to approach it. No one knew the way there. "Therefore," an old text tells us, "people consider this mountain a capital of immortals."[28]

Before Buddhists identified the mountain as the abode of Manjushri, the Bodhisattva of Wisdom, Taoists regarded Wu-t'ai Shan as the dwelling place of im-mortals and other heavenly spirits. According to a Taoist text:

The name of Wu-t'ai Shan is "Purple Palace." A purple haze constantly emanates from the mountain. Immortals live there.[29]

Along with the haze described in this passage, a number of unusual features attracted attention and set Wu-t'ai Shan apart as a sacred place inhabited by divine spirits. Early visitors remarked on the beauty of the flowering meadows—as Blofeld would centuries later—and the fact that blossoms appeared even in winter. Others noted a spring of marvelously clear water on top of the Central Terrace that always remained mysteriously full even though no one was able to discern its source. Like tracings of fine brocade, beautiful flowers grew among the stones that shone like crystals in the luminous depths of the spring's clear clean water.

Such features predisposed people to having visions of spiritual beings, whose presence confirmed the sanctity of the mountain. When Wu-t'ai Shan became a Buddhist peak, sometime around the sixth century A.D., these visions acquired the forms of various manifestations of Manjushri, the Bodhisattva of Wisdom. He would sometimes appear to pilgrims as an aged monk of great insight or as a youthful prince seated on a white lion, roaming the ethereal pathways of the sky. At other times he would assume the appearance of rainbow-colored clouds hovering about the summits of the sacred peak or manifest himself in the form of mysterious balls of light floating up the sides of ridges. He might also appear as a dragon playing among the clouds. Such visions became so commonplace that many pilgrims, lay people as well as monks, went to Wu-t'ai Shan fully expecting to see Manjushri and to receive blessings and teachings from the Bodhisattva himself. The writings of an eighth-century master give a vivid picture of the powerful impression that the mountain made on Buddhists of his day:

The splendid display of its resonant qualities fills the eyes and ears—and even so there is more of excellent nature. Dragon palaces in mountain pools open up at night to reveal a thousand moons. Fine and delicate grasses spread out in the mornings among hundreds of flowers. Sometimes there are ten thousand sages arrayed in space. Sometimes five-colored clouds settle in gaps between the hills. Globes of light shine against the serene and peaceful mountain. Auspicious birds soar in the hazy empyrean. One merely hears the name of Manjushri and no longer is beset by the cares of human existence.[30]

Monks in snow on O-mei Shan, one of the four Buddhist mountains of China. (Susan E. Thiele)

Manjushri and his mountain had such power because Buddhists regarded—and still regard—him as the embodiment of the highest transcendent wisdom and the blessings that flow from it. The sword that he brandishes in most of his forms represents the sharp awareness that slices through illusion to reveal the ultimate nature of reality. This direct and intuitive wisdom liberates people from ignorance and protects them from evil. Invoking the aid of the Bodhisattva as a protector of the state as well as the individual, the T'ang Dynasty emperor T'ai-tsung had a temple built in his capital called the Pavilion of the Great Holy Manjushri for the Protection of the Nation. The power of this protection extended even to the heavens. When a comet appeared in the sky, auguring ill for his rule, T'ai-tsung dispatched the tantric master Amoghavajra to Wu-t'ai Shan to perform rituals intended to dispel the evil omen. Amoghavajra's success in this undertaking reinforced yet another view of Manjushri as lord of the cosmos, able to liberate people from the baleful influence of the stars.[31]

According to texts from India translated into Chi-

Manjushri, Buddhist deity of Wu T'ai Shan, with the sword of wisdom symbolizing the spiritual insight that cuts off ignorance and illusion. (Edwin Bernbaum)

nese, the Buddha prophesied that in the future, when the force of his teachings would be nearly spent, Manjushri would appear as a youthful prince on a five-peaked mountain in China to reinvigorate Buddhism with the spiritual power of his wisdom. Drawing on such prophecies, Chinese Buddhists in the sixth century identified Wu-t'ai Shan as that mountain—the place in the world most conducive to the practice of meditation and the attainment of enlightenment. The first of the Buddhist sacred mountains to be established, it became in the eighth century a major center of Buddhism in China. Many famous teachers, such as the tantric master Amoghavajra, came to visit the mountain and meditate in the illuminating presence of Manjushri. During the T'ang Dynasty, between the seventh and tenth centuries, when Buddhism reached the height of its influence in China, no other peak came close to equaling the importance of Wu-t'ai Shan in the eyes of Chinese Buddhists. The great emperors of the period, T'ai-tsung in particular, dedicated vast amounts of money to sponsoring monasteries on the sacred mountain—a number of them built on the sites of visions that particular individuals had of magic cloisters inhabited by supernatural monks. Rulers of succeeding dynasties continued to contribute funds for the renovation and support of these monastic complexes. Although in later periods other Buddhist peaks, such as O-mei Shan and P'u-t'o Shan, became more popular as places of pilgrimage, none of them ever surpassed Wu-t'ai Shan in the magnificence of its monasteries and temples, some of which rivaled in splendor the imperial palaces of the capital itself.[32]

Because of its reputation as the prophesied abode of Manjushri, Wu-t'ai Shan became the Chinese mountain of greatest significance for Buddhists outside of China. Tibetans and Mongolians came to regard it as one of the Five Great Places of Pilgrimage, a group that included the most important place in the Buddhist world—Bodhgaya, the site of the Buddha's enlightenment in India. Before World War II, as a result of the interest taken by the Ch'ing or Manchu Dynasty in Tibetan Buddhism, the largest and most active monasteries at Wu-t'ai Shan were run by monks from Mongolia and Tibet. During the summer months great processions of lamas in ornate robes would issue from these monasteries, descending a great staircase of 108 steps to wind in lines of dazzling color across meadows of flowers spread between the sacred peaks. Surrounded by thousands of Mongolian and Tibetan pilgrims in native dress, an observer would have to think hard to remember that the pageant he or she was witnessing was taking place in China, rather than Tibet.

Wu-t'ai Shan has also occupied a place of great importance for the people of Japan. From the time it

A Buddhist monk stands on the Southern Terrace of Wu T'ai Shan, gazing out over the peaks and valleys of the sacred mountain. (Raoul Birnbaum)

became sacred for the Buddhists of China, a number of well-known Japanese pilgrims journeyed to the mountain in search of Buddhist teachings. In a famous episode enshrined in a Japanese No play, one of these pilgrims comes to the foot of Wu-t'ai Shan. A magic bridge that can be crossed only with divine assistance leads to the paradise of Manjushri. As the pilgrim watches in awe, the Bodhisattva appears, dancing through a field of flowers in the form of a lion with a mane of red hair forming an aura around his golden face.[33]

Ennin, a well-known Japanese monk and traveler whose visit to Wu-t'ai Shan in A.D. 840 inspired the No play, wrote a detailed description of the mountain in his diary. The following extract reveals the profound feelings of spiritual equality and respect for all beings that Wu-t'ai Shan inspired in Ennin and his fellow pilgrims:

When one enters this region of His Holiness [Manjushri], if one sees a very lowly man, one does not dare to feel contemptuous, and if one meets a donkey, one wonders if it might be a manifestation of the Bodhisattva. Everything before one's eyes raises thoughts of the manifestations of Manjushri. The holy land makes one have a spontaneous feeling of respect for the region.[34]

Most pilgrims who come to Wu-t'ai Shan today focus their attention on the monasteries and temples clustered in the central valley, but a hardy few also climb to the summits of the surrounding peaks, where they have a better chance of witnessing the mysterious phenomenon of globes of light drifting eerily through the night sky—similar in nature, perhaps, to the kind of electrical discharges seen in Saint Elmo's fire. The changeable weather of the sacred mountain, however, can make the ascent of these peaks uncomfortable and even dangerous. When Blofeld made the climb in 1938, a sudden gale hurling horizontal sheets of rain pinned his party down in a temple for two days until one of his Chinese companions who had neglected to make an offering to the dragon deity at the foot of the peak ran off in the downpour to slither down slopes of mud and rectify his mistake. More recently, in 1986, Raoul Birnbaum, an American scholar doing research on Wu-t'ai Shan, became lost in a deadly storm of rain and icy fog that enveloped the peak he was climbing. Only by chance did he and

Pines, rocks, ridges, and mist compose an evocative Chinese landscape on Huang Shan, a mountain favored by poets and painters. (Albert E. Dien)

his companion, an elderly monk, find the way to the weather station on the summit and avoid freezing to death in the middle of summer. The weather men on top told them that if they had arrived a few minutes earlier they might have been killed by enormous hailstones that had almost battered in the roof. The two were so soaked that the monk had to take off his robe and wring streams of water out of it. After they came down from the peak, the head of the Buddhist association at Wu-t'ai Shan told Birnbaum that the rain and storm had been Manjushri's way of purifying him by washing out his bad karma. Before this incident the Buddhist official had looked askance at the foreign scholar, but now he took a liking to Birnbaum and helped him with his research.[35]

As places of universal sanctity, rising above the reach of anyone's exclusive grasp, mountains have become associated with ideas and beliefs shared by different traditions. The principle of yin-yang, a concept basic to most systems of Chinese thought, derives from the image of a mountain. The terms *yin* and *yang*

originally referred, respectively, to the shaded and sunlit sides of a peak. Over time, through the ideas they called forth, beginning with the images of shadowy valleys and bright summits, these terms came to denote the complementary opposites whose union creates the world—darkness and light, moisture and dryness, female and male, nonexistence and existence. In Confucianism the way to bring order to society and in Taoism the way to put oneself in harmony with the Tao is to balance these opposites in every part of one's life—from the food one eats to the thoughts one thinks.

Mountains also play an important role in *feng-shui*, the Chinese system of geomancy that determines the orientation and nature of the place in which one lives. According to this system of belief and practice, which has spread to other countries in East Asia, the form of a hill or peak directs a flow of energy that can influence the atmosphere of the local region, shaping the character and views of its inhabitants. In Korea, for example, Buddhist monasteries have nearby "host" and "guest" mountains that determine the way the monasteries treat their visitors. Tong-do Sa near Pusan has a large host peak and a small guest mountain. The monastery, therefore, has a reputation for being closed to outsiders, regarding them with suspicion and even hostility. The host mountain of Hae-in Sa near the city of Taegu, on the other hand, lies off in the distance and appears very small in comparison with the nearby guest peak, which dominates the view. As a consequence, the monastery welcomes guests with great courtesy, and the monks even treat each other as transient visitors whose homes lie elsewhere.[36]

Despite official efforts to replace religious with secular values, traditional views of sacred mountains continue to influence the ways in which the Chinese people regard themselves and the land on which they live. When the government relaxed restrictions on the practice of religion at the end of the Cultural Revolution, peasants thronged by the thousands to T'ai Shan to give thanks to the mountain. In the winter of 1982 Japanese newspapers reported the bafflement of local officials who after years of Communist rule had never expected such a reaction and did not know how to handle it. According to these reports, the officials had turned to Beijing for advice on how to deal with the situation, which had become a political problem threatening to undermine confidence in the secular values of the state. The authorities' response seems to have been to try to cover up the resurgence of religious sentiments by encouraging the development of sacred mountains as places of tourism rather than pilgrimage. A cable car built by a Japanese firm now runs to the summit of T'ai Shan, making it a major attraction for foreign and local visitors, and the government is busily transforming Wu-t'ai Shan into a resort for the rest and recreation of workers.

Traditional views of sacred mountains have even influenced officially sanctioned thinking about modern mountaineering. When in 1960 Chinese mountaineers succeeded in reaching the summit of Mount Everest—the highest mountain in China, as well as the world—an official account declared, "Summing up our conquest of Everest, we must in the first place attribute our victory to the leadership of the Communist Party and the unrivalled superiority of the socialist system of our country. . . . " Along with Western ideas about the conquest of nature, these words recall the reason for which the ancient emperors climbed T'ai Shan, the most important peak in China—to perform sacrifices bearing witness to the success of their dynasties. Here, as in the past, the ascent of the mountain is an act of sacred politics, declaring to the world the triumph of the current regime.[37]

A passage from a government book describing a subsequent climb of Mount Everest in 1975 even makes mountain climbing a kind of sacred mission undertaken for the purpose of realizing the highest values and aspirations of the socialist system:

New China promotes mountaineering as a sport to serve proletarian politics, the interests of socialist economic construction and the building of national defense, to help improve the people's health, and to foster such fine qualities in them as wholehearted devotion to the people and the collective, and fearing neither hardship nor death. In the recent expedition to Qomolangma [Mount Everest], the climbers, united as one, helped each other and gave full play to their collective strength.[38]

The ascent of the mountain symbolizes the fulfillment of the people's dream: the transformation of China into a paradise of the proletariat.

Although the particular visions they inspire may have changed, mountains continue to awaken a sense of the sacred that for thousands of years has nourished the spirit of the Chinese people and put them in harmony with the world in which they live. The words that a Western scholar of Chinese art used to refer to the artist or poet of the past remain true of the worker today: "By climbing the hills and looking out over range upon range of peaks he discovers man's true place in the scheme of things."[39]

· 3 ·

CENTRAL ASIA
THE DISTANT RANGES

NORTH OF THE HIMALAYAS and the Tibetan Plateau, thousands of miles from the nearest ocean, rise the remote and mysterious mountains of Central Asia, shimmering like mirages on the distant horizon. The K'un-lun, the T'ien Shan, the Pamir, the Altai, ranges whose names conjure up visions of faraway places, stretch off in long ridges of snow peaks to waver and vanish in clouds of dust swept up from two of the harshest and most forbidding deserts on earth—the Gobi and the Taklamakan. Older than the Himalayas, and nearly as high, these little-known ranges form some of the most formidable barriers in the world, folded and squeezed up in great walls of metamorphic and sedimentary rock, topped here and there by impregnable towers of granite. Running from east to west across western China and the edge of Soviet Central Asia, the K'un-lun and the T'ien Shan enclose the desiccated heart of the continent—the Tarim Basin, an oval-shaped region filled with the golden dunes of the Taklamakan Desert.[1] The two ranges converge in the Pamir, a tangled complex of valleys and peaks that seals off the western end of this empty basin, on the Chinese border with Afghanistan and the Soviet Union. North and east of the T'ien Shan, the Altai and other, lesser mountains break the monotony of the Mongolian steppes.

Across the middle of Central Asia, skirting the northern and southern edges of the Tarim Basin, runs the Silk Route, a system of ancient caravan tracks linking China to India and the West. For more than two thousand years, tenuous lines of oases strung along the feet of the K'un-lun and the T'ien Shan have made it possible for merchants, pilgrims, and explorers to undertake some of the longest and most difficult

journeys on earth. In 139 B.C. the Chinese emperor Han Wu-ti sent an envoy named Chang Ch'ien west to negotiate an alliance with a Central Asian people called the Yüeh-chih against the Hsiung-nu, nomadic warriors of Mongolia whose attacks had prompted the construction of the Great Wall of China. Although Chang Ch'ien failed in his diplomatic mission, he did manage to cross the Pamir Mountains and open the Silk Route. Chinese historians speak of him as "having made the road." Merchants who followed the path of Chang Ch'ien's epic journey initiated the exchange of silk and other goods between China and the empires of India, Persia, and Rome.[2]

Carried by traders and pilgrims, religious and cultural ideas also traveled along the Silk Route. Starting in the first century A.D., Buddhism entered China by way of the oases of Central Asia, populated at the time by Indo-Europeans, Chinese, and others. During the following centuries merchants and missionaries spread the teachings of Nestorian Christianity and Manicheism—a religion from Iran—eastward on the Silk Route. Major centers of learning, many of them organized around meditation caves carved out of cliffs, developed in oases such as Kashgar, Khotan, Turfan, and Tun-huang.[3] Paintings and sculpture found in Central Asian ruins of the first millennium A.D. show a mix of Indian, Greek, Iranian, and Chinese influences. In the seventh century the most famous of all Chinese travelers, the monk Hsüan-tsang, followed the Silk Route west on a journey of sixteen years to obtain Buddhist teachings from India. In the account he wrote of the places he visited, he describes some of the uncanny impressions made by the awesome deserts and mountains of Central Asia:

Kirghiz silhouetted in front of a snow peak in the Pamir Mountains. (Galen Rowell/Mountain Light)

Rappelling into ancient Buddhist caves carved in a cliff on the Silk Route outside the oasis of Kashgar. (Edwin Bernbaum)

At times sad and plaintive notes are heard and piteous cries, so that between the sights and sounds of this desert men get confused and know not whither they go. Hence there are so many who perish in the journey. But it is all the work of demons and evil spirits.[4]

Impressed by the divine as well as the demonic aspects of the landscape, the Chinese named the great mountain range that runs along the northern rim of the Tarim Basin the T'ien Shan or "Mountains of Heaven." There they envisioned a heavenly abode of Taoist immortals.

Conquests and migrations sent entire peoples surging in waves across Central Asia. Around the beginning of the first millennium A.D., a branch of the Hsiung-nu who had threatened the Chinese empire under Han Wu-ti vanished toward the west to reappear in the fourth century as the Huns in Europe. During this period the Chinese established settlements and constructed beacon towers along the Silk Route. After the end of Hsiung-nu dominance of Central Asia, various groups speaking Turkish languages succeeded one another in Mongolia. In the ninth century the Kirghiz, one of these groups, drove the

Uighurs, another Turkic people, out of the region, forcing a number of them to move to the Tarim Basin, where their descendants form the largest group living in the area today. During the eighth century the Tibetans came out of their mountains to seize control of the Silk Route from the Chinese and hold such oases as Tun-huang for almost a hundred years. In the tenth century the Kirghiz were pushed out of Mongolia and drifted slowly through the mountains of Central Asia, reaching the Pamir perhaps four hundred years ago. Between the eighth and twelfth centuries, Muslim invaders from the west destroyed Buddhist monasteries along the Silk Route and converted most of the people to Islam. In the thirteenth century the Mongols under Genghis Khan[5] overran the oases of the Tarim Basin on their way to establishing the greatest empire in history.

During the rule of Kublai Khan, a grandson of Genghis, the Italian merchant Marco Polo followed the Silk Route to China. His description of the deserts and mountains he traversed in the thirteenth century reads remarkably like that of Hsüan-tsang, written six hundred years earlier. Following in the footsteps of Marco Polo and Hsüan-tsang, European explorers around the beginning of the twentieth century crossed the great ranges of Central Asia to make some of the most important archaeological discoveries of recent times. In 1907 the British-Hungarian archaeologist Sir Aurel Stein came across a cave at the oasis of Tun-huang that yielded the oldest printed book discovered up to that period—a Buddhist text on the philosophy of emptiness pressed from a wood block in A.D 868. Tracking down legends of cities magically engulfed by the dunes of the Taklamakan Desert, he and the Swedish explorer Sven Hedin unearthed sand-buried ruins of settlements nearly two thousand years old. Watered by the melting snows of the K'un-lun, these dwellings had been abandoned as mountain glaciers receded and the rivers flowing from these glaciers gradually dried up. Hedin described the excitement and magic he felt at discovering one of these sites:

No explorer had an inkling, hitherto, of the existence of this ancient city. Here I stand, like the prince in the enchanted wood, having wakened to new life the city which has slumbered for a thousand years.[6]

Suspended in haze at the limits of the imagination, rising over the mysterious traces of ancient civilizations, the mountain ranges of Central Asia lend themselves naturally to myth and legend. For nearly a thousand years, the people of Tibet have looked north in the direction of these distant ranges for the mythical kingdom of Shambhala, an earthly paradise hidden, like the Tarim Basin, behind a ring of snow mountains. There, in a setting of unmatched splendor and beauty, a line of divine kings is said to be guarding the highest Buddhist teachings for a time in the future when wars will ravage the earth and destroy Buddhism outside of Shambhala. Then, according to ancient texts, a great king, much like the Messiah of Judeo-Christian prophecy, will emerge from behind the veil of peaks hiding his sanctuary to defeat the forces of evil and establish a golden age throughout the world. Guidebooks written in the Tibetan language describe the way to Shambhala, a long and perilous journey that only yogis endowed with supernatural power and spiritual insight can hope to accomplish. Those who manage to reach the hidden sanctuary will find what they need to attain the ultimate goal of enlightenment for the sake of all.[7]

The great civilizations of India and China have also turned their gaze toward the mysterious ranges of Central Asia, where their ancient texts fix the location of two of their most important mythical mountains. According to the *Mahabharata,* the great epic of Indian literature composed between the fourth century B.C. and the fourth century A.D., Mount Meru lies somewhere in that direction, far to the north of India, at the center of the world, directly beneath the immovable point of the North Star. Gleaming with gold and precious jewels, the mountain soars to unimaginable heights, forcing the sun and moon to circle around it in homage to its undisputed splendor. On its lofty summit, high above four sacred rivers that issue from its foot, rests the divine seat of Brahma, creator of the universe. Looking north from the Himalayas, a sage in the *Mahabharata* points toward the mountain and says to his disciple, a king who embodies the virtues of righteousness, "Behold the pure land, the superb peak of Meru, where the Grandfather dwells with the Gods, who are content with their souls."[8] There, he tells his royal listener, runs the path of those who seek the ultimate truth and attain the highest goal of the Hindu religion—release from bondage to the painful round of life and death. Vanishing into the dazzling brilliance of the supreme deity, becoming one with the essence of all that exists, they never return.

Some Indian scholars have attempted, without much success, to identify Meru with one of the ranges of Central Asia, in particular the Pamir, from whose central knot of ridges and valleys the ranges of the K'un-lun, Karakoram, Hindu Kush, and T'ien Shan radiate out like the petals of an enormous flower. Such a configuration of mountains would accord with the traditional conception of Meru as the center of a mandala in the form of a lotus blossom embracing the world. As a consequence of this conception, paintings of the cosmic mountain show it widening toward the summit, like a flower. Because of a preoccupation with symbolism, descriptions of Meru found in traditional texts make its precise location difficult, if not impossible, to ascertain; indeed, the peak functions in both Hindu and Buddhist thought primarily as a cosmic axis outside space and time—the still point around which everything turns and takes its place in the order of existence.[9]

Looking west from China, the ancient Chinese saw another mythical mountain in the place of Mount Meru. According to Taoist texts, in the desert regions of the Silk Route, hidden on the awesome heights of the K'un-lun, lay a paradise ruled by a goddess of immortality. There, close to heaven, near the source of the Yellow River, the immortals of Taoist mythology were said to dwell in perpetual bliss, free from human cares and concerns. Only those who had mastered the secret of immortality could penetrate the desert wastes that concealed the gleaming peak of this lofty paradise.

K'un-lun Mountains

Of all the mountains in Central Asia, the most remote, desolate, and forbidding, as well as the least known, are those of the K'un-lun. A system of ranges longer

than the Himalayas and nearly as high, its windswept ridges, devoid of all but the most tenacious shreds of desiccated vegetation, run between the icy expanses of the Tibetan Plateau on the south and the arid dunes of the Taklamakan Desert to the north. So barren and hostile are the mountains that even nomads avoid their cold and sterile valleys. The caravan routes that come close to the range skirt its northern edge, jumping from the safety of one oasis to another. Only a few explorers and scientists have ventured into the inner reaches of the K'un-lun, where no one lives but the wind—and wild yaks, asses, and other animals hardy enough to survive its icy blast. The highest peak of the range, Ulugh Muztagh, 23,923 feet, was not climbed until 1985, when a Sino-American expedition made the first ascent, backed by the resources of the Chinese army.

According to ancient Chinese tradition, perched on a mythical mountain that rises above the barren heights of the K'un-lun lies a magnificent palace of jade surrounded by ramparts of gold. Around the base of this mountain, cutting off all access to ordinary mortals, flows a river with magic waters so insubstantial that they cannot support the weight of a feather. There, high among heavenly peaks hung with gardens of fragrant pine, Hsi wang-mu, the Queen Mother of the West, dwells at ease in the company of Taoist immortals. Beyond the reach of worldly concerns, they pass their days in endless delight, enjoying the purest pleasures of body and spirit. The Chinese *Classic of Mountains and Seas* describes the paradise of Hsi wang-mu in the following terms:

There is the country of satisfaction, which satisfies its people. In this place are the Fields of Satisfaction. Phoenix eggs are their food and sweet dew is their drink; everything that they desire is always ready for them.[10]

Beneath the palace, near a magic fountain made of precious stones, grows a tree with the peaches of longevity. Every six thousand years, when its flowers bear fruit, the immortals gather beside it on the shore of a jeweled lake to celebrate the birthday of Hsi wang-mu. Regaled with celestial music and song, they dine on such delicacies as bear's paws, dragon's liver, and phoenix marrow and partake of the peaches that enable them to live forever.[11]

In *The Journey to the West*, a famous sixteenth-century Chinese novel inspired by the travels of Hsüan-tsang to India, the peach tree and banquet are transposed to the court of the Jade Emperor in hea-

ven. In an attempt to quell the main character, a rambunctious monkey who has attained immortality through Taoist practices, the heavenly officials make him custodian of the peach tree of longevity. When he discovers that he has not been invited to the party along with the other immortals, he gobbles up all the peaches and ruins the banquet, much to the dismay of Hsi wang-mu, who has been waiting six thousand years to celebrate her birthday. Unable to control him with the forces of heaven, the Jade Emperor finally appeals to the Buddha himself, who imprisons the monkey under a mountain for five hundred years. As penance for his sins, he must wait for the Chinese pilgrim Hsüan-tsang to appear and then help him follow the Silk Route to India in order to bring the Buddhist teachings back to China.[12]

In the earliest Chinese texts Hsi wang-mu has nothing to do with the palace of the immortals or the peaches of longevity. She dwells instead on a jade mountain north of the K'un-lun and west of the Moving Sands, where she appears in a terrifying human form with a leopard's tail, the fangs of a tiger, and wildly disheveled hair. Rather than bestowing eternal life, she sends epidemics and other calamities to punish people for transgressions against heaven. Later sources, beginning in the Han Dynasty around the second century B.C., transform her from a hybrid hag into a gracious lady with a white head and a jade comb placed neatly in her hair. A queen of considerable refinement, she entertains visiting emperors with song and conversation, offering them the secret of long life. The transformation of Hsi wang-mu from an ogress of disease into a goddess of immortality coincides with a shift in the location of her dwelling place from a cave in a mountain to the north, perhaps in the T'ien Shan, to a palace on the heights of the K'un-lun to the south.[13]

Under the influence of Buddhism, which entered China in the first century A.D., the paradise of Hsi wang-mu on the K'un-lun underwent yet another transformation. The mythical mountain, which the ancient Chinese had tended to regard as a cosmic axis, took on the form of a palace with nine levels and became identified with Sumeru, the peak at the center of the Buddhist universe. Four rivers corresponding to the four rivers issuing from Sumeru in Indian mythology flowed out from the base of the K'un-lun to water the four points of the compass. Partly as a result of interest aroused by the opening of the Silk Route, the mysterious region of Central Asia to the

Mount Sumeru opens up like a flower, surrounded by concentric rings of mountains and oceans with the palace of Indra on its summit. Heavens representing states of meditation form colored bands at the top. Our world appears as the centermost of three red triangular shapes at the bottom. Mural at Paro Dzong, Bhutan. (Edwin Bernbaum)

west of China and to the north of India became the exotic setting for a cosmic mountain shared by both civilizations. There, far from the world known to either the Chinese or the Indians, in a mysterious region inhabited by divine and demonic beings, lay the center of the universe and the paradise of the immortals.

Before the introduction of Buddhism, myths about the K'un-lun had played a part in inspiring the Chinese to look west and open the Silk Route in the second century B.C. Han Wu-ti, the emperor who sent Chang Ch'ien out "to make the road," was obsessed with discovering the secret of immortality. He climbed T'ai Shan, the sacred mountain of ancient China, in quest of that secret, and his biography claims that Hsi

Chinese immortals enjoy themselves on the heights of the K'un-lun. The palace of the goddess Hsi Wang-mu, the Queen Mother of the West, appears through clouds in the upper left corner of the painting. (Reproduced by courtesy of the Trustees of the British Museum)

wang-mu came to his palace to present him with the peaches of longevity from her garden on the heights of the K'un-lun. Aware of Wu-ti's interest, Chang Ch'ien and other envoys dispatched on diplomatic missions to the west went looking for the abode of the goddess, as well as an alliance against the enemies of the Chinese empire.[14]

Because of the extreme difficulty in seeing, much less reaching, the K'un-lun, James Hilton chose this range, rather than the Himalayas, as the setting for *Lost Horizon*, his famous novel about an idyllic monastery hidden in a remote valley where, like the immortals of Hsi wang-mu's paradise, people can live for hundreds of years without growing old. Like the kingdom of Shambhala in Tibetan mythology, the purpose of this monastery, called Shangri-La, is to preserve the highest spiritual and cultural treasures for a time in the future when wars will destroy everything of value in the world outside. Then, when the strong have devoured each other, out of this hidden sanctuary will come what is needed to build a new and better world.

Toward the end of the novel one of the characters chases after the hero, who left Shangri-La and is now trying to find his way back to the earthly paradise. He traces him as far as the K'un-lun Mountains, which he describes to the narrator of the book in the following conversation:

"The Government people were quite right—all the passports in the world couldn't have got me over the Kuen-Luns [K'un-lun]. I actually went as far as seeing them in the distance, on a very clear day—perhaps fifty miles off. Not many Europeans can claim even that."

"Are they so very forbidding?"

"They looked just like a white frieze on the horizon, that was all."[15]

In 1986 I visited Khotan, a major oasis on the southern branch of the Silk Route, right at the foot of the K'un-lun Mountains. The haze of dust churned up from the dunes of the Taklamakan Desert was so thick that I could just barely make out brown hills at the base of the range; of the high snow peaks, where Eastern myth and Western fiction envision an earthly paradise, there was not the slightest trace. The mountains remained as mysterious as they had ever been in the musings of my imagination.

Amnye Machen

At the eastern end of the K'un-lun, near the source of the Yellow River, rise the peaks of Amnye Machen. A mysterious massif often hidden in clouds, it lies in a region so remote and difficult to reach that rumors developed in the middle of the twentieth century that

Snowy wastes of the K'un-lun Range on the road linking the Silk Route of western China to Tibet. (Edwin Bernbaum)

it might be higher than Everest. For centuries a fierce group of nomads who preyed on passing caravans kept outsiders away from the sacred peaks of Amnye Machen. Their name, the Golok, means "Those with Their Heads Turned Backwards," reflecting their contrary and rebellious nature. Descended from marauding Tibetan warriors of the eighth century, they refused to acknowledge the sovereignty of either the Tibetan or Chinese governments until finally forced to do so by Mao Tse-tung long after the Communist Revolution in China in 1949.

One of the first Westerners to enter the territory of the Goloks, a hapless Frenchman named Dutreuil de Rhins, was sewn inside a yak skin bag and dumped in the Yellow River to drown. Some thirty years later, in the 1920s, a couple of British and American explorers got close enough to see the mountain in the distance and estimated its summit to be more than 29,000 feet above sea level. During World War II pilots blown off course while flying from Burma to China reported sighting a peak more than 30,000 feet high in the vicinity of Amnye Machen. For a brief period of time after the war, sensational accounts in newspapers and magazines convinced many in the West

that it was, in fact, the highest mountain in the world. More accurate measurements made by Chinese scientists after the Goloks had been pacified by the Communist government reduced its altitude to a modest 20,610 feet. The Japanese expedition that succeeded in making the first ascent of Amnye Machen in 1981 found the mountain much harder to reach than climb.[16]

Only occasionally visible in short breaks of clear weather, Amnye Machen hovers as an invisible presence over a green plateau of rolling grassland where enormous herds of wild yaks and asses used to roam in the company of bears and snow leopards. Here and there, tucked in the shelter of a ridge or gully, appear the black tents of the Goloks, whose modern rifles, supplied by the Chinese, have decimated most of the local wildlife. When the winds blow away the bluish gray clouds that usually envelop the sacred massif, a magnificent view of three peaks composed of pure white snow is revealed. The most dramatic, a pyramid to the south, is named Chenrezig, after the Bodhisattva of Compassion and patron deity of Tibet. The highest, a great dome to the north, appears to float like a cloud above the glaciers that issue from its foot. The central peak, which rests serenely between its

higher and more impressive neighbors, houses the actual deity of the sacred range—the warrior god Machen Pomra. Tibetan texts describe his mountain abode as a giant reliquary made of crystal with its square base buried deep in the earth, its round body girded with rain clouds, and its spire reaching into the ethereal zone of the sun and moon.

Paintings on the murals of Tibetan monasteries depict Machen Pomra as a horseman arrayed in golden armor with a spear in one hand and a vessel of jewels in the other. He is accompanied by 360 deities who reside on peaks surrounding his lofty fortress. The Goloks regard him as their principal god and protector. Machen Pomra also acts as the protective deity of Ganden, a major monastery just outside Lhasa, more than five hundred miles away from the mountain itself. Tsongkhapa, the fourteenth-century lama who founded Ganden and the Gelugpa, or Yellow Hat, sect of Tibetan Buddhism that ruled Tibet under the Dalai Lamas, was born not far from Amnye Machen. When he went to Lhasa, he took with him the worship of Machen Pomra. Until the Chinese destroyed Ganden in the 1960s, the monks of the monastery performed elaborate rituals to the god of the distant sacred peak.[17]

Despite the suppression of their religion after the Communist Revolution in 1949, the Goloks still practice ritual circumambulations of the sacred range, much like the pilgrimages made around Mount Kailas in western Tibet. The circuit, which is here much longer, takes at least a week. Ten thousand people a year used to make the pilgrimage, all of them on foot, even the highest chiefs and lamas. In doing so they not only acquired religious merit, which would help them in this life and the next, but also paid honor to Machen Pomra, the divine lord of the region.

Amnye Machen is also linked with Gesar of Ling, the supernatural hero of the Gesar Epic of Tibetan literature. Many of the legendary events recorded in this national epic, which assumed its present form around the seventeenth century, are supposed to have taken place in the vicinity of the sacred range. Son of a mountain god and a lake goddess, Gesar is born in this region of eastern Tibet to rid the country of demons threatening the Buddhist teachings. After winning a horse race against an evil uncle, he regains the kingdom that is rightly his and goes on to subjugate enemies at the four quarters of the compass, establishing himself as a universal monarch and protector of the Buddhist religion.

More than an epic hero, Gesar is a deity invoked by bards in trances, worshipped in rituals, and depicted in religious paintings. The Goloks call Amnye Machen the Palace of Gesar and identify various features of the pilgrimage route around the sacred mountain with his activities. One distinctive rock, for example, is both the place where Gesar tied his horse and an embodiment of his younger brother. They believe, moreover, that his sword lies hidden inside the mountain, waiting for him to return in a future rebirth as the King of Shambhala, come to defeat the forces of evil and establish a golden age throughout the world. His name, Gesar, reflects the influence of ideas traveling along the Silk Route: the name comes, in fact, from Kaiser or Caesar, the title of the emperors of Rome and Byzantium. Passed from ruler to ruler along the caravan routes from the West, this title eventually reached the far end of the Tibetan Plateau, where it came to designate the hero-king of the Tibetan national epic.[18]

Muztagh Ata

Far to the west, beyond the other end of the K'un-lun range, rises the great white dome of Muztagh Ata, the "Father of Ice Mountains," 24,757 feet high. Swelling up from the broad valleys of the Chinese Pamir, it stands alone, like a colossal monument of rock and ice left to commemorate the glories of some forgotten god. Like Mount Kailas in Tibet, Muztagh Ata dominates the surrounding landscape, drawing the eye inexorably to its smooth rounded summit, set above cliffs of dark stone. The Kirghiz nomads who graze their yaks and camels within sight of it see in its symmetric form the shape of paradise—and the tomb of their holiest saints. When I first saw the mountain, it was in moonlight. A thin veil of silver mist was flowing off its summit so that its highest glaciers appeared to be streaming into the black sky to merge with the stars. Dissolving into the infinite reaches of space, the mountain seemed to belong more to heaven than earth.

When Hsüan-tsang passed through the Pamir, following the Silk Route back from India in A.D. 644, he recorded a Buddhist legend that, in the judgment of Aurel Stein, was inspired by the mysterious and impressive appearance of Muztagh Ata:

Two hundred li *or so to the west of the city we come to a great mountain. This mountain is covered with brooding vapors, which hang like clouds above the rocks. The crags rise one above another, and seem as if about to fall where they are suspended. On the mountain top is erected a* stupa *of a wonderful and mysterious character.*[19]

Climbing toward the hidden summit of Amnye Machen. (Brock A. Wagstaff)

Above: The great dome of Muztagh Ata floats above Little Karakul Lake. The Kirghiz revere the mountain as the tomb of Muslim saints. (Edwin Bernbaum) Below: Kirghiz women in a yurt at the foot of Muztagh Ata. (Edwin Bernbaum)

According to the story heard by Hsüan-tsang, one day the mountain opened up to reveal an ancient Buddhist monk seated within it, immersed in meditation. A hunter happened to see him and told the king about the extraordinary sight. When the people came to awaken the monk from his trance, he rose up in the air and passed into nirvana, leaving his bones behind. The king and his monks built the *stupa* over them — a hemispherical monument used to honor the relics of Buddhist saints. Linking the story to Muztagh Ata, Stein points out that "it is certain that the remarkable shape of the huge dome of ice rising above all other mountains must have vividly suggested to Buddhist eyes the idea of a gigantic Stupa."[20]

Even though the Kirghiz came to the region in the sixteenth or seventeenth century, long after Buddhism had succumbed to Islam, they seem to have picked up a survival of this pre-Islamic tradition and adapted it to their own beliefs, inspired by feelings of reverence for the sacred peak. When Stein visited the mountain in 1900, the Kirghiz told him what he took to be a Muslim version of the original Buddhist legend: a story of hunters who long ago beheld a Muslim sage residing on the inaccessible summit of Muztagh Ata. Local beliefs collected around the same time by Sven Hedin reinforce Stein's conclusions. According to the Swedish explorer, the Kirghiz regard the mountain as the gigantic tomb or shrine of seventy saints, the two most important being Moses and Ali, the son-in-law of Muhammad, the founder of Islam. They believe that the soul of Moses dwells within Muztagh Ata, which they also call, in his honor, Hazrett-i-Musa, the "[Shrine of] Moses." As for Ali, the legend goes that when he lay dying, murdered by his enemies, he prophesied that a white camel would descend from heaven to carry him away. As soon as he breathed his last, the camel appeared and flew with him on its back to the snows of Muztagh Ata, where his soul now rests in peace.[21]

Like many other Muslim shrines, Muztagh Ata derives its sanctity from the holiness of the saints associated with it. The Kirghiz do not worship the peak itself, for that would be idolatry in the Muslim tradition. They look to the mountain, instead, for the kind of power and blessings they would expect to receive at the shrine of a sage, whose sanctity lingers after his death. According to Hedin, whenever his Kirghiz companions first saw Muztagh Ata after being away on a journey, they would fall on their knees and pray, as they would to Mecca, the holy center of the Muslim world where Muhammad founded the religion of Islam in the seventh century A.D. The Communist Revolution in China in 1949 put an end to such visible expressions of "superstitious beliefs," but with the loosening of restrictions on the practice of religion, the old myths have revived.

The Kirghiz also revered the mountain as a symbol of paradise. According to one of their legends, on the summit of Muztagh Ata, hidden in its snows, lies the ancient city of Janaidar, built thousands of years ago in a time of universal peace and happiness. When strife and misery broke out below, the people of Janaidar withdrew from all contact with the outside world. There, they have preserved the peace and happiness that humankind knew at the beginning of creation. Delicious fruit of all kinds grows all year round, and flowers do not fade; death, cold, and darkness have been banished forever. The paradise that the Kirghiz see on the summit of their sacred mountain carries echoes of the golden age described in the biblical legend of the Garden of Eden, preserved for them in the Muslim scriptures of the holy Koran.

A story the Kirghiz told Hedin may reflect the distant influence of the ancient Chinese myth of the peach tree of longevity on the heights of the K'un-lun to the east. According to this account, an aged Muslim holy man wandered by himself up the mountain. High on its slopes, beyond where ordinary people can go, he came across a beautiful lake with a white camel grazing on its shore. Near it was a garden of plum trees, where old men dressed in white were walking to and fro. The holy man reached into one of the trees and plucked a plum, which he ate with relish. One of the old men immediately came over to congratulate him for having eaten the fruit. "If you had despised the plums, as they all did," he said, pointing to his aged companions, "you would have been condemned to stay on this mountain like them, walking to and fro, until the end of time." A rider suddenly appeared on a white horse, picked up the holy man, and carried him down the mountain. When the sage recovered his senses in the valley below, he could only vaguely remember what had happened to him. Whereas in the Chinese myth of the K'un-lun, the magic fruit grants the Taoist sages immortality, here it saves the Muslim holy man from a pointless life of eternal boredom. The story also reflects the Kirghiz view of Muztagh Ata as a tomb of the dead: white, the color of the camel and the garments worn by the old men, is the color of death and mourning in Islam.[22]

Because of its isolated position and striking appearance, Muztagh Ata attracted the attention of a number of well-known Western explorers and mountaineers in the nineteenth and twentieth centuries. The mountain so impressed them that they ignored or overlooked the higher summit of Mount Kongur, 25,325 feet, clearly visible a few miles to the northeast across the dark waters of Little Karakul Lake. Each one of these travelers, whether or not a mountaineer, tried without success to climb Muztagh Ata. Sven

A camel caravan on a mountaineering expedition follows a remote river bed between the Karakoram and Aghil ranges near the western end of the K'un-lun Mountains. (Edwin Bernbaum)

Hedin, the most daring and flamboyant of these Westerners, was the first to try. In 1894, in the course of one of his many harrowing expeditions across the deserts of Central Asia, he made four attempts on the peak. With the help of the Kirghiz who lived at the foot of the mountain, he reached its glaciers only to succumb to snow blindness. Figuring the best way to deal with the problems of high altitude was to avoid exertion, he tried on his next two attempts to ride yaks to the summit, but the poor beasts got bogged down in soft snow and crevasses, forcing him to abandon that tactic. Obsessed with a longing to reach the summit, he came back yet again, but this time the mountain simply brushed him off with a flick of wind and snow.

More than any other Westerner, Hedin was fascinated by Muztagh Ata. In fact, the mountain assumed for him the role of a religious shrine, embodying the god of exploration he sought, wittingly or unwittingly, in the most remote and forbidding reaches of Central Asia. In his book *Through Asia* he concludes the account of his attempts to climb Muztagh Ata with a paean of praise in the form of a prayer addressed to the mountain itself:

Like the holy Dalai Lama, thou permitest none but thy chosen children to approach the sacred precincts of thy temple. Shed then thy saving light as from a lofty beacon-tower across the desert ocean, which stretches to a boundless distance from thy eastern flank. Let the gleam of thy silver brow scatter the dust-haze of the desert hurricane— let the cool refreshing airs of thy palace of etenal snows be wafted towards the weary traveler toiling through the burning heats of sun and sand—let the life-giving streams, which flow from thy mighty heart, abound in strength for thousands of years to come, and for thousands of years to come still maintain their fight against the all-devouring all-devastating sands! Among the lights of Asia thou art, and always wilt be, one of the brightest, as thou art amongst the mountains of the earth one of the noblest, one of the most sublime![23]

In 1900, when Aurel Stein passed by Muztagh Ata on one of his many archaeological expeditions along the Silk Route, he also tried to climb the mountain. Following Hedin's lead, he attempted to ride a yak up to the summit but soon realized the folly of the enterprise: "More and more frequently we had to dismount and drag the stubborn animals out of the deep snow-drifts into which they had plunged."[24] Leaving the unhappy yaks behind, he proceeded on foot to his high point, the buttress of a ridge at 20,000 feet, still almost 5000 feet from the top. Like Hedin he remarked on the religious awe in which the Kirghiz held the sacred mountain.

Nearly fifty years later, in 1947, two experienced mountaineers tried to climb Muztagh Ata: Eric Shipton and H. W. Tilman, both well-known veterans of British expeditions to Nanda Devi and Mount Everest. At the time, Shipton was posted to the British consulate in Kashgar, not far from the mountain itself. Although Muztagh Ata presented no technical difficulties, the endless rise of its upper slopes, combined with wind and extreme cold, forced him to give up not far from the summit, which still seemed infinitely remote. The mountain was finally climbed in 1956 by a Sino-Soviet expedition that put thirty-one climbers on top, all at the same time, to demonstrate the supremacy of the Communist doctrine. A more unusual climb took place in 1980, when three Americans—Galen Rowell, Jan Reynolds, and Ned Gillette—reached the summit and then made a heavenly descent on skis, soaring down fields of golden powder, floating above a darkened world in the last light of the setting sun.[25]

More than other mountains, the distant ranges of Central Asia by their very nature and location inspire a sense of the sacred as the mysterious other, remote and apart from the familiar world of everyday life. Cultures of both East and West have turned to these little-known ranges in quest of earthly paradises impossible to reach or attain by ordinary means—the hidden kingdom of Shambhala, the cosmic axis of Mount Meru, the palace of Hsi wang-mu, the monastery of Shangri-La. The mountains that conceal such sanctuaries—the K'un-lun, the T'ien Shan, the Pamir, and others—represent the extreme limits of the physical world, the borders between the possible and the impossible, the known and the unknown, the imaginable and the unimaginable. Their remoteness and inaccessibility are a measure of their power to evoke the ultimate mystery of the sacred.

Despite the fact that it lies beyond our reach, or perhaps because it does, something deep within us yearns to experience that mystery, to find ourselves transported outside the boundaries of the world we know. We sense that in the strange and distant place symbolized by the palace of the immortals or the monastery of Shangri-La is hidden the answer to our deepest longings—a secret sanctuary so far and so close that we have great difficulty recognizing it, much less reaching it. Yet seek to reach it we must—or spiritually wither and die.

A Tibetan story tells of a young man who sets off across the ranges of Central Asia in search of Shambhala. After traversing many deserts and mountains, he comes upon an old hermit in a cave. The sage asks him, "Where are you going across these wastes of snow?"

"To find Shambhala," the youth replies.

"Ah, well then, you need not travel far," the hermit says, "The kingdom of Shambhala is in your own heart."[26]

· 4 ·

JAPAN

MOUNTAINS OF THE RISING SUN

EMERGING FROM THE EASTERN RIM of the Pacific Ocean, the islands of Japan rise up to form one of the most mountainous countries in the world. The prominent outline of a peak or hill dominates the view from almost any spot on the islands' rugged surface. Japan abounds with mountains of all kinds: elegantly shaped volcanoes sweeping up from low-lying plains, gnarled crags of rock lunging out of the sea, ranges of alpine peaks serenely tipped with crystal snows, long massifs of twisting ridges softened with covers of green forest. The subterranean forces that created these mountains, giving the country its distinctive character, continue to shape and shake the islands. Steam and gases seethe out of the cauldrons of volcanoes, and the land frequently quivers in response to the unseen collision of tectonic plates. The overwhelming power of the earth, made visibly manifest in the mountainous landscape, imbues human life with a fragile beauty reflected in the delicacy of Japanese art and culture.

Living in close proximity to their mountains, knowing them almost as members of the family, the Japanese have developed a special affection for them. A passage from a well-known scripture by Dogen, the founder of an important school of Zen Buddhism in Japan, reveals the depth of this affection and the sense of intimate relationship it expresses:

Although we say that mountains belong to the country, actually they belong to those who love them. When the mountains love their master, the wise and the virtuous inevitably enter the mountains. And when sages and wise men live in the mountains, because the mountains belong to them, trees and rocks flourish and abound, and the birds and beasts take on a supernatural excellence.[1]

Of all the features of the natural landscape, the Japanese have tended to regard mountains as the most sacred—the places most intimately associated with the gods. Intricate networks of shrines and pilgrimage routes bear witness to the devotion the Japanese have lavished on an incredible number of hills and peaks. According to one authority, Japan has 354 major sacred mountains—to say nothing of the minor ones, which remain uncounted.[2]

The Japanese reverence for mountains has its origins in a loose collection of beliefs and rituals practiced by prehistoric hunters, woodcutters, and farmers whose ancestors migrated to the islands from the Asian mainland as early as 10,000 B.C. When Shinto, the indigenous religion of Japan, emerged sometime before the Japanese began writing down their history in the sixth century A.D., it pulled these practices together in the veneration of prominent peaks as major abodes of the *kami,* or spirits believed to animate rocks, trees, and other features of the natural world. The observance by the general populace of these ancient beliefs and practices led them to revere mountains in various ways. Farmers and villagers worshipped local hills and peaks, such as the ones around Kyoto, as sources of fertility. They regarded them as the abodes of the *yama no kami,* the "mountain spirits" who would descend during the spring and summer to become the rice *kami* of the fields, bringing with them the life-giving streams that issued from the moist and cloudy heights. Hunters and woodcutters who

A shrine painting shows pilgrims climbing Mount Fuji with three deities depicted on the triple summit of the mountain. (Fujisan Hongu Sengen Taisha, photograph courtesy of Japan Society)

roamed the forested slopes of these mountains revered them as the dwellings of somewhat wilder *kami* on whose favor they depended for a plentiful supply of game and wood.

Certain mountains, such as Gassan in northern Honshu, were viewed as hallowed places of the dead. There, at the entrance to the other world, the souls of the deceased underwent a process of purification that transformed them into the mountain *kami* on whom their offspring depended for sustenance in this life. In accordance with such beliefs, the burial mounds of ancient emperors assumed the suggestive form of hills. The mausoleums of later Japanese rulers were actually called *yama*, or "mountains." The person in charge of building such a tomb bore the title of "the official who erects the mountain." Even today, in parts of rural Japan, people call a coffin a "mountain box" and the process of digging a grave "mountain work." At the start of a funeral procession, the leader cries out, "*Yama-yuki! Yama-yuki!*"—"We go to the mountain!"[3] The summit of Tateyama, one of the highest peaks in the Japanese Alps, is regarded as the doorway to hell, from which the dead occasionally emerge for a brief escape from their sufferings. A *torii*, or gate, on the approach to the temple on the highest point of the mountain marks the passage from this world to the next.

The Japanese venerated other mountains as the dwellings of gods who descended from heaven. Many of the peaks in this category, such as Mount Fuji, possess the distinctive form of isolated cones, both symmetric and irregular. The people believed that mountains of such a shape caught the attention of heavenly deities, who would step down from the sky to alight on the mountain summits. Over the centuries, hunters, shamans, and hermits were drawn to these peaks to make contact with the gods and spirits who could give them the supernatural powers needed to hunt, heal, and meditate. In a modern continuation of such practices, many artists and actors from Tokyo go to Mount Ontake to put themselves in trances as a means of obtaining divine inspiration for the creative aspects of their work.[4]

In the prehistoric period before the sixth century A.D., the Japanese did not climb their sacred mountains, which were regarded as a realm apart from the ordinary world, too holy for human presence. The people built shrines at their feet and worshipped them from a respectful distance. With the introduction of Buddhism from China in the sixth century came the

A climber approaches a shrine on the summit of Tateyama, the mountain of hell, Japanese Alps. (Galen Rowell/Mountain Light)

practice of climbing the sacred peaks all the way to their summits, there to commune directly with the gods and the divine reality they embodied. The development of this favorable attitude toward the heights also drew inspiration from Chinese Taoist practices of seeking freedom and immortality in the highest and wildest reaches of the mountains—the craggy ridges and cloud-hung summits where immortals took delight in living in harmony with the Tao.

Buddhists, in particular, regarded the higher parts of mountains as the perfect setting for the practice of meditation. There in the peaceful solitude of forest and stream, monks might develop the tranquility of mind needed to penetrate the mysterious nature of ultimate reality. The two great monastic centers responsible for the early propagation of Buddhism in Japan, especially within imperial circles, were founded on top of sacred mountains: the center of the Tendai sect on Mount Hiei, just outside Kyoto, and the seat of the Shingon school on Mount Koya, several days' walk southeast of the city. Monasteries of the meditation school of Zen Buddhism, which became popular later, were called "mountains," titles that they have retained to this day. Dogen, the thirteenth-century founder of the Soto sect of Zen, composed an influential work extolling mountains and rivers in which he wrote, "From time immemorial the mountains have been the dwelling place of the great sages; wise men and sages have all made the mountains their own chambers, their own body and mind."[5]

Whereas the orthodox schools of Buddhism found the mountains a congenial environment for pursuing meditation, a new and fascinatingly eclectic sect arose that made mountain climbing itself the focus of its doctrine and ritual practice. Shugendo, which emerged in the period between the ninth and twelfth centuries,

Yamabushi practitioners of the mountain-climbing sect of Shugendo are greeted by their leader at Kotaku-ji, temple of retreat on Mount Haguro. (H. Byron Earhart)

blended shamanistic practices of Shinto with magical doctrines of Buddhism and Taoism to produce a tradition unique in the religious landscape of the world. The name *shugendo* means "the way of mastering ascetic powers." The practitioners of this sect were called *yamabushi*, "those who lie down or sleep in the mountains." They looked to mountain peaks as places to obtain supernatural powers that they could bring down to the world below. There, on the heights, the *yamabushi* practiced austerities and rituals designed to purify themselves and make contact with the deities who could grant them such powers—the old *kami* of the Shinto tradition transformed into the Buddhas and Bodhisattvas of esoteric Buddhism.

Shugendo practices, which continue today, center on the ritual ascent of sacred mountains. As a means of purifying and strengthening themselves for the influx of supernatural powers, *yamabushi* begin their climb by standing under icy waterfalls at the foot of the peak they intend to ascend. Then, wearing sandals and carrying staffs, they follow a leader up the mountain, stopping at various shrines to perform fire sacrifices and other rituals that involve the recitation of mantras, or sacred spells, to call forth the powers of various Buddhist deities. As they climb toward the summit, they visualize themselves passing through the stages leading to enlightenment. At one place on the ascent of Mount Omine, a major center for modern practitioners of Shugendo, the *yamabushi* will hang each member of the party by his heels, head down, over a cliff to contemplate the transient nature of all things and repent the evil he has done in his life. In the old days, if the unfortunate person had committed too many sins, they would let him fall to his death on the rocks below.

The ritual ascent of Mount Haguro, another peak important to modern-day Shugendo, involves the powerful symbolism of death and rebirth. On reaching a hut high in an isolated ravine or valley on the side of the peak, the *yamabushi* sequester themselves inside its dark interior and imagine themselves dying and entering the womb of the mountain itself. Red and white cords hanging from the beams symbolize arteries and veins running through the new body each will receive. On the last night of the ritual, the *yamabushi* burn logs representing the bones of their old bodies, turning to ashes all remnants of their passions and illusions. The next morning as they descend the peak, they squat in fetal positions and jump up with a sharp cry, symbolizing the ecstatic moment of rebirth from the mountain—and the entrance into a new and more spiritual life leading to the ultimate goal of enlightenment itself.

The *yamabushi* who acquired ascetic powers in this manner on mountains sacred to Shugendo—mountains such as Haguro in the north and Omine in the south—traveled to villages throughout Japan, bringing blessings to the people, often in the concrete form of amulets charged with magic forces believed capable of improving the conditions of life in this world. During the medieval period, between the fourteenth and sixteenth centuries, many of these wandering ascetics settled down to become village priests devoted to helping the community worship a particular mountain in the local vicinity. Spreading in this manner, Shugendo became the predominant form of religion for the common people of Japan, an underground network of folk beliefs and practices anchored to the mountains as the holy source of sacred power and blessings.

The practice of Shugendo came to an abrupt end in 1868 when the nationalistic Meiji Restoration overthrew the feudal rule of the preceding Tokugawa period and forcibly separated Buddhism from Shinto to make the latter the state religion of Japan, free from the contamination of foreign influences. Too much an inextricable blend of Buddhist and Shinto elements, Shugendo fit neither category and was banned, its temples given into the hands of Shinto priests who scorned its mongrel background. Only after the end of World War II, when Japanese nationalism fell into disgrace and freedom of religion was established as a constitutional right, did its practice regain some of its former importance. Today small bands of *yamabushi* continue to climb Haguro, Omine, and a few other sacred peaks of Shugendo, but the worshippers represent only a ghostly remnant of a vast and pervasive sect that once spread the power and authority of holy mountains throughout the islands of Japan.[6]

As a consequence of their deep reverence for sacred mountains, the Japanese have one of the oldest traditions of mountain climbing in the world. Ascents of major peaks were recorded as early as the ninth century A.D., long before the sport of mountaineering developed in the European Alps in the eighteenth and nineteenth centuries. Over the succeeding years countless pilgrims climbed such peaks as Mount Fuji, seeking the spiritual exhilaration and power of the heights. Those who returned from pilgrimages to sacred summits were treated with special reverence. As one scholar of Japanese religions has noted, "It is well known that pilgrims coming back from sacred spaces [in the mountains] were regarded with awe: common people saluted them, made offerings, even tried to touch them."[7]

The power attributed to mountains and those who climb them helps to explain the keen interest that the Japanese take in the modern sport of mountaineering. On weekends during the summer and fall, thousands of climbers dressed in boots and knickers take commuter trains out of Tokyo to scramble over every peak and ridge of the Japanese Alps. A national passion for mountains and mountaineering has inspired an increasing number of Japanese to seek out more distant ranges, such as the Himalayas. During one season shortly after Nepal reopened its peaks to climbing in 1969, the Nepali government reported that out of thirteen expeditions applying for climbing permits, eleven were from Japan. In 1975 Junko Tabei, an interpreter and teacher from a suburb of Tokyo, became the first woman to reach the summit of Mount Everest. On a visit to Japan in 1978, I saw a billboard in the Ginza, the commercial center of downtown Tokyo, announcing a successful Japanese ascent of K2, the second-highest mountain in the world. It would be hard to imagine such a billboard in Picadilly Circus or Times Square.

Mount Fuji

The mountain that most represents Japan in the eyes of the world is, of course, Mount Fuji. No peak more beautifully embodies the spirit of a nation. The elegant simplicity of its lines, sweeping up into the graceful shape of an inverted fan painted with delicate patterns of pure white snow, symbolizes the quest for beauty and perfection that has shaped so much of Japanese culture, both secular and sacred. Suspended between heaven and earth, neither rock nor cloud, the volcano appears as a cone of crystallized sky, floating above a vast landscape of fields, villages, lakes, and sea. On a clear day, when smog clears and the world seems fresh and new, the mountain's symmetric outline can be seen from the cramped and polluted city of Tokyo, sixty miles away. The very perfection of its form, startling in its incredible simplicity, suggests the mystery of the infinite.

Innumerable poets and painters have attempted to depict the divine beauty of Mount Fuji. A famous poem from the *Manyoshu*, the oldest collection of Japanese poetry, compiled in the seventh century A.D., describes the mountain in the following words:

Lo! There towers the lofty peak of Fuji
From between Kai and wave-washed Suruga,
The clouds of heaven dare not cross it,
Nor the birds of the air soar above it.
The snows quench the burning fires,
The fires consume the falling snow.
It baffles the tongue, it cannot be named
It is a god mysterious.[8]

Although many writers and artists have tried to capture the spirit of Fuji, no one has ever completely succeeded. Something about the mountain, the mystery of its sublime perfection, evades the stroke of pen and brush. As one modern Japanese critic has said, "The reason why there are curiously few fine poems in Japanese or Chinese, or fine paintings about Fuji, is that the subject is too overpoweringly splendid."[9]

Although the sight of Fuji may inspire thoughts of eternal beauty, the mountain itself was created quite recently. Much younger than most Japanese mountains, it burst forth from the earth only 25,000 years ago when an eruption buried the surrounding plain beneath ten feet of volcanic ash. The cone we see today assumed its general form about 8000 B.C., but the mountain continued erupting over the succeeding millennia, growing to its present altitude of 12,388 feet. A composite volcano, Fuji owes its graceful shape to alternating layers of ash and lava that have given its smoothly rising slopes the internal structure and strength to withstand the forces of upheaval and erosion. Nine eruptions convulsed the peak between A.D. 781 and 1083 alone. The mountain last shot forth fire in 1707 when ash swirled up to drift down on Edo—modern-day Tokyo. Patches of sand on the crater rim hot enough to cook eggs indicate that Fuji could erupt

Konohana Sakuya Hime, goddess of Mount Fuji, depicted on a hanging scroll. (Edwin Bernbaum)

again. Although the volcano's eruptions have brought destruction and death through the centuries, they have also created the beautiful landscape of forests and lakes that surrounds the sacred mountain like the garden of an earthly paradise. Some of the most unusual features of this landscape, now a national park interspersed with small towns, are networks of hollow tubes formed by flows of lava. Religious tradition has associated these tunnels and caves with the womb of the mountain, giving the volcano a distinctly female character.

Japanese myths about the origin of Fuji reflect the deep impression made by the fiery activity of the sacred peak. One account, published in 1680, holds that the ancient deities declared through an oracle the sanctity of the ground on which the volcano would stand: "Here the Gods and Buddhas will manifest themselves and protect our land. Witness now a great wonder!" Immense clouds billowed over the earth, bringing the blackness of a solar eclipse. When they cleared, Fuji appeared shining in all its splendor, created in a single night.[10]

The name Fuji probably comes from an Ainu word meaning "fire" or "deity of fire." Obviously a god of great power, the mountain had to be placated. In A.D. 806 a local official built a shrine near the foot of the volcano to keep it from erupting. The priests assigned the task of pacifying the mountain apparently neglected their duties because Fuji erupted with great violence in 864, causing much damage in a nearby province. The governor of that province blamed the priests for failing to perform the proper rites and constructed another shrine in his own territory, where he could make sure everything was done correctly. The fiery god of the mountain became at a later date the more peaceful Shinto goddess of Mount Fuji—Konohana Sakuya Hime, the "Goddess of Flowering Trees." Today she is worshipped at the shrine originally built for the older deity.[11]

A romantic story, attributed to the fifth century A.D., explains the eruptions of the volcano after its emergence. One day an old couple with no children found a beautiful baby girl in a bamboo grove near the foot of Mount Fuji. They named her Kaguya Hime, or "She Who Lights Up the Area." The girl grew up to become the most beautiful woman in the region, attracting the attention of the local governor, who made her his wife. After a number of years of blissful marriage, she told her husband that she was, in fact, the Immortal Lady of Mount Fuji and that she was returning to the Palace of the Immortals on top of the sacred mountain. To alleviate his distress, she gave him a magic mirror in which to see her image and then she vanished. Unable to bear life without her, he followed her up to the crater of Fuji. Finding no trace of her

BOY AND MOUNT FUJI. *Painting by Katsushika Hokusai. (Courtesy of the Freer Gallery of Art, Smithsonian Institution, Washington, D.C.)*

and having nowhere else to go, he clutched the mirror to his breast and leapt off a precipice. The love that burst from his heart set fire to the mirror, and the smoke that rose from it is the smoke that used to issue from the summit of Fuji itself. A famous verse from the twelfth century commemorates the sublimity of the official's love for the goddess of the sacred mountain:

Trailing down the wind
The smoke of Fuji
vanishes into space
till nothing lingers
of my love's deep fires.[12]

Gracefully walking through the mists, Kaguya Hime appears from time to time as an elusive representative of the immortals believed to dwell on the summit of Fuji. The Japanese long ago identified their sacred volcano with the mountainous island of P'eng-lai, the eastern Isle of the Blest in the Taoist mythology of China. Shih-huang-ti, the Chinese emperor who climbed T'ai Shan in the third century B.C. and left

behind the buried army at Xi'an, sent an expedition of five hundred youths and five hundred maidens on a voyage to bring back the elixir of immortality from this earthly paradise. They disappeared into the eastern ocean, never to return. According to Japanese legend, they reached Japan and their leader climbed Fuji to obtain the elixir from the summit, but the deity of the peak refused to give it to him. Fearing the wrath of the tyrannical emperor, who had executed many scholars, the youths and maidens decided to stay at the foot of the mountain, and there their descendants have lived to this day.[13]

Regarding the peak from a different perspective, Buddhists came to view Fuji as the abode of a Buddhist deity who embodied the divine light of spiritual wisdom. According to a story recorded in a text of esoteric Buddhism, Shotoku Taishi, a famous prince of the sixth century, went to the summit and descended through the crater into the mountain itself. There, in a vast cavern deep within the volcano, he spied a fire-breathing dragon coiled around a rock in the middle of a pond. When he addressed the supernatural creature, it turned into Dainichi Nyorai, the Buddha of All-Illuminating Wisdom, and said, "I have come from the empty and limitless realm of ultimate reality to live forever in the cave palace of this peak in order to save all sentient beings."[14] This cosmic Buddha, whose glorious form reflects the blazing light of the sun, embodies the highest teachings of Shingon or esoteric Buddhism.

In the twelfth century a Buddhist priest named Matsudai Shonin climbed the mountain and built on its summit a temple dedicated to Dainichi Nyorai—henceforth identified with Sengen Dainichi or Sengen Daibosatsu, the Buddhist deity of Mount Fuji. Matsudai's pilgrimage marks the first historically attested ascent of the volcano. He climbed Fuji several hundred times, inspiring many cults devoted to the mountain.

Impressed by its purity of form and extraordinary height, Buddhists found in Fuji a sublime symbol of meditation. The word they used to describe its summit, *zenjo*, is a Buddhist term for the flawless state of perfect concentration. Just as the peak of a mountain soars above the mists that gather in the valleys below, so a person in meditation rises above the passions and illusions that obscure the vision of ordinary people. The Japanese say that the clouds that cover the tops of other peaks only curl around the foot of Fuji. Its summit, a lofty place of contemplation, provides an attractive sanctuary for the gods, who dwell there free from the sorrows that trouble the world below.

In the Kamakura period, between the twelfth and fourteenth centuries, the sublime state of meditation symbolized in the summit of Fuji took on the complex and concrete form of the mandala or sacred circle of Dainichi Nyorai. Buddhists identified eight subsidiary summits along the crater rim as the earthly manifestations of eight petals of the lotus forming the divine abode of this cosmic Buddha. Those inclined toward the practice of esoteric Buddhism would visualize various deities standing on these petals, each in his or her own paradise of meditation, thus striving to transform their perceptions of the ordinary world into the luminous realm of the glorious Buddha embodying the empty nature of ultimate reality.[15]

According to the hybrid tradition of Shugendo, the legendary founder of the sect, En no Gyoja, a wild ascetic who possessed the nature and appearance of a mountain wizard, made the first ascent of Mount Fuji around A.D. 700. Having offended important officials with the indiscriminate display of his magic powers, he was exiled to the Izu Peninsula. Legend tells us that every night he would step across the sea to climb to the summit of Fuji. Regardless of whether En no Gyoja actually climbed the mountain, sometime around the fourteenth century Shugendo prac-

titioners who counted themselves as his followers established a climbing route up the southern flank of Fuji, the side with the mildest weather. These *yamabushi* would lead parties of pilgrims up to the summit, stopping to engage in ritual baths of icy water on the way. In the fifteenth century they began to build huts high up on the mountain for the use of religious mountaineers. A record dated 1518 tells us that "thirteen persons met their deaths at the summit during a storm and a bear killed three climbers."[16]

The *yamabushi* effectively controlled all access to the summit of the peak until the beginning of the seventeenth century, when Fuji-ko or Fuji devotional societies began climbing the mountain from the north. Kakugyo Hasegawa, an ascetic revered as the founder of Fuji-ko, was drawn to Fuji around 1560 by a visionary dream in which En no Gyoja commanded him to go to the mountain. There he took up residence in a sacred cave, where he spent most of his life practicing austerities, most notably splashing himself with icy water and standing for hours without moving on a tiny square of wood. Out of such practices and the ascent to the summit of the mountain, he gained spiritual powers and insights that allowed him to formulate a new religious doctrine centered around his own conception of Sengen Dainichi, the Buddhist deity of Mount Fuji, as the one supreme god, creator and sustainer of all things. Although the Japanese had worshipped sacred mountains for centuries, no one had ever elevated a particular peak to such a rank of divine preeminence over all others.

During his lifetime Kakugyo attracted only a small band of followers devoted to the worship of Fuji. One of his successors, Jikigyo Miroku, however, transformed the relatively minor cult of Fuji-ko into a major religious movement. Born in 1671, about forty years after the death of Kakugyo, Jikigyo became a devotee of Fuji at an early age and made it a practice to climb the peak at least once a year. On one of these ascents, at the very summit, he had an overpowering vision of Sengen, whom he saw as a deity no longer identified with Buddhism. The experience so moved him that he abandoned his old life and became an inspired teacher, proclaiming the advent of a messianic era of peace and plenty in which people would abandon the worship of false deities and devote themselves instead to helping each other.

According to Jikigyo's teaching, in the beginning only Mount Fuji existed in a sea of mud: two primordial gods, the Precious Parents, emerged from the womb of the sacred mountain to give birth to Sengen, a supreme deity who transcended all established religions, including Buddhism. Unfortunately the people of Japan turned away from this deity to worship Buddhas and other illusory deities who were only creations of their own minds. Jikigyo prophesied, however, that Sengen was coming to dispel these false gods and establish himself as the divine ruler of the world. Jikigyo taught that simple faith in the god of Fuji would make people honest, diligent, happy, and prosperous. Drawing on another name for the peak as Kokushuzan, the "Mountain of Heaped-up Grain," he proclaimed Sengen to be the divine embodiment of food, the universal deity who would finally rid the world of starvation.

In another vision, the god of Fuji conferred upon Jikigyo the title of Miroku, the Japanese name of Maitreya, the Buddha prophesied to bring about a messianic era in the future. Feeling that through his devotions to the sacred mountain he had become one with Sengen, Jikigyo decided that by offering his body as a physical manifestation of the deity he could feed the world and initiate the golden age to come. In 1733, in an effort to end a famine wasting Japan, he climbed up the peak, entered a portable shrine beneath the summit, and fasted to death. His dramatic self-sacrifice inspired thousands to respond to his call and begin climbing Fuji. The Fuji-ko cult swelled into a large movement that appealed, in particular, to tradespeople and laborers. Over a hundred devotional societies appeared in Edo alone, and as many more in surrounding areas. By casting aspersions on the present time and looking forward to a golden age in the future, the messianic teaching these societies espoused potentially criticized the established feudal order, and the authorities from the ruling nobility tried, with varying success, to suppress the movement.

When Kakugyo, the founder of Fuji-ko, had stood immobile in his cave inside Mount Fuji, he had been trying through his meditation to restore stability to a nation rocked by political and social unrest. He hoped in this way to transfer the calm serenity of the mountain as the axis of the universe to the people of Japan, who had been fighting among themselves in an endless series of civil wars. Legend has it that he conferred this stabilizing power on Tokugawa Ieyasu, the military ruler who finally unified the warring factions of the country in 1600. Drawing on Kakugyo's action as a precedent, Shibata Hanamori, a nineteenth-century Fuji-ko leader, combined the emerging cult of the emperor with the practice of worshipping Fuji. Shibata taught that the peak was, in fact, the very foundation of national security and the primeval womb and brain of the world. Through devotion to emperor and mountain, a new age of salvation would come in which Japan would reign supreme over all the earth. A Japanese scholar of art wrote at the beginning of World War II: "The soaring of Mount Fuji above other volcanic mountains and its majesty characterizes the potential power of the Japanese people as a nation."[17] Disillusionment with the disastrous consequences of such nationalistic pretensions led to a backlash of hostility against the mountain after the defeat of Japan. Some Japanese even felt personally betrayed by Fuji because its prominent outline had helped to guide Allied war planes to their targets in the terrible firebombing of Tokyo during World War II.[18]

The people of Edo, modern-day Tokyo, had felt a special fondness for Fuji. They even attempted to bring the mountain to the city. In 1765, in accordance with the final instructions of Jikigyo Miroku, a disciple of the Fuji-ko martyr initiated the practice of building replicas of Mount Fuji in Tokyo. Devotees of the sacred mountain responded with great enthusiasm: between the end of the eighteenth century and the beginning of the twentieth, more than fifty of these models were constructed. One of the largest and most impressive was a cone over thirty feet high, constructed on top of an ancient burial mound and capped with volcanic rocks brought back from Fuji itself. A zigzag path with nine switchbacks mimicked the ninety-nine turns of the actual ascent. Little monuments marked sacred sites, such as the place of Jikigyo's fast to the death. Fuji-ko members, especially women and the young, as well as the old and sick, would ritually climb these replicas, imagining themselves climbing the peak itself. To provide an authentic experience of the ascent, the models had to evoke the inner feeling, rather than the external appearance, of the sacred mountain. Many of these Fuji replicas still exist in Shinto shrines scattered throughout Tokyo, and every summer Fuji-ko members ritually climb the more important ones before going on pilgrimages to the actual peak.[19]

Echoes of this practice are found in the festival that marks the end of the official climbing season on Mount Fuji. Each year, on August 26, the people of Fuji-Yoshida, a town at the base of the mountain, form a religious procession to carry around two brightly lacquered models of the sacred peak. The larger of the two, borne on wooden beams supported by men with muscular calves, weighs 2475 pounds. Laughing and shouting, enthusiastic schoolboys bear the smaller replica through the crowded streets of the village. Priests in smooth white robes and peaked black caps initiate the ceremonies at the Shinto shrine situated at the foot of Fuji. At night people light a line of straw towers that runs up the main street of the town. The resulting bonfires burn like red-and-orange geysers of molten lava spewing out of the ground. Beneath the mysterious light of the stars, the gods appear to emerge from the mountain to dance in a line of wraithlike flames.

The fire ceremony has its origins in the earliest known myth about Konohana Sakuya Hime, the prin-

Members of a Fuji devotional sect perform a fire ceremony before setting out to climb the sacred mountain. (Henry D. Smith II)

cipal goddess of Mount Fuji. According to the *Kojiki*, the great eighth-century A.D. compilation of Japanese mythology, she married a god who grew suspicious of her when she became pregnant shortly after their wedding. To prove her fidelity to her husband, she entered a burning bower and miraculously gave birth to a son, unscathed by the surrounding flames. The ceremony at Fuji-Yoshida recalls this story as a means of protecting the town from fire and promoting easy childbirth among women.

Konohana Sakuya Hime originally had little or no connection with Mount Fuji. Sometime between the fourteenth and sixteenth centuries, the belief arose among the people of the region that she would protect them from the eruptions of the volcano as she had her newborn son from the flames of the burning bower. During the Tokugawa period, between 1600 and 1868, the Fuji-ko movement confirmed Konohana Sakuya Hime as the principal goddess of the sacred mountain. She is now the central deity in major shrines at the base of the volcano and on the rim of its crater. Fuji-ko members worship her at altars in their homes, and each group lights a torch in her honor at the fire ceremony of Fuji-Yoshida.

During the nineteenth century, when the movement reached the height of its popularity, there were said to be 108 Fuji-ko groups, each with its own leader and distinctive symbol. Today only 10 or 11 remain active, and every summer they make pilgrimages up the mountain. The ritual ascent of the most active group, the Maruto Miyamoto-ko, begins with a ceremony at the leader's home in Tokyo. After making offerings to the deities of Fuji, the members proceed to the mountain, stopping in Fuji-Yoshida at an inn run by an *oshi*, the descendant of a family of religious climbing guides. The current *oshi* of this particular inn, Shiro Tanabe, runs a local museum devoted to preserving the lore of the sacred peak. His father was the head priest of the Shinto shrine where the fire ceremony of Fuji-Yoshida begins each year.

The members of Maruto Miyamoto-ko used to start the actual climb from this shrine, set in a grove of tall cedars at the base of Mount Fuji, but today they take a bus halfway up the volcano. As they ascend the reddish brown ash slopes above tree line, moving together in a congenial party of white-clad pilgrims, they sing a traditional song of purification: "May the six sense organs be pure and may the weather on

The moon and torii gates guide lines of climbers toward the summit of Mount Fuji. (Henry D. Smith II)

the honorable mountain be fine." The whole group stops to pray at the rock where Jikigyo Miroku, the saint of Fuji-ko, fasted to death in 1733. Having attained an important objective of their pilgrimage, the more tired members return from this point while the others proceed to the summit, there to worship at the shrine of Konohana Sakuya Hime and ritually circumambulate the crater rim.

When I met Ida Kiyoshige, the leader of the Maruto Miyamoto-ko, in 1987, he was seventy-one years old and had climbed the mountain 146 times. Mr. Ida, or Ida-san in the familiar form of address used by his followers, comes from a long line of Fuji-ko devotees. He made his first ascent of Fuji at the age of seven with his father, the previous leader of the group. A stone statue of the latter, dressed in pilgrim's garb, stands beside the Shinto shrine that marks the beginning of the traditional climbing route from Fuji-Yoshida. The statue was erected to commemorate his 150th ascent of the mountain. One of Ida-san's ancestors was the first person to construct a Fuji replica, back in 1765. The owner of a small factory in Tokyo, Ida-san has devoted much of his time and wealth to the Fuji-ko movement, reconstructing a *torii*, or cere-

monial gateway, at the Fuji-Yoshida shrine and sponsoring expensive ceremonies. When I asked if his son would succeed him as leader, he smiled and said, "There is a saying that if the leadership of the group stays in a family for three generations, that family will go bankrupt."

A quiet, gentle man, Ida-san is deeply devoted to preserving the teachings and memory of Jikigyo Miroku. On the yearly ascent of the volcano, he carries up a ritual mirror inscribed with the word *Sengen*, the name of the deity of Mount Fuji, and places it in front of the rock where the saint martyred himself. He regards this place as the most sacred feature on the entire mountain, a focal point of the pilgrimage. The teachings that Jikigyo dictated there as he was fasting to death for the sake of others encourage people to care for each other. As Ida-san puts it, "The most important thing in climbing is the inner strength to help each other, so that not just the strongest but all the members of Fuji-ko reach the goal. Mount Fuji, from our religious point of view, is a mountain that accepts all comers."[21]

Like other religious phenomena in Japan, pilgrimage has undergone a process of secularizing, taking on the modern forms of mountaineering and tourism—when the latter have not simply replaced it. On Fuji, the Japanese peak most often climbed, hikers and tourists follow routes established long ago by practitioners of Shugendo and devotees of Fuji-ko. In the height of the official climbing season, during the months of July and August, more than a million people swarm up the sides of the volcano to gather like insects around the summit. So many have made the ascent, scattering refuse in every direction, that Fuji has the dubious distinction of possessing the world's first polluted snowfield. Here and there, sprinkled like grains of rice among the crowds, appear groups of pilgrims dressed in white, wearing straw hats and carrying staffs, still seeking the sacred where the mountain meets the sky.

Four major routes zigzag up the gritty ash slopes of Fuji, converging on the crater rim from different sides of the volcano. The long lines of climbers, representing all ages from smooth-faced children to wizened grandmothers, pass through ten stations, equipped with huts and such amenities as food and bedding. The hut keepers, descended from the religious guides and shrine attendants of the past, burn characters into each climber's staff, indicating the station reached. Whether religiously motivated or not, the Japanese take these staffs home as treasured mementos of their climb of Japan's most famous landmark. Most spend the night at the eighth station to reach the summit the next morning at dawn.

Moving through the eerie silence of the final hours of night, pebbles crunching beneath their feet, the climbers come at last to the crater rim, poised on the edge of darkness. As they wait on the summit, arms crossed over their chests to hold in the warmth of their bodies, the sky to the east gradually brightens and the sun rises from the sea. Beneath their feet, above a darkened world, a reddish gold light ignites the tops of clouds to leap like fire from one to the next, spreading to the limits of the far horizon. Off to the west, in the shadow of the mountain, the dark forms of hills and ridges emerge in graceful curves from a sea of soft blue mist. Immersed in light that flows through the air like water, tourist and pilgrim alike stand in wonder, bathed in splendor beyond the world of ordinary experience.

Mount Koya

Two hundred miles southwest of Mount Fuji, hidden in the forested depths of the Kii Peninsula, lies another sacred mountain that has played a major role in the religious and cultural history of Japan. Quite different from lofty Fuji, whose isolated cone can be seen from a great distance, Mount Koya rises to a modest altitude of only 3320 feet in a tangled range of densely wooded mountains concealing its secret beauties from the eyes of the outside world. A mountain sanctuary like Wu-t'ai Shan in China, Koya consists of an elevated plateau surrounded by eight peaks, whose soft forms enclose the central area in a womb-like space. Only by ascending to this plateau by means of a steep cableway can the visitor distinguish Mount Koya from the confusing maze of its surroundings. There, secluded from the bustle of the outside world, appears a gentle valley of giant cedars dotted with the sharp spires and broad pagoda roofs of Buddhist monasteries and temples. Over the occasional honking of automobiles, muted here compared to most other places in Japan, the murmured sound of chanting punctuated by the occasional ringing of a bell rises to hover like smoke among the trees. From this sacred area emanates the atmosphere of another world where people live in harmony with the underlying reality of their surroundings.

More than any other mountain in Japan, Mount Koya is intimately involved with the life of one person—Kukai, the founder of Shingon Buddhism and one of the most remarkable figures in Japanese history. A brilliant man of many accomplishments, Kobo Daishi—as Kukai is also known—was a gifted philosopher, artist, poet, engineer, and writer, as well as a great religious leader. Born into an aristocratic family in A.D. 773, he abandoned a worldly career to become a wandering monk whose life and work profoundly influenced the subsequent course of Japanese civilization. Tradition credits him, among other accomplishments, with devising the phonetic script used in conjunction with Chinese characters to write the Japanese language, and many Japanese today regard him as the patron saint of learning and calligraphy in Japan.[21]

In the course of his early wanderings, Kobo Daishi came across Mount Koya and spent time meditating in its wild but peaceful setting. He regarded mountains as the ideal environment for spiritual practices leading to enlightenment. The natural freedom of stream and forest, far from the irksome constraints of society, especially appealed to him, as his own description of a wandering ascetic similar to himself so clearly attests:

The blue sky was the ceiling of his hut and the clouds hanging over the mountains were his curtains; he did not need to worry about where he lived or where he slept. In summer he opened his neck band in a relaxed mood and delighted in the gentle breezes as though he were a great king. . . . Not being obliged to his father or elder brother and having no contact with his relatives, he wandered throughout the country like duckweed floating on water or dry grass blown by the wind.[22]

His words express exactly the kind of sentiments that make mountains attractive to so many people today.

From Mount Koya, Kobo Daishi's wanderings took him eventually to China, where he studied with various Buddhist masters. According to legend, when it came time to return, he went to the shore of the ocean and hurled a three-pronged diamond thunderbolt, the symbol of esoteric Buddhism, toward Japan, praying that it might land at a place suited to become the center of the secret teaching in his native land. The *vajra*, as this thunderbolt is called in the Sanskrit language of India, dwindled into the sky to vanish over the ocean. Years later, when he went to look for the thunderbolt, he met a hunter with two dogs, one black, one white. The man told him that the *vajra* was on the sacred plateau of Mount Koya and offered to lead him there. Then, revealing himself as the god of the hunting field, the man mysteriously vanished. Guided by the hunter's dogs, Kobo Daishi climbed Mount Koya and found the *vajra* hanging in an unusual pine tree with clusters of three needles corresponding to the three prongs of the thunderbolt. The goddess of the mountain, Nibutsu Hime, none other than the mother of the hunting god himself, emerged from the forest and gave him permission to construct a monastery. In return he built a shrine in honor of the two deities as guardians of Shingon and its sanctuary on Mount Koya. In this way Kobo Daishi incorporated the older Shinto *kami* or spirit of the sacred mountain into the new Buddhist tradition he had imported from China and India.[23]

At the time Kobo Daishi established his center, Koya was a remote mountain wilderness, frequented only by hunters and hermits. Whether or not he found the *vajra*, as in the story, in 816 he did write a letter to the emperor of Japan asking for permission to build a monastery on the sacred site:

According to the meditation sutras, meditation should be practiced preferably on a flat area deep in the mountains. When young, I, Kukai, often walked through mountainous areas and crossed many rivers. There is a quiet, open place called Koya located two days' walk to the west from a point that is one day's walk south from Yoshino. . . . High peaks surround Koya in all four directions; no human tracks, still less trails, are to be seen there. I should like to clear the wilderness in order to build a monastery there for the practice of meditation, for the benefit of the nation and of those who desire to discipline themselves.[24]

Kobo Daishi was drawing here on older Shinto beliefs concerning the role of mountain deities in protecting the local community and, by extension, the nation itself. As a further justification, he spoke of the important role of mountains as settings for many of the Buddha's sermons and pointed out that in China "students of meditation fill the five Buddhist temples on Wu-t'ai Shan."[25]

The emperor granted him permission to establish a monastic center on Mount Koya, and Kobo Daishi immediately dispatched his disciples to build meditation huts there. Tied up with commitments in the capital city of Kyoto, he himself was only able to come to the sacred mountain two years later, in 818. The following year he performed the rituals of consecration to sanctify the site and clear it of all demonic influences. The actual construction of temples, lecture halls, and other buildings took a long time — too long for Kobo Daishi to see them completed in his lifetime. However, he came as often as he could to meditate in the place where he felt most at home in the universe. The beautiful poetry he composed there reflects his deep feelings for nature and the joy he experienced in his mountain retreat:

Spring flowers and autumn chrysanthemums smile upon me,
The moon at dawn and the breezes at morn cleanse my heart.[26]

At the age of sixty-two, his mind focused in meditation, Kobo Daishi died on his beloved mountain, surrounded by his disciples. At his request, instead of cremating him, they buried his body on the eastern side of Mount Koya, in a beautiful forest of cedars that he loved to frequent. According to traditional accounts, however, he did not die, but instead entered the mountain, where, suspended in a state of meditation, he awaits the coming of the next Buddha, Maitreya. A vast cemetery grew up around this site so that people could have their remains interred in the living presence of Kobo Daishi, whose spirit, they believed, still remained in this world, invisibly working for the liberation of all beings. Over the succeeding centuries monks journeyed forth from Koya to spread the teachings of Shingon throughout Japan and to bring back to the holy mountain the ashes of all those who saw it as the gateway to paradise and the ultimate goal of enlightenment itself.

In keeping with its remote and sequestered setting, Mount Koya is one of the most important centers of esoteric Buddhism in Japan — the physical and spiritual home of Shingon, the Buddhist school that teaches the secret path leading to the swift attainment of enlightenment in this very life. For more than a thousand years the priests and holy men of this mountain sanctuary have preserved a rich and complex tradition centered on the visualization and evocation of Dainichi Nyorai, the Buddha of All-Illuminating Wisdom. Like the summit of Fuji, the entire plateau of Mount Koya is regarded as the divine abode of Dainichi, whose name in Japanese means the "Great

Sun," referring to the blazing light of enlightenment itself. Over a hundred monasteries and temples, some of them wooden pagodas covered with moss, enshrine his gilded image, serenely seated in the midst of peaceful and wrathful deities who symbolize the spiritual forces that must be awakened on the way to realizing the nature of ultimate reality.

Many of these temples also function as inns for the million or so pilgrims and visitors who come each year from all parts of Japan. There, in a setting of beautiful gardens, surrounded by magnificent works of art, they absorb the contemplative atmosphere of the sacred mountain. Some come to study the deepest points of Buddhist doctrine with priests and professors at the Shingon university located in Koya. Others pass their time in a more leisurely way, visiting the numerous shrines and monuments scattered around the plateau. A small town complete with shops and restaurants caters to their material needs so that they may make of their pilgrimage a spiritual holiday. During the summer large groups of schoolchildren, each

A Buddhist image adorned with brightly colored cloth stands in the graveyard of Mount Koya. (Edwin Bernbaum)

group marked with caps of a distinctive color, tour the sacred sites, sketching the temples and learning about the religious heritage of their country.

Most of the visitors to Mount Koya focus their attention, and devotion, on its graveyard, the largest and most impressive in Japan. Over a mile long, it runs through an ancient grove of gigantic cedars, whose straight trunks stand like columns of silver gray marble rising into a cloud of foliage. On either side of the main path, rows of stone monuments of all sizes, many of them gone green with moss and lichen, vanish into the shadows of the forest. Some of these monuments are in the shape of jovial little dolls, almost primitive in the open simplicity of their expressions, and are draped in brightly colored clothes to honor the dead whose memory they commemorate. Gray wisps of smoke, rising from incense burners, drift up through beams of sunlight, slowly dissipating among the trees. A scent of holiness hangs in the air.

When I came to Koya in the summer of 1978, the graveyard with its trees had a particularly haunting effect. Walking through the cedars, watching pilgrims make offerings to images of silent Buddhas, I was strongly reminded of hikes I had taken through redwood groves in California, such as the Muir Woods outside San Francisco. Here in Koya the cedars rose as straight and tall as California redwoods, with the same aura of primeval simplicity, but the additional presence of shrines and incense accentuated the natural sanctity of the forest, producing an overwhelming atmosphere of spiritual devotion in which the living could commune with the dead. A sense of another reality, another world, deeper and more mysterious than the one I knew, hovered on the edge of awareness, palpable and evanescent as the gray mist of incense floating around me, ascending toward the sky.

The sense of the sacred issuing from peaks like Koya and Fuji gradually spread to other parts of Japan. Kobo Daishi spent much of his life traveling between mountain and plain, the sacred and the profane. In the pure and peaceful space of Mount Koya, he found the spiritual sustenance needed to further his work in the corrupt and congested atmosphere of Kyoto, where he built another center for the propagation of esoteric Buddhism. As a Japanese commentator on his life and works has written, "In time Kukai came to be regarded as a prototype of the holy ascetic of magical power who from time to time descended from the sacred mountain into the world to help save people."[27] In this way he provided a powerful model for the *yamabushi* practitioners of Shugendo, who incorporated many of the ideas and practices of esoteric Buddhism in their mountain religion.

The *yamabushi* drew, in particular, on a doctrine basic to all of Shingon philosophy — the idea that one can attain enlightenment in this very body and life. Practitioners of esoteric Buddhism put this doctrine into practice through rituals and meditation based on the visualization of two mandalas or sacred circles representing the universe as the realm of Dainichi Nyorai, the Buddha of All-Illuminating Wisdom, who stands at the center of all things. The first, called the womb mandala, embodies the principle of enlightenment inherent in all beings. It has the basic form of an eight-petaled lotus. Impregnated by the compassion of Dainichi, this womb gives birth to the Buddha each person is destined to become. The second mandala, called the diamond realm, symbolizes the awakened wisdom of enlightenment, no longer latent but fully manifest in the person of Dainichi himself. Through offerings of fire and recitations of mantras, the practitioners of Shingon visualize themselves passing through these mandalas in order to reach the ultimate goal of Buddhism right here in this world.[28]

Seeking to make such practices as concrete as possible, a matter of experience rather than imagination, Kobo Daishi imposed the womb mandala on Mount Koya itself and viewed the eight peaks of the mountain as the eight petals of the lotus representing the sacred circle. In the plateau at the center of this schematic blossom, he laid out the main temple complex in the form of the diamond mandala, with an image of Dainichi enshrined in the central pagoda. The unity of the two mandalas, one within the other, represented for him the transcendence of all differences, in particular the difference between the beginning and the end of the path to Buddhahood. In this way he sought to transform his disciples' experience of Mount Koya and the world around it into that of the divine realm of the Buddha who embodies the ultimate nature of reality. One who could see what was truly there could walk through the two mandalas to the goal of enlightenment, shining through the streams and forest of the mountain itself.

This process of mandalization — as one scholar of Japanese religion, Allan Grapard, has termed the way in which geographical places take on the character of mandalas — spread to other mountains, deepening and enhancing their sacred nature. By the twelfth century, practitioners of Shugendo to the east of Koya had projected the womb and diamond realms of Shingon onto an immense mountainous region stretching from Mount Yoshino in the north of the Kii Peninsula to Kumano in the south. The two mandalas overlapped in the middle, at Mount Omine, a peak with special significance, as the following Japanese text reveals:

This peak is the pure temple of the two realms: it is the original, noncreated mandala; the summits covered with trees are the perfect altars of the nine parts of the diamond mandala, and the caverns filled with fragrant herbs are the eight petals of the lotus in the womb mandala. Mountains and rivers, trees and plants are the true body of the Buddha Dainichi. . . .[29]

Yamabushi adopted the practice of going from mountain to mountain, following pilgrimage routes laid out on the pattern of painted womb and diamond mandalas found hanging on temple walls. At each place where a deity would be invoked in a particular mandala, they would perform the corresponding ritual on an actual peak. By entering the mountains in this way, they would plunge into the divine realm of the Buddha, seen as this very world.

Unfolding from the summits of sacred peaks, mandalas blossomed like giant flowers, spreading over the landscape of Japan. Overlapping with each other, they eventually covered the entire country, filling the atmosphere with the fragrance of their sanctity. The mandalas that emerged from mountains as the original seed beds of sacred power and space, embodied from ancient times in the *kami* deities of the Shinto tradition, transformed Japan itself into a divine nation in the eyes of its people, a land protected and blessed by the gods. It makes perfect sense, therefore, that in the modern period, following the Meiji Restoration in 1868, one of the most important symbols of the country became Mount Fuji, the highest and most beautiful mountain in Japan. In its inspiring form the Japanese saw a divine crystallization of all that they revered in their land and their society.

The divinization of the country came to an end when the disastrous defeat of World War II provoked a profound disillusionment with everything that had contributed to the growth of Japanese imperialism. Japan's present-day infatuation with technology and economic development has added to the process of disenchantment by shifting attention away from the realm of the sacred. However, at a deeper level reflected in the maintenance of traditional beliefs and customs, old attitudes toward the *kami* and the places they inhabit still influence Japanese views of the world around them. Mountains, the most impressive abodes of the *kami*, have reverted to their natural role as metaphors of spiritual power and freedom. Although modernization has masked their influence in the guise of such pursuits as mountaineering and tourism, mountains continue to inspire a sense of the sacred that unites the Japanese people in a love for the land on which they live and for everything connected with it — from the unearthly heights of heavenly peaks down to the mundane realities of everyday life.

SOUTH AND SOUTHEAST ASIA
COSMIC CENTERS

Dry hills of ancient rock, green ranges of jungle peaks, slick summits of wet granite, fiery volcanoes of fresh ash—a great diversity characterizes the mountains of South and Southeast Asia, reflecting the variety of cultures that fill the region. Aside from the Himalayas (covered in a separate chapter), South Asia comprises the Indian subcontinent and the island of Sri Lanka and includes crystalline mountains formed of some of the oldest bedrock in the world, remnants of the ancient continent of Gondwanaland. Southeast Asia covers the Asian peninsula east of India and south of China, along with its continuation in a line of Pacific islands that curves eastward toward Australia and New Guinea. Its mountains range from the Himalayan outliers of northern Burma through the hills of Thailand and Cambodia to the volcanoes of Indonesia. The region also includes the mountainous islands of Borneo and the Philippines.

For many cultures of South and Southeast Asia, a single idea of great power brings together these diverse and far-flung peaks—the concept of a cosmic center that gives order and stability to the universe around it. This idea finds its principal expression in two major images of sacred mountains: Mount Kailas as the abode of Shiva and Mount Meru as the axis of the cosmos. Throughout South Asia numerous hills and temples dedicated to Shiva as Lord of the Universe are regarded as replicas of his Himalayan peak; they serve as centers of sacred power, drawing pilgrims from all directions. South Asians also revere a number of peaks as fragments of Mount Meru given them by

the gods to bring them closer to the center of the cosmos as the source of blessings and stability. With the spread of Hinduism and Buddhism from India through Southeast Asia, images of Kailas and Meru have adapted to local beliefs and passed from peak to peak, traveling as far as the distant islands of Indonesia. Along the way the idea of a cosmic center enshrined in these mountains has attached itself to kings and religious leaders, giving them the stature and power of gods.

SOUTH ASIA

South of the Himalayas and the Gangetic Plain, low mountain ranges border the Deccan Plateau, which makes up the well-worn heart of the Indian subcontinent. The Vindhya Range forms the northern rim of this ancient plateau, while the Eastern and Western Ghats slant down its sides to converge in the Nilgiri Hills to the south. Aside from these ranges, smaller hills and ridges ripple across the Indian landscape, shimmering green and brown beneath the tropical sun. Across from the southern tip of India, a fragment of Gondwanaland rises out of the Indian Ocean to form the mountainous island of Sri Lanka, with its steep escarpments and jungle-covered peaks of gneiss and schist.

Since at least 1000 B.C., hills and mountains throughout South Asia have attracted attention as objects of religious veneration. After the Buddha attained enlightenment in the sixth century B.C., he delivered

Worshippers stream up and down the stairway of Besakih during the Eka Dasa Rudra festival, an elaborate ceremony of cosmic exorcism that takes place once a century. Gunung Agung, the sacred mountain of Bali, appears behind the split gate of the temple. (Fred B. Eiseman, Jr.)

many of his most important sermons on Vulture Peak, a rocky hill outside the North Indian village of Rajgir, not far from the holy city of Benares. Imitating the model of Vulture Peak, Buddhist communities elsewhere in India and Sri Lanka claimed local hills and mountains as sacred places blessed by their founder's teaching activities. The Jains, followers of a nonviolent religion arising during the same period, also revered mountain peaks as sites of important acts performed by their various Tirthankaras, or savior teachers, including Mahavira, a near contemporary of the Buddha. Two principal places of Jain pilgrimage today are sacred mountains: Mount Abu in the north, noted for the intricate delicacy of its temple sculpture, and Shravana Belagola in South India, with its enormous statue of Bahubali, son of the first Tirthankara.[1]

Following the rise of Hindu devotional sects around the third century B.C., numerous mountains became identified with the abode of Shiva, modeled on mythic images of Mount Kailas in the Himalayas. One of the most important today is the sacred hill of Arunachala in South India. Revealing its role as a cosmic center, the *Skanda Purana* says of it:

That is the holy place! Of all Arunachala is the most sacred. It is the heart of the world. Know it to be the secret and sacred heart-center of Shiva. In that place He always abides as the glorious Arunachala.[2]

Each year in November thousands of pilgrims come to Arunachala, the "Mountain of Light," to worship Shiva as a phallus of fire — a cosmic pillar of creative power. Drawn by the spiritual magnetism of the sacred hill, which he viewed as a manifestation of the god himself, Ramana Maharshi, one of the greatest Indian sages of the twentieth century, spent his life meditating and teaching at its foot.

Followers of Shiva also constructed many of their temples as man-made replicas of his Himalayan abode. East of Bombay a famous rock-hewn shrine carved out of the cliffs of Ellora in the eighth century A.D. is dedicated to Shiva as Kailasanatha, "Lord of Mount Kailas." In South India, near Madras, another well-known temple of the period, built at Kanchipuram, bears the same name. Like its counterpart at Ellora, it has the general form of a mountain covered with deities, portrayed in beautifully sculpted images, but there is little evidence that it was modeled on the particular shape of Mount Kailas. Hindus of the period knew Kailas from idealized descriptions in religious texts, not from direct observation.[3] Like the Himalayan peak it vaguely resembles, the temple serves as

an abode of Shiva, evoking his presence in the hearts of those who came to worship him.

The symbolism of Mount Meru as the axis of the universe and dwelling place of the gods lies imbedded in the architecture of many Hindu temples. The pointed superstructure rising directly over the sanctuary holding the main deity is called a *shikhara*, the Sanskrit word for a mountain peak. Like a mandala with Meru at the center, the temple renews the profane world of ordinary experience, transforming it into the sacred realm of a god. A number of well-known Hindu shrines in South India, such as Tirupatti near Madras, actually rest on hills regarded as pieces of Meru, reinforcing their association with the cosmic axis symbolized in the mountain. In so doing they express a concern for order and unity fundamental to Indian thought. Through the experience of the cosmic center embodied in the temple as a sacred peak, what seems haphazard and chaotic becomes ordered and meaningful. Organized around the image of Mount Meru, the world makes sense. Through the symbolism of the cosmic mountain linking earth to heaven, a vision opens of the ultimate path leading to release from the suffering of death and rebirth.[4]

At an early period Buddhist and Hindu ideas concerning sacred mountains spread to Sri Lanka. In the fifth century B.C., Indo-Europeans from North India sailed south to colonize the island, and their descendants became the Sinhalese, the people who dominate Sri Lanka today. Between the third and second centuries B.C., Buddhist missionaries from India converted the island to the Buddha's teaching. After Muslim invasions forced Indians to abandon Buddhism in the thirteenth century A.D., the Sinhalese maintained the religion in Sri Lanka, making it a stronghold of Theravada, or Southern, Buddhism. During the preceding millennium, Tamils from South India had migrated to the island; more came in the nineteenth century to work as laborers in the British Empire. A Dravidian people who preserved a mixture of Aryan and pre-Aryan traditions, they brought with them the sect of devotional Hinduism focused on the worship of Shiva and his son Karttikeya. Bloody clashes between Hindu Tamils and Buddhist Sinhalese have created the political turmoil that agitates Sri Lanka today. Despite their differences, both groups revere a mountain that brings together ideas associated with hills and peaks throughout South Asia. Because of its ecumenical nature, having attracted the attention of Christians and Muslims as well as Hindus and Buddhists, this mountain, Adam's Peak, has become one of the most famous sacred mountains in the world.

A Jain temple on Mount Abu with a mandala carved in its ceiling depicting a circular arrangement of deities. A priest sits within the shrine room. (©Robert Holmes)

Adam's Peak

Among the peaks set like precious stones upon the heights of Sri Lanka, known to the ancients as the "Isle of Jewels," one has the unique distinction of being regarded as sacred by the adherents of more major world religions—Buddhism, Hinduism, Islam, and Christianity—than any other mountain on earth: Adam's Peak. A marvelous cone of polished gneiss, its beautifully sculpted summit rises cleanly out of the green waves of jungle that swirl about its flanks. From the heights of its summit on the southwestern part of the island, long ridges of undulating hills ripple away to the gleaming edge of the Indian Ocean, a blue band on the horizon. Viewed from a ship at sea, the solitary situation of the mountain presents such a striking appearance that early Western travelers greatly overestimated its altitude. One of the first Europeans to see Adam's Peak was so impressed that he wrote:

It hath a pinnacle of surpassing height, which on account of clouds can rarely be seen; but it lighted up one morning just before the sun rose, so that [we] beheld it glowing like the brightest flame. It is the highest mountain on the face of the earth.[5]

The mountain is, in fact, only 7360 feet high, less than half the height of the Matterhorn, the archetypal Alpine peak to which other European travelers have tended to compare it.

On the very summit, high above the surrounding jungle, is a strange depression carved out of the rock in the rough shape of an enormous foot. This footprint makes the mountain sacred in the eyes of millions of Buddhists, Hindus, Muslims, and Christians, who come on pilgrimage each year by the tens of thousands to venerate it as a holy relic. According to the ancient chronicles of Sri Lanka, the Buddha, who lived in India during the sixth and fifth centuries B.C., visited the island three times to rid the country of demons and preach sermons to various divine beings. On his third visit he "left the trace of his footsteps on Sumanakuta"—an early Buddhist name for Adam's Peak.[6] According to a later account recorded by Fahsien, a famous Chinese pilgrim who came to India to obtain Buddhist teachings around A.D. 400, the Buddha used his magic powers to plant one foot near the capital of Sri Lanka and the other on top of the mountain. Local tradition maintains that his actual footprint lies engraved on a huge sapphire imbedded in the rock beneath the depression visible on the surface today.[7]

The followers of Islam regard the hollow as the footprint of Adam, whom they revere as the first prophet and patriarch of the human race. The high esteem in which they hold this biblical figure derives, in part, from earlier beliefs of Christian Gnostics, who venerated him as the primordial man, created from the breath of God. A fourth-century Coptic manuscript from the Middle East contains a cryptic reference to a footprint left by Adam, but fails to mention where the sacred relic might be. After the founding of Islam in the seventh century, a belief arose among Muslims that after Adam's expulsion from the Garden of Eden, he spent a number of years doing penance on a mountain in India before being reunited with Eve on Mount Arafat near the holy city of Mecca. When Arab seafarers in the ninth century brought back reports of the Buddhist relic on the summit of Adam's Peak, their countrymen concluded that it must be the footprint left by Adam in the course of expiating his sins.[8]

The adherents of two other major religions also revere the depression in the rock as a holy relic. Hindus worship it as the footstep of the great god Shiva, whose abode lies far to the north on the summit of Mount Kailas. Shiva seems to have come here to establish himself as an object of devotion at a fairly late date. The first evidence we have for a cult of his footprint on the summit of Adam's Peak occurs in the sixteenth century, when the chronicles record that a king of Sri Lanka repudiated the dominant religion of Buddhism and put control of the shrine in the hands of Shaivite priests devoted to the worship of the Hindu god. Before that time Hindus probably venerated the peak as the abode of Agastya, a noted sage of Hindu mythology, and Dharma, the divine embodiment of righteousness. Finally, not to be outdone by other traditions, the Christians of India and Sri Lanka have long claimed the relic as the footprint of Saint Thomas, the apostle who is said to have come to India to preach the Gospels after the death of Jesus. The Portuguese, who controlled parts of Sri Lanka for a number of years, encouraged this belief, although they were the ones who gave the mountain the name by which it is most widely known—Pico de Adam, or Adam's Peak.[9]

The name Adam's Peak comes from Muslim beliefs concerning the legend of Adam. The people of Sri Lanka, who are mostly Buddhist and Hindu, have other names for the mountain. In memory of the Buddha's visit to the peak, they call it Siripada, the "Glorious Footprint." Another name refers to the ancient Buddhist deity of the mountain itself. According to the early chronicles of Sri Lanka, written in Pali, when the Buddha came to the island, one of the deities who

The rock cone of Adam's Peak rises over the jungles of Sri Lanka. (Jeffrey Falt)

heard his sermons and made great progress on the path to enlightenment was Sumana, the god of Sumanakuta, or "Sumana's Peak." In the modern language of Sinhalese, Sumana has become Saman and the name of the mountain Samanala Kanda, the "Peak of Saman." Saman plays an important role in the Buddhist religion of Sri Lanka as one of the four deities charged with protecting the island from evil. He has given his name to the Samanalaya butterflies that come each spring to swarm like yellow clouds about the cliffs of the sacred peak. The Hindus of Sri Lanka call the mountain Shivan Adipadam, the "Primordial Step of Shiva." This name evokes images of the cosmic dance of Shiva in which, as supreme deity, he creates, preserves, and destroys the entire universe, all at the same time.[10]

For more than a thousand years, Adam's Peak has fascinated countless pilgrims, drawing them to its summit regardless of their different traditions and beliefs. The first recorded ascent of the mountain—undertaken for religious reasons—took place in the last decade of the seventh century. A Chinese biography of Vajrabodhi, an Indian master of esoteric Buddhist teachings, describes his climb of Adam's Peak in a matter-of-fact, almost contemporary, fashion:

When at last, he reached the foot of the mountain, he found the country wild, inhabited by wild beasts and extraordinarily rich in precious stones. After long waiting, he was able to climb to the summit and contemplate the impression of Buddha's foot. From the summit, he saw on the north-west the kingdom of Ceylon, and on all other sides the ocean.[11]

The Chinese account does not make clear whether Vajrabodhi was following an established route of pilgrimage or just exploring the area on his own. The earliest historical evidence for an actual Buddhist practice of ritually climbing the mountain to venerate the footprint on the summit occurs in rock-cut inscriptions dating from the eleventh century. They indicate that a king named Vijayabahu (1055–1110) ascended Adam's Peak and made arrangements for villages along the way to help other pilgrims in their attempts to reach the top. Ibn Batuta, an Arab traveler who visited Sri Lanka in 1340, wrote that an Islamic sage by the name of Imam Abu-Abd-Allah had already initiated a Muslim pilgrimage to the mountain in the tenth century.[12] Arabic inscriptions carved on the wall of a cave near Adam's Peak indicate that such pilgrimages had become established practice by the thirteenth century: one reads, "Muhammad, may god bless him [the father of Man]. . . ."[13]

The classic pilgrimage route begins in Ratnapura, the "City of Jewels," near a beautiful temple dedicated to Saman, the Buddhist god of Adam's Peak. Pilgrims who take the difficult route of ancient times and travel by foot leave behind the reassuring world of civilization to plunge into the wilderness of the peak—a maze of jungle-covered hills and ravines arranged like a series of ramparts and moats guarding the approaches to the sacred mountain. The hot sun of the open sky vanishes behind a dense canopy of overarching foliage, replaced by a cooler world of hanging vines, gently twisting in a greenish light that floats like mist between the trees. Watered by streams trickling through the lush undergrowth, flowers of various shapes and sizes appear here and there in stands of bamboo lining the trail, their bright colors muted in the soft shadows of the forest. The occasional cry of a wild creature or the musical call of a bird brings to sharp attention a sudden awareness of the invisible life concealed in the dark recesses of the surrounding jungle. As the path gains height and climbs toward the peak hidden in the sky above the trees, it winds through an obstacle course of enormous boulders draped with moss and hanging ferns.

As they move through this eerie world of mysterious beauty, the pilgrims chant from the religious texts they carry in their hands and perform the rituals that give their journey the added dimension of a spiritual quest. They stop at a river to bathe in its smoothly flowing waters and put on a clean set of clothes, purifying themselves for entry into the sacred presence of the holy relic above. Where the peak first comes into view, they pause to add a pebble to a cairn of stones. At another spot Buddhist pilgrims hurl a needle with a piece of thread into a bush to commemorate the place where legend says that helpful spirits mended a tear in the Buddha's robe. Tradition also dictates that those who are climbing the peak for the first time must wind a turban of white cloth about their heads.

In the old days pilgrims would emerge from the jungle to confront the final, terrifying obstacle of the journey: a smooth cliff of bulging precipices that they would have to surmount by means of ancient chains fastened to the polished rock, some say, by Alexander the Great when he came to India in the fourth century B.C. The more sober historical chronicles of Sri Lanka, however, state that these chains were installed by the minister of a local king at a much later date, in the thirteenth century A.D. Since the chains hung free with only their upper ends attached to the cliff, they would sway over the void as the terrified pilgrim lurched from rung to rung, anxiously seeking a firmer purchase on top. Stories tell of people watching in horror as an entire family, caught on the chains in a sudden windstorm, would be whipped about and flung, one by one, to their deaths on the jungle-covered rocks below.[14]

Today most pilgrims ascend the final cliffs by means

Pilgrims warm themselves by a fire, awaiting sunrise on the summit of Adam's Peak. (Jeffrey Falt)

of a staircase constructed on the other, easier side of the mountain. In 1947 the minister of transport made a vow to Saman, the Buddhist god of Adam's Peak, that in exchange for the deity's help in completing a hydroelectric project to provide power for the city of Colombo, the government would illuminate the pilgrimage route to the summit. In 1950, after the project had been completed, apparently with divine assistance, the minister fulfilled his side of the bargain by having electric lights installed along the staircase leading up the final peak. Seen from the distance at twilight, they appear as a string of bright pearls draped in a graceful curve across the mountain, a dark silhouette against a dark blue sky.[15]

Most pilgrims make use of these lights to climb the peak at night to reach the summit before dawn. As they huddle about fires, reciting prayers and warming themselves in the sharp chill of the early hour, a marvelous spectacle unfolds to the west. The long tapering shadow of Adam's Peak materializes on a layer of clouds and ridges, looking like an elongated finger pointing toward some ultimate mystery concealed behind a blue-gray mist on the edge of the world. As the light brightens, the delicate tip of this ethereal pyramid glides swiftly back from the horizon, contracting into the mountain from which it came, bringing with it a touch of the infinite that leaves the entranced pilgrims with the overwhelming sense of standing poised for one brief moment at the unimaginable center of the universe. Then, with the rising of the sun, they turn to pay homage to the sacred footprint beside them.

Whether Buddhist or Hindu, Muslim or Christian, the pilgrims join in a spirit of harmonious devotion, each worshipping the appropriate deity in his or her own traditional way. Some recite prayers and make offerings; others sit in quiet meditation. Those who have already made the pilgrimage ring a bell, once for each previous ascent. Many collect the rainwater that has accumulated in the depression of the holy relic or in a nearby well: they will take it back for its healing powers. Some offer a coil of silver wire as long as their bodies in a ritual meant to help them recover from illness. Whatever form their devotion takes, the pilgrims all seek the spiritual blessing or power left by the holy person or deity who, they believe, made the sacred footprint on the summit of Adam's Peak.

SOUTHEAST ASIA

Separated by the valleys of such rivers as the Irawaddy, the Salween, and the Mekong, mountain ranges curve off the eastern end of the Tibetan Plateau and meander down the peninsula of Southeast Asia to peter out in jungles and plains along the coast. Beyond the southern tip of the Malaysian mainland, mountains surge back up in a line of island peaks created by some of the most intense volcanic activity on earth—an activity made dramatically manifest in the great explosion of Krakatoa in 1883, which darkened skies around the world with clouds of dust and ash. North of this line of fire, traced by Indonesian islands such as Java and Bali, rises the highest mountain outside of Burma—the 13,455-foot jungle peak of Kinabalu in northern Borneo. For the Dusun people who live near Kinabalu, the granite slabs that cap its summit are the haunted abode of ancestral spirits.

Before traders introduced Indian culture in the beginning of the first millennium A.D., the peoples of Southeast Asia, whose ancestors had come mostly from southeastern Tibet and southern China, revered mountains as abodes of the dead and sources of water and fertility. Devoted to the worship of sacred hills as ancestral shrines of important chieftains, they responded with enthusiasm to Buddhist and Hindu ideas of Meru and Kailas as centers of cosmic power. Beginning with the ancient Cambodian kingdom of Funan in the second century A.D., Southeast Asians made these conceptions the basis for systems of divine kingship in which kings and emperors were treated as deities enthroned on the summit of one or the other of the two mountains of Indian mythology. The name *Funan* comes from an expression meaning either "Sacred Mountain" or "King of the Mountain," the title for the ruler of the country in his role as an incarnation of Shiva. The Khmer, who took over Cambodia in the ninth century and dominated much of Southeast Asia until the fifteenth century, built their capital cities as models of the universe centered around architectural representations of Mount Meru, such as the famous temple of Angkor Wat. Cambodians even called their temples *giris*, meaning "mountains."

Beginning with the kingdom of Pyu in the seventh century A.D., Buddhist rulers of Burma constructed their palaces as replicas of the residence of Indra on the summit of Meru. Whoever occupied the palace did so as the representative of the king of the gods, from whom he derived his authority to rule. According to this model of government, the wives and ministers of the ruler had to correspond in number and rank to those surrounding Indra on top of his mountain. Burma itself was a microcosm of the universe governed by the god of the sacred peak. The sym-bolism of Mount Meru as the center of cosmic power made it relatively easy to usurp the throne: all one had to do was to take the palace—success was proof positive that one had received the authority needed to rule the entire country in the name of Indra. The last king of Burma, who reigned in the nineteenth century, refrained from venturing out of his capital for fear that one of his relatives might overthrow him by moving into his residence as an uninvited guest. A few years later, after the British took over, a group of rebels armed only with swords tried to take back the kingdom by storming the palace, which the English had converted into a social club. Some officers there for drinks stopped the rebels with hunting rifles.[16]

Sometime before the fifth century A.D., the influence of Buddhism and Hinduism spread from the mainland of Southeast Asia to the islands of Indonesia. During the eighth and ninth centuries, the Sailendra Dynasty of central Java constructed the magnificent monument of Borobudur, the largest Buddhist structure in the world—and one of the most enigmatic. Hidden beneath the ashes of a volcanic eruption and the subsequent growth of jungle, Borobudur was discovered by Europeans only in the nineteenth century. Much of its symbolism remains unknown. Laid out in the form of mandala covered with statues and smaller monuments, called *stupas*, Borobudur appears to represent a model of the spiritual universe organized around the axis of Mount Meru. In climbing a series of galleries and terraces that rise toward its central spire, one follows a symbolic path to the ultimate peak of perfect enlightenment. Significantly, *Sailendra*, the name of the dynasty that built the mountain-shaped monument, means "Lord of the Mountains"—a reference to Meru itself.[17]

As Buddhist and Hindu dynasties vied for control of Indonesia, the merging of different traditions that held mountains in the highest regard transformed volcanoes of distinctive size and shape into sacred centers of great power and significance. Peaks that originally had only local importance as abodes of ancestral spirits became awesome replicas of Mount Meru, the axis of the universe. The highest peak in Java, 12,060-foot Mount Semeru, was named after Sumeru, the cosmic mountain of Indian Buddhist mythology. According to myths that developed during this period, Java and Bali were originally unstable, bobbing like boats on the surface of the ocean. To stop them from wobbling, the gods reached far to the north of India and brought Meru down to fix the islands to the earth. The body of the mythical mountain they drove into the interior of eastern Java like a nail to create the magnificent volcano of Semeru. The summit they set down not far from there, to become Mount Penanggungan, a perfectly shaped

The sun silhouettes the topmost spire of the stupa of Borobudur, Java, in golden red light. The Buddhist monument probably represents a vision of the cosmos organized around the axis of Mount Sumeru. (Eric Oey)

cone surrounded, like Meru, by four subsidiary peaks. Penanggungan played an important role in the ceremonial life of Java's most glorious empire, ruled by the great Majapahit Dynasty, which unified most of Indonesia in the fourteenth century A.D. Between 1935 and 1940 the ruins of more than eighty temples and shrines were discovered on the slopes of the sacred mountain. Carvings at one of these structures show that its builders regarded the spring feeding its bathing pool as the source of the elixir of immortality created when the gods used Meru for a stick to churn the ocean.[18]

When Islam took over Java in the sixteenth century, the mountains of that island lost much of their sanctity, rejected as sites of pagan worship. Bali, however, remained Hindu, ruled by refugees from the Majapahit Dynasty. According to Balinese myth, after Java fell to Islam, the Hindu gods moved to Bali, where they created peaks at the four points of the compass to have dwelling places of suitable height and grandeur. The highest mountain, Gunung Agung, they set at the east, the place of honor that receives the first light of the rising sun. In another version of the legend, the gods placed Gunung Agung at the very center, again

indicating its status as the most important peak in Bali.[19] Today the Balinese volcano reigns supreme as the most actively worshipped mountain in all the islands of Indonesia—and most of Southeast Asia.

Gunung Agung

Many of the peaks that form the dramatic highlands of Bali, an island noted for the beauty and prominence of its sacred mountains, are active volcanoes that run from east to west like a line of watchtowers set high above the deep blue sea. The ominous plumes of white smoke and steam that rise from their craters to twist and billow in the wind lend a visible dimension of physical power to the aura of sanctity that hangs, and sometimes rumbles, about their lofty summits. When the volcanoes erupt, as they often do in terrifying signs of divine displeasure, they bring destruction and death to the tranquil villages that cluster about their feet. Yet the very ash that sears the landscape produces in time the fertile soil that has made the islands a green paradise of plentiful food and pleasant life.

The people of Bali have a special reverence for the

sacred mountains that loom so visibly over their land and lives. There, on blue summits above the clouds, dwell the high gods and ancestral spirits who have the power to reward them for their virtue and punish them for their sins. As the divine link between heaven and earth, the mountains embody all that is holy and good, the source of innumerable blessings made manifest in the waters of life that flow down from their heights. Every hill and peak in Bali has a shrine; the higher the hill or peak, the more sacred the shrine. The sea, on the other hand, is associated with the underworld and all that is demonic and evil. The people of Bali regard the ocean with its turbulent waves and treacherous currents as a dangerous realm of terrifying monsters and malevolent spirits who threaten to disrupt the divine order of existence. Living in the intermediate zone between the mountains and sea, the Balinese seek to balance within themselves the system of opposites that pervades their worldview: high and low, gods and demons, good and evil, life and death.

According to Balinese tradition, after the gods had stabilized Java, they took a piece of the original Meru from Mount Semeru and used it to pin down the island of Bali. This piece, which the Balinese regard as a form of Meru itself, became the highest and most sacred mountain in Bali—a magnificent volcano called Gunung Agung. Situated in the northeast corner of the island, the peak rises in unrivaled majesty to an altitude of 10,308 feet. Its summit, visible from most parts of Bali, often floats above a layer of white clouds, appearing to belong to another higher world beyond the reach of ordinary mortals. As one gazes up toward it, looking into the light of the morning sun, the patterns that may appear in the sky can easily take the form of bright gods hovering about the highest peak. Steep slopes of gray ash laid down by recent eruptions make the ascent to the crater a difficult task, especially for those laden with food and other offerings to give to Mahadewa, the Great God of Gunung Agung. Beneath the forbidding reaches of the upper cone, the base of the volcano fans out to form a warm and inviting region of green terraces glowing with rice and tranquil villages shaded by tropical trees.

The people of Bali, in fact, speak of Gunung Agung as the "Navel of the World." This term comes from its association with Meru, the mythical mountain at the center of the Hindu universe. The name Gunung Agung means in the Balinese language the "Great Mountain," a direct reference to the peak as Mahameru or Great Meru, the title by which the cosmic axis of Indian mythology is known in Bali. The conception of Gunung Agung as the Navel of the World is no mere abstraction created to delight the mind of a scholar. The mountain functions for the Balinese in a very real and tangible way as the principal center toward which they orient themselves in the course of their daily lives. Each night when they lie down to sleep, they must point their heads toward Gunung Agung: to align themselves the other way, pointed toward the sea, would be equivalent to dying and assuming the position of a corpse. Without a sense of their relationship to the mountain, they become ill with feelings of dizziness and nausea. Since most Balinese live south of Gunung Agung, the peak defines for them *kaja*, the sacred direction of the north, in opposition to *kelod*, the southern direction of death, represented by the sea.[20]

Every temple on the island of Bali, no matter how small, contains a shrine dedicated to the deity of Gunung Agung. They resemble little pagodas in the shape of miniature peaks with a single roof made of sugar palm fiber. Known as *merus*, these pagodas—and their full-size counterparts with multiple roofs at major temples, also termed *merus*—call to mind Mount Meru, the cosmic mountain embodied in Gunung Agung. The shrines link the island's many temples to the largest and holiest of all—Pura Besakih, the mother temple of Bali, perched high on the southwest slope of the sacred peak itself. There, beneath an impressive view of the massive cone wreathed in streaming clouds that suggest the latent power brooding in its crater, a great complex of nearly two hundred pagodas, shrines, and courtyards cascades in graceful steps down a series of terraces cut into the side of the mountain. A ceremonial gate shaped like a temple spire split vertically down the middle guides the eye up to the sacred summit of Gunung Agung as it leads the body into the spiritual heart of Besakih, an elaborate shrine of three altars dedicated to Sanghyang Widhi Wasa, the supreme deity in his manifestation as Brahma, Vishnu, and Shiva, the Hindu gods responsible for the creation, preservation, and destruction of the universe.

Besakih with its awesome backdrop of Gunung Agung is the site of one of the largest and most impressive ceremonies in the world—the great Eka Dasa Rudra festival, so magnificent and powerful that tradition dictates that it be performed only once a century. Over the course of several months, in which everyone on the island visits the temple, the high priests of Bali summon all the deities of the Hindu pantheon in a series of elaborate rituals intended to dispel the forces of evil and destruction that threaten the order of existence. In a great ceremony of exorcism and purification, they seek to restore harmony and balance to the entire universe, from the lowest depths of hell to the highest reaches of Heaven. Eka Dasa Rudra means literally the "Eleven of Rudra," indicating the all-encompassing nature of the ritual.

"Eleven" stands for the eleven directions that account for the totality of space: east, south, west, north, southeast, southwest, northwest, northeast, the nadir, the zenith, and the center. "Rudra" refers to the one god Sanghyang Widhi Wasa in his wrathful manifestation as Siwa or Shiva, the Hindu deity of destruction invoked to pacify or destroy the forces of evil.[21]

Tradition allows one exception to the rule that Eka Dasa Rudra take place only once a century; in times of exceptional trouble, the ceremony can be performed to purify the world of malevolent influences. In 1963 Indonesia was experiencing famine and political turmoil that eventually led in 1965 to the overthrow of President Sukarno and a bloodbath against Communists in which 300,000 people lost their lives. The priests made a controversial decision to conduct the ceremony to restore harmony to the nation. As they began preparations in February 1963, Gunung Agung, which had remained dormant for 136 years, stirred with ominous rumblings and began to give off erratic puffs of turgid smoke and noxious fumes. On March 17, just nine days after a major sacrifice to the god of the mountain, the volcano erupted with a tremendous roar, spewing a black cloud thousands of feet into the sky. Waves of superheated water mixed with searing ash and poisonous gases swept down the slopes of the peak, wiping out villages and killing more than 1000 people. A strange dry rain of smoking pebbles fell from the sky, clattering on the ground and turning the bright green terraces into a gray wasteland of ruined fields and crops.

A villager who just managed to escape from his villege temple before it burned described his terrifying experience:

Suddenly it was dark again. I ran out of the temple, but a cloud that was very hot came toward us, and I went back and prayed with the others.

There was no noise at first, but then the duk-duk-duk-duk-duk *of falling stones. Some people in the temple seemed to be sleeping. I tried to wake them, but they would not answer—they were dead. There were children, too, but they could not cry. They made strange wailing noises, because they had ashes in their mouths.*[22]

In May the volcano erupted again, killing 100 more people in villages along its northern slopes. Most of these and the earlier victims died in avalanches of hot mud and lava accompanied by searing gases. In all, more than 1500 people perished. Eighty-five thousand had to flee their homes, half of them never to return. The bitter ash that fell miles from the volcano rendered large areas of Bali uninhabitable for a generation, forcing 40,000 refugees to be resettled on other islands in Indonesia. The Balinese interpreted the disaster as the deities' chastisement for sins they had committed. One person told me that many of the people blamed it on the government for having skimped on offerings to the gods.[23]

Throughout this period, even during the eruptions themselves, the ceremonies of Eka Dasa Rudra continued on the slopes of the volcano itself. Although ash fell on Besakih, covering it in a layer six inches deep and flattening weaker structures made of thatch, most of the temple survived intact, its pagodas and shrines left standing like black tombstones in a gray cemetary. The people of Bali took the sparing of the temple as a sign of the great sanctity of their holiest shrine. After the first eruption, the government tried to keep worshippers away from the temple, but their faith in the inherent goodness of their sacred mountain was so strong that they continued to come in great numbers. The ceremony itself, however, was cut short.

Sixteen years later, in 1979, when the time was judged correct, Eka Dasa Rudra was performed without incident. This time the festival took place in an atmosphere of peace and rejoicing, accompanied by great pageantry. Each day, beginning at the end of February, thousands of Balinese came to Besakih in long processions bearing gifts for the gods—flowers, fruit, animals, and grains. Using all the artistic skills for

A ceremonial dancer performing in the Eka Dasa Rudra festival at Besakih Temple on Gunung Agung. (Eric Oey)

A rice dough offering at Eka Dasa Rudra festival depicts Bhoma, god of the middle world and symbol of growth and vegetation.
(Eric Oey)

which the people of Bali are famous, hundreds of them spent countless hours sculpting rice flour on structures of woven palm leaves to create ornate and colorful offerings depicting the highest deities in the midst of the crowded realms of existence over which they rule. Like the fruits of the earth from which they sprang, these magnificent works of art flowered for a brief moment before being consumed in ceremonies symbolizing the great sacrifice that constitutes the eternal round of life and death. In honor of the deities being invoked in these rituals, troops of men and women wearing graceful costumes with fanlike head-dresses came to make their offerings with the sacred steps and gestures of traditional Balinese dances. By the time the multiple ceremonies of the festival drew to an end in early May, more than a million people had come to worship the highest gods on the slopes of the holy mountain of Gunung Agung.

At an important juncture several weeks after the opening rituals, the high priests invited the deities of other sacred mountains to come to Besakih and honor the festival with their exalted presence. Emissaries were dispatched to Mount Semeru in Java and Mount Rinjani on the neighboring island of Lombok. They returned with bamboo tubes filled with water ritually drawn from springs high on the slopes of the sacred peaks. These tubes, along with others gathered from holy places in Bali, most notably the volcanic lake of nearby Mount Batur, embodied the essence of the deities themselves. The waters of Semeru had a particular importance for the Balinese, who regard the deity of that mountain as the father of the gods of Gunung Agung, their most sacred peak, and Mount Rinjani, the volcano whose silhouette looms across the straits in Lombok. In an elaborate ceremony the deities of Besakih were ritually installed in images and ceremonially carried to the entrance of the temple to welcome their distinguished guests.

On March 25, as the festival approached the main sacrifice of Eka Dasa Rudra, a procession set out from a village east of Besakih in the early hours before dawn to climb to the summit of Gunung Agung. With them they had a buffalo, a goat, and a duck, along with other sacrificial offerings for the god of the sacred mountain. Half asleep, they stumbled in darkness across the treacherous ground of a recent ash flow and up a path winding through scrub and wasteland. As the sun emerged from behind the glowing outline of Mount

Rinjani, visible across a lavender sea of water and cloud, they reached the site of Pura Pasar Agung, the highest temple on Gunung Agung, destroyed in the eruption of 1963. After stopping at a nearby spring to pray and sprinkle themselves with holy water, they proceeded toward the crater itself. Hauling and pushing the reluctant buffalo, perspiration mixing in their eyes with the gritty dust, they struggled up steep slopes of frustrating rock and ash. Collapsing to rest every few minutes, their hearts pounding and their breath coming in staccato gasps, they broke up into random groups, zigzagging across the final cone.

As they approached the crater, mist blew in, cutting them off from their usual surroundings and isolating them in an unreal world of blurred forms and deceptive shadows. Carefully watching out for a misstep that would plunge them to their own deaths, the men pulled the buffalo up to a crevice between two outcrops of rock on the rim itself. Gaps in the blowing cloud revealed wisps of smoke rising from the crater floor hundreds of feet below; changes in the direction of the wind brought the pungent smell of sulfur. Each person in the party approached the rim and paid homage to the god of Gunung Agung, whose presence they could feel swirling in the mists around them. While the officiating priest and his wife looked on, sanctifying the sacrifice, the men quickly shoved the gently mooing buffalo over the edge and threw the goat and duck after it, along with the other offerings. Then, their task completed, everyone turned and scrabbled back down the mountain to the familiar realm of ordinary mortals.

The ceremony on the summit of Gunung Agung, along with eight similar rituals conducted at various other sacred places in Bali, purified the environment of oceans, lakes, mountains, and heavens in preparation for the performance of the major sacrifice of Eka Dasa Rudra, held at Besakih itself on March 28. On the appointed day two hundred thousand people massed on the slopes of the sacred mountain to participate in the greatest religious ritual they would ever have the opportunity to witness. Although not a Hindu himself, President Suharto came with an entourage of high officials to express his government's tolerance and respect for the sanctity of all the diverse religions practiced in Indonesia. For this occasion the shrines and staircases of the temple glowed with ceremonial umbrellas and streams of brightly colored cloth, made all the more striking by the black pagoda towers rising like somber mountains above them.

Intoning mantras and chanting hymns of praise, twenty-three high priests—the most ever to gather in one place—performed rituals over the offerings of numerous animals sacrificed the previous day. They called on Sanghyang Widhi Wasa, the supreme deity in all his myriad manifestations, including his lofty form as the god of the sacred mountain, to bestow his blessings and restore peace and harmony throughout the universe. Invoking him as Shiva, the wrathful lord of destructive forces, they offered the sacrifices spread before them as gifts to placate and purify the demons hidden within every human being that relentlessly afflict the world with misery and suffering. This time the god of Gunung Agung seemed to hearken to their entreaties: the volcano did not erupt.[24]

Images of Kailas and Meru play important roles in various systems of meditation found in South and Southeast Asia. As a means of attaining unity with the underlying reality of the cosmos, practitioners of Hindu and Buddhist yoga often visualize the universe as one with their bodies. In their visualizations they identify the central axis of Mount Meru with a channel that rises from the base of the spine to the top of the head. Spaced along this channel lie centers of energy called chakras or wheels. Symbolized by lotus blossoms with different numbers of petals, these chakras control the various emotional and spiritual states that people experience in daily life as well as meditation. By activating the chakras one after the other in the practice of yoga, a Hindu climbs the symbolic mountain within his body to merge with Shiva on the summit of Kailas, visualized as a thousand-petaled lotus on the crown of his head. Buddhists also visualize the central channel as Mount Meru, but they attain the heights of spiritual realization through union with a Buddhist, rather than a Hindu, deity. For Hindus the goal of the inner ascent is to become one with the true self, symbolized by Shiva; for Buddhists it is to awaken to the empty nature of reality, embodied in a Buddha.[25]

The tantric systems of meditation that make use of these visualizations are esoteric and difficult to comprehend. Although we cannot follow the intricacies of their practice, they suggest the more general significance of mountains in South and Southeast Asia. According to these systems, the deities we see enshrined on sacred peaks actually reside within ourselves as expressions of our own true nature, however we may conceive of it—as the true self, God, or emptiness. The inner symbolism of the footprint on Adam's Peak and the cosmic axis in Gunung Agung lead us to the realization that what we seek, the center of meaning in our lives, lies right here in the place where we stand. If we can know that place as it truly is, if we can recognize the sacred mountain within ourselves, we can, as in the Eka Dasa Rudra ceremony, exorcise the inner demons of our illusions and find the harmony we need to live together in peace.

· 6 ·

THE MIDDLE EAST
HEIGHTS OF REVELATION

As SUPREME SITES OF DIVINE POWER and revelation, the mountains of the Middle East have influenced the course of Western civilization, shaping the values of millions of people throughout the world. One mountain, Mount Sinai, is inextricably associated with the Ten Commandments, the basis of law and ethics in modern society. Another, Mount Zion, embodies the messianic vision that has inspired the highest ideals and aspirations of the Western world. Events on the heights of other mountains have played a major role in the lives of prophets and founders of three major religions—Judaism, Christianity, and Islam. Although many of these mountains serve as places of pilgrimage, they assume an even greater importance in works of religious literature, where they stand out as sublime symbols of the mind and spirit.

In their material forms the mountains of the Middle East range from the lofty volcanoes of Iran and Turkey in the west and north to the harsh desert peaks of Sinai and Saudi Arabia in the east and south. The northern end of the Great Rift, a fracture in the earth's crust that runs down to southern Africa, divides the well-watered hills of Israel and Lebanon from the drier ranges of Jordan and Syria. Scattered among these mountains—on their heights and in the flood plains between them—lie the ruins of some of the most ancient civilizations known to history—those of Sumer, Babylon, and Canaan. Discoveries of clay tablets dating back to the third millennium B.C. show that since the earliest recorded times sacred hills and peaks have played important roles in the religion and culture of the Middle East.

The Epic of Gilgamesh, the oldest surviving work of epic literature, speaks of a journey through an awesome mountain guarding the approach to the garden of the sun and the secret of immortality. Although the Sumerians and Babylonians who composed this work lived on the flat plains of Mesopotamia, far from mountainous regions, the hazy images of distant peaks beyond the horizon influenced these peoples' conceptions of the world around them. They built great temples of bricks, called ziggurats, in the form of stepped pyramids that mimicked the role of sacred mountains, linking the earth to the sky and providing a place for mortals to communicate with the gods. The names of some of these temples—Assur, "House, the Mountain of the Universe," and Larsa, "House, Link of Heaven and Earth"—suggest an underlying identification with a mythic peak at the center of the cosmos. Modern scholars disagree over whether ziggurats were actually viewed as mountains, but most of them agree that these temples did function as cosmic centers connecting various spheres of existence. In any case, the Tower of Babel in the well-known story of the Bible appears as an enormous ziggurat in the shape of a man-made mountain.[1]

Unlike the low-lying plains of the Tigris and Euphrates, the land of ancient Canaan to the west, where the modern states of Israel, Lebanon, Syria, and Jordan now lie, is mountainous. Here, where the people lived among hills and valleys, the influence of sacred mountains on religious traditions is much more apparent. Clay tablets inscribed in the second millennium B.C. and found near the ancient city of Ugarit

Jesus being transfigured on the summit of Mount Tabor, flanked by Elijah and Moses—a major event in which he is revealed, for Christians, as the Son of God. THE TRANSFIGURATION. *Painting by Raphael, Vatican Pinacoteca. (Scala/Art Resource, N.Y.)*

in 1929 speak of a number of Canaanite gods closely associated with prominent peaks. The two most important gods, El and Baal, had their own mountains, on whose summits they dwelled and where shrines were built to worship them. Seated on the throne of his peak, as a king in the splendor of his royal tent, El would call the gods together to make decisions and issue divine decrees. The waters flowing from the paradise on the summit of his mountain brought life and fertility to the world below. Baal, the other great deity, resided on an awesome peak where, wrapped in thunder and lightning, he fought great battles and celebrated his victories over the lesser gods of the Canaanite pantheon. The name of Baal's mountain appears in the Old Testament as Zaphon, which became the Hebrew word for north, the direction in which it lay in relation to Jerusalem. A younger, more virile god of war and fertility, Baal eventually wrested sovereignty from El and reigned supreme over the hills of Canaan, the promised land to which Moses led the Children of Israel. Although the prophets who followed Moses attacked the worship of these deities, modern scholars have shown how the mountains of Baal and El influenced biblical conceptions of Sinai and Zion.[2]

The land that the Hebrews took from the Canaanites around 1200 B.C. lay for the most part on mountainous heights overlooking the Mediterranean to the west and the Dead Sea to the east. There, like an oasis suspended above the shimmering heat of the surrounding lowlands, the Hebrews found a country of pleasant hills and valleys—cool, green, and well watered. After years of wandering in the harsh and sterile wilderness of Sinai, the mountains of Israel seemed an earthly paradise, a land flowing with milk and honey. Clusters of moist grapes, shade trees heavy with olives, meadows of smooth grass, all that they had longed for in the desert sands lay waiting for them as gifts from heaven, where the hills received the blessings of God in the form of soft rain from the sky. Rugged enough to block the passage of war chariots from the plains, but not too rugged to graze and cultivate, the mountains offered a natural fortress and sanctuary where the Hebrews could live, at last, in peace and plenty. Moreover, the sight of massive summits set firmly around the horizon provided a constant reminder of the eternal presence of God, watching over his people and protecting them from their enemies. As the 125th Psalm so beautifully puts it:

They that trust in the Lord
Are as mount Zion, which cannot be moved, but
  abideth for ever.
As the mountains are round about Jerusalem,
So the Lord is round about His people from this time
  forth and for ever.[3]

From the beginning mountains have been central to the experience of the holy land in Judaism. The Old Testament abounds with passages extolling their goodness and beauty. In Deuteronomy Moses implores God to grant him a view of the land denied to him: "Let me go over, I pray Thee, and see the good land that is beyond the Jordan, that goodly hill-country, and [Mount] Lebanon."[4] The Psalms, in particular, express deep feelings of affection for mountainous heights and the virtues that flow from them, such as peace and righteousness:

Let the mountains bear peace to the people
And the little hills, through righteousness.[5]

Something about the long smooth sweep of their ridges, the way their crests mark the edge of the sky, gives the mountains of Israel an impression of height that far exceeds their actual altitude. The sight of them leads the mind naturally to thoughts of the Creator, who dwells in the heights of heaven above and beyond them. In the words of the 121st Psalm:

I will lift up mine eyes unto the mountains:
From whence shall my help come?
My help cometh from the Lord,
Who made heaven and earth.[6]

Many of the most important events in the Bible and the history of Judaism are closely associated with mountains. As the source of four rivers, including the Tigris and the Euphrates, the Garden of Eden lies, by implication, on the heights of a hill or mountain. In fact, the prophet Ezekiel refers to Eden as "the holy mountain of God." In the story of the flood, Noah's ark comes to rest on Mount Ararat, the first point of land to emerge from the waters that have covered the earth. Here God makes the first of three covenants made on mountains: he promises never to destroy the world by water again, no matter how evil its creatures may become.[7]

The second covenant on a mountain takes place on the summit of Mount Moriah: there, in one of the most powerful and poignant scenes in the Bible, God commands Abraham, the patriarch of the Jewish people, to build an altar and offer up his only son in sacrifice. When Abraham raises up his knife to slay Isaac, an angel stays his hand and says:

By myself have I sworn, saith the Lord, for because thou
hast done this thing, and hast not withheld thy son, thine
only son, that in blessing I will bless thee, and in multiply-
ing I will multiply thy seed as the stars of the heaven, and
as the sand which is upon the sea shore; and thy seed shall
possess the gate of his enemies; and in thy seed shall all
the nations of the earth be blessed; because thou hast
obeyed My voice.[8]

The successful passage of this terrible test on top of a mountain forms a pivotal event in the sacred history of Judaism: it confirms the promise God has made with Abraham and his descendents, thereby assuring the future of the Jewish people and their religion. It is significant that when God appears to Abraham he refers to himself as El Shaddai, the "One of the Mountain."[9]

The third covenant, the most important single event in the history of Judaism, takes place on Mount Sinai. There Moses ascends the mountain to converse with God and receive the Torah, the divine law that makes of his people a true nation and marks the actual beginning of the Jewish religion. The earlier incident of the Burning Bush, in which God first reveals himself to Moses and commands him to deliver the people from bondage in Egypt, occurs on the slopes of Mount Sinai. In a later part of the Bible, the prophet Elijah comes to this mountain to hear God speak to him in a still, small voice — a tiny echo of the thunder with which he had spoken to Moses.

Other mountains also play important roles in the lives of Elijah and Moses. In a famous episode that reflects the triumph of Judaism over the Canaanite worship of mountains gods, Elijah defeats the priests of Baal in a contest on Mount Carmel. A broad ridge rising over the Mediterranean, Carmel served as a popular retreat for hermits and inspired, at a much later date, the Carmelite Order of the Catholic Church. At the end of Deuteronomy, after years of wandering in the wilderness, Moses climbs to the top of a mountain to see the Promised Land before passing away, alone with God on the heights of Mount Nebo.

In the later sections of the Bible, Mount Zion replaces Sinai as the mountain where God makes his will known to the people of Israel. David establishes his city on a hill next to it, and his son Solomon builds the temple on Mount Moriah, which, along with the city of Jerusalem, becomes identified with Zion. The great tragedies of ancient Jewish history, culminating in the destruction of the Second Temple and the Diaspora of the Jewish people, almost all take place on hills in the vicinity of this mountain. Prophets such as Isaiah foretell the coming of the Messiah and the establishment of the Kingdom of God on the summit of Mount Zion. Finally, in recent times, the Zionist movement, which resulted in the establishment of the modern state of Israel in 1948, took its name and inspiration from the sacred mountain that embodies the ideals and hopes of the Jewish people.

The mountains of the Middle East have also played an important role in Christianity. Most of the major events in the life of Jesus in the New Testament take place on hills and mountaintops. His birthplace, Bethlehem, lies high on a ridge south of Jerusalem. The star that guides the three wise men to his manger resembles the polestar that often appears in other traditions suspended over the summit of a sacred mountain, such as Mount Meru in the Hindu and Buddhist religions. At the beginning of his ministry, right after his baptism by John the Baptist, Jesus passes the final test of his spirituality, the last and greatest of his temptations, on top of a mountain:

Again, the devil taketh him up into an exceeding high mountain, and sheweth him all the kingdoms of the world, and the glory of them; And saith unto him, All these things will I give thee, if thou wilt fall down and worship me. Then saith Jesus unto him, Get thee hence, Satan. . . .[10]

After he has begun to teach, Jesus climbs a hill near the Sea of Galilee to deliver the most important and influential teaching in the New Testament, the Sermon on the Mount. This discourse, which begins with the well-known words "Blessed are the poor in spirit: for theirs is the Kingdom of Heaven," has as its implicit model the revelation of the Ten Commandments on Mount Sinai — for Jesus has come to fulfill the old law and institute a new one.

Another significant event, the transfiguration that confirms Jesus as the Son of God in the Christian tradition, also takes place on a mountain — traditionally Mount Tabor near Nazareth. There, just as Moses was transformed on Sinai so that his face shone with a divine light, so is Jesus. The older prophet even appears in person to bear witness to this affinity between himself and his divine successor:

And after six days Jesus taketh Peter, James, and John his brother, and bringeth them up into an high mountain apart, And was transfigured before them: and his face did shine as the sun, and his raiment was white as the light. And, behold, there appeared unto them Moses and Elias talking with him.[11]

God then speaks out of a cloud, as he did on Mount Sinai, to announce: "This is my beloved Son, in whom I am well pleased."

The final, crucial events of Jesus' life and death center on hills and mountains. His crucifixion takes place on the hilltop of Mount Calvary or Golgotha, the Place of the Skull. Christian tradition makes this place the center of the world and the summit of a cosmic mountain linking earth to heaven. After his resurrection, at the very end of the Gospel of Matthew, Jesus appears to eleven of his disciples on a mountain in Galilee. There, in his final act, he bids them to spread his teachings to all the nations throughout the world. Although the Gospels do not specify this, traditional accounts maintain that he ascended to heaven from the Mount of Olives.[12]

Although later Christian theologians in western Europe cast aspersions on mountains as demonic disfigurations of the earth, the early Christians of the New Testament clearly regarded them in a positive light, as places of miracles and divine revelation. This view continued among the monks and hermits of the Eastern Orthodox Church, who sought to find God in the solitude of mountain hermitages. The ninth-century writings of Theodore, abbot of the Monastery of Studios in Constantinople, express his high regard for mountains and reveal the importance of their symbolism for him and others of his faith:

It seems to me that a mountain is an image of the soul as it lifts itself up in contemplation. For in the same manner as the mountain towers above the valleys and lowlands at its foot, so does the soul of him who prays mount into the higher regions up to God. . . . Jesus himself, our divine king, ascended the mountain, that he might send up his prayer after the manner of men.[13]

Sacred mountains also appear in Islam, the other major religion of Middle Eastern origin, but they do not occupy as prominent a place as they do in Judaism and Christianity. In the Muslim tradition they play a most significant role in mystical writings and practices having to do with stages of the path leading to spiritual absorption in Allah, or God. However, mountains are also associated with events of great importance in the life of the Prophet Muhammad, who founded Islam in the seventh century A.D. A caravan merchant who lived in the vicinity of Mecca, he would go alone to seek spiritual solace in a rocky cave on Mount Hira, a nearby hill of sun-scorched sand and stone. There, on a fateful night in the fortieth year of Muhammad's life, known thereafter as the Night of Glory, the Archangel Gabriel appeared before him and said:

Recite: In the name of thy Lord who created,
created man from a clot of blood.
Recite: and thy Lord is the Most Bountiful,
He who taught by the pen,
taught man what he knew not.[14]

With these commanding words came the first of an extraordinary series of revelations that became in time the Koran, the holy scripture of Islam. Fearing that he might be possessed by an evil spirit, Muhammad fled down the mountain, but the angel stopped him and told him that he had indeed been designated the Messenger of God. The religion that he founded, based on revelations that began on Mount Hira, swept swiftly across Asia to reach the far-off islands of the Pacific and to become the major rival of Christianity as the faith of millions throughout the world.

Muhammad viewed Islam as the culmination of the Jewish and Christian religions, from which he drew much of his inspiration. Regarded by Muslims as the seal, or last, of the prophets, he took a great interest in his biblical predecessors. A number of passages in the Koran refer to Moses and the revelation of the Ten Commandments on Mount Sinai. One that shows the influence of later Jewish legend and commentary tells how Allah suspended the mountain over the Children of Israel and commanded them to hold fast to the teachings he had given them through Moses. In the rabbinic account God holds Mount Sinai over their heads and threatens to bury them under its massive bulk if they do not accept the Torah and obey its commandments. In both versions the image of the mountain conveys the awesome power and glory of the Lord before whom all must submit. And, indeed, Islam, the name of the Muslim religion, means submission—to the merciful will of Allah.[15]

The summit of Mount Moriah, the site of the ancient Temple of Solomon and the place where according to tradition God commanded Abraham to sacrifice Isaac, served as the point of earthly departure for Muhammad's celebrated night journey to heaven. According to accounts of the Prophet's life, he traveled with the Archangel Gabriel in a single night from Mecca to Jerusalem, where he met Abraham, Moses, Jesus, and other prophets. Then, mounting on a winged steed, he flew up through seven heavens to a tree that marks the limit of all knowledge, beyond which lies the inscrutable mystery of God. There, he received the revelation that contains the central creed of Islam. The golden dome of the Mosque of Omar, a striking symbol of modern-day Jerusalem, stands over the rock on the summit of the Temple Mount from which Muhammad soared to the heights of heaven.[16]

On his final pilgrimage, shortly before his death, Muhammad went to Jabal al-Rahmah, the Mount of Mercy, just outside Mecca, and according to tradition told his followers, "Truly, all Muslims are brothers." There he is said to have received the revelation of the last verse of the Koran recited by him: "This day I have perfected your religion for you and have chosen for you Islam as your religion."[17] As part of the pilgrimage to Mecca, all Muslims go to stand at the Mount of Mercy, commemorating the brotherhood of their faith and their commitment to the teachings of Islam. According to Islamic legend, after their expulsion from the Garden of Eden, Adam and Eve, the parents of the human race, came together and "knew" each other on the plain of Arafat beneath the sacred mountain.[18]

Al-Ghazali, one of the greatest theologians and mystics of the Islamic faith, used the image of mountains to symbolize the relationship between man and the realm of the spirit. Around the end of the eleventh century, long after Islam had become a well-established religion, he wrote:

If among the objects of the world of the spirit there is something fixed and unalterable, great and illimitable, something from which the beams of revelation, the streams of knowledge, pour into the mind like water into a valley, it is to be symbolized by a mountain. *If the beings who receive these revelations are of differing ranks, they are to be symbolized by a* valley; *and if these beams of revelation reach the minds of men, and pass on from mind to mind, then these minds are likewise to be symbolized by* valleys.[19]

The tradition of Sufi mysticism to which Al-Ghazali turned in the latter part of his life speaks of Kaf, a mythical range of mountains that surrounds the earth like a great ring. In the writings and visions of Muslim mystics, this mountain range symbolizes the impassable barrier that stands between the world of matter and the inconceivable realm of the spirit. There, beyond its transcendent heights lies the throne of the most high, the glorious presence of God that only those who abandon all worldly attachments and take the Sufi path to its very end can ever hope to attain. *The Conference of the Birds,* a poetic allegory composed by the Persian poet Farid ud-Din Attar in the twelfth century and often compared to Dante's *Divine Comedy,* describes the journey of a group of birds who follow this path over the mountains of Kaf, seeking a king who turns out to be a divine reflection of their own true nature. Some have sought to identify the mythical range of Sufi mysticism with the Caucasus Mountains, the boundary of the known world for the ancient Greeks.[20]

The Muslims appear to have borrowed both the name of Kaf and the idea of a ring of mountains surrounding the world from the ancient cosmology of Iran. Texts incorporating material dating from the time of Zarathustra, the Iranian prophet who founded the religion of Zoroastrianism around 1000 B.C., speak of Hara Berezaiti or Elburz, the first mountain to stand upon the flat disc of the earth. Actually a range of mountains, it encircles the rim of the world and rises up to the heights of heaven. From the roots that it extends underground, one of its peaks, the Peak of Hara, has grown up like a plant to form the mountain at the center of the earth. The stars, moon, and sun circle around this peak as they do around Mount Meru, the cosmic axis of Indian mythology. Indeed, the two mountains probably have a common ancestor in the myths of the Indo-European peoples who settled both Iran and India during the second millennium B.C. One of the subranges that grows out of Elburz is called in Zoroastrian texts the "Mountain of Kaf," leading some later authors to identify Kaf with Elburz itself.[21]

From the summit of the Peak of Hara, the highest point on earth, the Chinvat Bridge arches up to heaven. According to Zoroastrian teachings, over the bridge's narrow span, suspended above a terrifying abyss, the souls of the dead must pass. There, they receive their final judgment. Those who have sinned lose their balance and plunge into the depths of hell, to be tormented by demons and ghosts who remind them of the evil they have wrought. Those who have been virtuous are greeted by a beautiful maiden, who embodies all the good deeds they have done. Drawn by the sweet scent of paradise, they follow her up into the realm of the infinite lights of heaven.[22] Zoroastrian preoccupations with the struggle between good and evil—and the eventual triumph of the former over the latter—strongly influenced the development of messianic ideas in Judaism, Christianity, and Islam.

Demavend and Ararat

The two highest mountains of the Middle East, Demavend and Ararat, have drawn to themselves, like mist and clouds, the stories and myths of the ancient past. Extinct volcanoes, they rise in soaring cones of white and gray to float like visions from another realm poised high above the surrounding world. Streaked with lines of snow, the tapered summit of Demavend hangs in the sky at 18,606 feet, visible from Tehran, the capital of Iran, forty miles away. Towering 4000 feet higher than its companion cone of Little Ararat in the Armenian region of eastern Turkey, Ararat reaches an altitude of 16,945 feet. With nothing close to them in height, no other summits to distract the attention, Ararat and Demavend possess an overwhelming presence that compels the eye to regard them and only them. Seeking the mountains of their myths, the ancients naturally turned to these, the highest peaks in the world they knew.

Demavend, and more particularly the range of which it forms the most impressive peak, became the Elburz or Hara Berezaiti of Zoroastrian mythology—the cosmic mountain surrounding the earth and rising from its center to surpass the stars. There, somewhere on the heights of the Elburz Mountains grew the white haoma plant that produced the elixir of immortality. There, too, beyond the reach of darkness, cold, and night, lay the hidden source of the divine river bringing the waters of life to the world below. With the waning of Zoroastrianism and the coming of Islam, Demavend became the setting of more mundane stories and legends, many of which have their origins in much older myths of ancient Iran. According to the one most commonly told in recent times, an evil king named Zohak required the sacrifice of young men to feed snakes growing out of his shoulders. A

Mount Ararat, traditional landing place of Noah's ark in eastern Turkey. (John Elk III)

hero named Feridun, who had grown up in the safety of the Elburz, came down from his alpine refuge to defeat the tyrant and imprison him in a cavern on Mount Demavend, where he chained him to the living rock. According to local legend, whenever Zohak groans and writhes within his mountain prison, the volcano rumbles and shakes as if about to erupt.[23]

Like a ship seeking a place to land, the biblical story of Noah's ark came to rest on the summit of Mount Ararat. Although some have looked to Demavend or the high peaks of western Iran, most traditional authorities have identified the lofty volcano of eastern Turkey as the mountain that first emerged from the receding waters of the flood. According to the Book of Genesis:

And the waters returned from off the earth continually: and after the end of a hundred and fifty days the waters decreased. And the ark rested in the seventh month, on the seventeenth day of the month, upon the mountains of Ararat.[24]

The word *Ararat* refers to a country or region and comes, in fact, from Urartu, the name of an ancient civilization that thrived during biblical times in the vicinity of the mountain now traditionally identified as the place where the ark landed. The Bible only mentions Ararat in two other passages, where it makes it clear that it is speaking of a land and kingdom.[25]

Noah and his fellow passengers descended from the heights to repopulate the world, making Ararat the site of the second creation. The theme of a high peak as the first piece of land to emerge from a flood, the place from which life begins again, has a peculiar power and fascination: it appears in myths from regions as diverse and scattered as the Middle East, Europe, South Asia, and North America. On one level the story of the biblical deluge probably has its origins in memories of floods that inundated centers of Mesopotamian civilization in the low-lying areas between the Tigris and Euphrates rivers. On another, deeper level it reflects a longing to sweep away the past and begin again, to cleanse the world and make it new. Because of its deep and widespread appeal, the story of Noah's ark has made Mount Ararat one of the most famous, and intriguing, mountains in the world — despite the fact that the Bible gives it only the briefest mention.

Mount Ararat has also become, for the Armenian

people, a symbol of their homeland and national identity. For centuries they have viewed it not only as the resting place of Noah's ark, but also as the center of the world, around which they were privileged to dwell. Dispersed in various parts of the world, many of them still remember with longing the sacred mountain of Armenia, called Masis in the language of their fathers. A number of villages and other features around Mount Ararat bear names relating them to events in the story of the flood: the site where Noah built an altar to make sacrifices after descending from the mountain, the vineyard where he planted the first grapes, the place where he buried his wife. The Armenians of the past regarded the mountain with such veneration that they believed no one could ever reach its holy summit.[26]

An old story tells of an Armenian monk who so desired to venerate the ark that he tried to climb Mount Ararat three times. Each time, after struggling up to a great height, he would fall inexorably asleep, only to wake up and find himself back at the foot of the mountain. On his third attempt an angel entered his dreams and told him that God had decreed that no mortal should ever step on the summit where Noah had alighted to renew the human race. However, in recognition of the monk's devotion, the angel brought down a piece of the ark and placed it on his breast. A relic preserved at the ancient monastery of Echmiadzin in the thirteenth century was said to be that piece of wood. A small chapel built on the site of the monk's dream was destroyed in an earthquake that buried it and other monastic buildings under an avalanche of rock in 1840.[27]

Visible from great distances, looming over intervening ridges and plains, the summits of Demavend and Ararat appear at first sight impossibly high and remote. Until modern times the idea of climbing them seemed to the local people the height of folly. But aside from the thin air of high altitude and the drudgery of struggling up endless slopes of loose rock, sliding ash, and soft snow, the ascent actually poses little difficulty, especially for mountaineers accustomed to climbing more precipitous peaks in the Alps. An Englishman named W. T. Thomson made the first ascent of Demavend in 1837. Frederic Parrot, a Russian doctor, succeeded in reaching the summit of Ararat in 1829, but the local people, and even his European companions, refused to believe that he had climbed the sacred mountain. When James Bryce reported his own ascent of Ararat nearly fifty years later to the Archimandrite of Echmiadzin, the monastery with the reputed relic of Noah's ark, the Armenian prelate replied, with a gracious smile, "No, that cannot be. No one has ever been there. It is impossible."[28]

A number of expeditions have climbed Ararat in search of Noah's ark. Some have found pieces of wood high on the mountain, others have reported seeing the boat itself, buried in ice or perched in a ravine near the summit. Pilots flying over the peak have reported sighting the outline of an enormous ship nearly a thousand feet long. But none of these reports have withstood subsequent scrutiny. Aside from the improbability of their having survived for thousands of years, the pieces of wood discovered high on Ararat prove little: they could have been taken up by local people or used in the ancient shrines of other religions. Pictures reportedly taken of the ark itself have vanished under mysterious circumstances, leading one to question if they ever existed. But for those who believe, and those who wonder, it matters little whether physical evidence of the biblical story will ever be found: it is enough to gaze on the mountain and think of a time when the world was made fresh and clean.

Mount Sinai

The two most important sacred mountains in the Middle East are neither the highest nor the most impressive. One, Mount Sinai, rises to an altitude of less than 9000 feet. The other, Mount Zion, is scarcely more than a rounded hill. But despite their lack of physical height and imposing demeanor, the two tower spiritually over all other mountains of the Middle East. The image of God's descent in fire and cloud on Mount Sinai is burned into the minds of millions throughout the world as *the* paradigm of man's confrontation with the awesome power and majesty of the sacred and occupies a place of distinction as one of the most impressive and influential images in all of religious history and literature.[29]

In the Jewish tradition Mount Sinai marks the site of the most important covenant made between God and the people of Israel. The Ten Commandments issued as part of this covenant also form the basis for much of Western law and civilization. What happened on Mount Sinai over three thousand years ago had a power that keeps it alive even today, making it central to the ongoing history and practice of Judaism. The following passage from a biblical commentary used in many synagogues reveals the significance for contemporary Jews of the covenant made on the sacred mountain of the distant past:

The arrival at the foot of Mt. Sinai marks the beginning of Israel's spiritual history. We reach what was the kernel and core of the nation's life, the Covenant by which all the tribes were united in allegiance to One God, the Covenant by which a priest-people was created, and a Kingdom of God on earth inaugurated among the children of men.[30]

The Bible itself provides a vivid picture of the sacred mountain and what transpired on its awesome heights. After leading the Children of Israel out of bondage in Egypt, Moses brought them through the wilderness to the foot of Mount Sinai. There they camped to await the will of God. The Book of Exodus continues:

And it came to pass on the third day, when it was morning, that there were thunders and lightnings, and a thick cloud upon the mount, and the voice of a horn exceeding loud; and all the people that were in the camp trembled. And Moses brought forth the people out of the camp to meet God; and they stood at the nether part of the mount. Now mount Sinai was altogether on a smoke, because the Lord descended upon it in fire; and the smoke thereof ascended as the smoke of a furnace, and the whole mount quaked greatly. And when the voice of the horn waxed louder and louder, Moses spoke, and God answered him by a voice. And the Lord came down upon mount Sinai, to the top of the mount: and the Lord called Moses to the top of the mount; and Moses went up.[31]

The passage provides a dramatic setting to highlight two movements in opposite directions. God comes down and Moses goes up. They meet at the top of Mount Sinai. The mountain reveals the presence of the Lord and serves as the meeting place between God and man, the sacred and the profane. In the powerful image of Mount Sinai, we find the resolution of two profound, but opposite, inclinations that define the poles of much of religious thought and practice throughout the world: the impulse to rely on personal effort and the need to depend on divine grace. On the summit of the sacred mountain, the two come together as one. God calls and man responds.

Moses ascended Mount Sinai once again, this time to receive stone tablets engraved with the Ten Commandments:

And the glory of the Lord abode upon mount Sinai, and the cloud covered it six days; and the seventh day he called unto Moses out of the midst of the cloud. And the appearance of the glory of the Lord was like devouring fire on the top of the mount in the eyes of the children of Israel. And Moses entered into the midst of the cloud, and went up into the mount; and Moses was in the mount forty days and forty nights.[32]

If we look closely at both of these passages, we find that neither describes the physical appearance of the mountain itself—the shape of its ridges or the texture of its rock. They focus instead on its transformation: the fire and smoke, the thunder and cloud that envelop the peak to reveal the awesome presence

of God. The mountain becomes charged with the electricity of a divine power that renders it lethal to all but Moses and a select few judged worthy to set foot on its lower slopes. Some have attempted to explain the biblical passages as a description of a volcano erupting, but a thunderstorm would just as easily fit the picture of a peak enshrouded in fire and cloud. In any case, whatever the physical explanation, we have here a classic example of a hierophany—an eruption of the sacred, or the experience of the sacred, into the profane world of ordinary reality.

Moses first experiences such a hierophany in the famous episode of the burning bush, which also takes place on Mount Sinai. Grazing his flocks on the side of Horeb, the Mountain of God, Moses comes across a bush that burns but is not consumed. When he turns aside to see this wonder, the voice of God speaks out of the fire and commands him to deliver his people from bondage in Egypt. The Bible implicitly identifies Horeb with Sinai since the passage adds, "When thou has brought forth the people out of Egypt, ye shall serve God upon this mountain"—the place where Moses will receive the Ten Commandments.[33] Like Mount Sinai, the bush burns with a miraculous fire that reveals the presence of the sacred. In both cases a feature of the natural landscape is transformed into a conduit of supernatural power and glory, a means by which God may communicate with man.

For all its emphasis on what happened on Mount Sinai, the Bible leaves no clear indication of where the mountain lies. It tells us only that the Children of Israel went into the wilderness of Sinai and that "it is eleven days' journey from Horeb by the way of Mount Seir to Kadesh-barnea"[34]—but no one has been able to identify Mount Seir. As a result, various religious traditions and scholarly conjectures have sprung up, each proposing its own candidate for the sacred mountain. Some believe Moses met God not far from the crossing of the Red Sea, on Jebel Helal in the northern part of the Sinai. Early Christians looked, instead, to the more dramatic peaks of Jebel Serbal and Jebel Musa near the southern tip of the peninsula. One archaeologist has recently deduced evidence pointing to a low butte in the Negev, where numerous ancient artifacts have been found, indicating a long history of cultic activity. The Jewish tradition itself has shown remarkably little interest in the actual site of the sacred mountain.[35]

The story of the Exodus reveals part of the reason for this lack of interest—why Mount Sinai never became an established place of pilgrimage for Jews. God

Moses on Mount Sinai: witnessing the Burning Bush in the upper illumination, receiving the Tablets of the Law in the lower. Psalter of Ingeburg of Denmark. (Giraudon/Art Resource, N.Y.)

Si come moysses uit dieu ou buisson.

Si come moysses a la gerre les tables de la loi

commanded the Children of Israel to build a tabernacle for him so that he might come down from the mountain and travel with them to the promised land. As Flavius Josephus, a Jewish historian, wrote in the first century, "God also desired that a tent should be built for Him, in which He would condescend to meet them, and which could also be carried with them on their journey, so that in the future it would no longer be necessary to ascend Mount Sinai, since He himself would descend into the tent and in that very place would hear their prayers."[36] From that time forward, the Jews never felt a need to return to the mountain, for God had left it to come with them to the holy land of Israel. Unlike mountains such as Fuji, which are sacred in their own right, Sinai derived its sanctity from the divine presence that descended on its summit. When that presence departed, the physical mountain lost its significance, overwhelmed by the power of what had transpired on its hallowed heights.

As the site of the most important event in the history of Judaism, the image of Sinai, however, lived on in the memory of the Jewish people as a powerful symbol of God's covenant with them. Moses himself impressed upon his followers the importance of remembering the sacred mountain and what it signified:

Only take heed to thyself, and keep thy soul diligently, lest thou forget the things which thine eyes saw, and lest they depart from thy heart all the days of thy life; but make them known unto thy children, and thy children's children; the day that thou stoodest before the Lord thy God in Horeb, when the Lord said unto me: 'Assemble Me the people, and I will make them hear My words, that they may learn to fear Me all the days that they live upon the earth, and that they may teach their children.' And ye came near and stood under the mountain; and the mountain burned with fire unto the heart of heaven, with darkness, cloud, and thick darkness. And the Lord spoke unto you out of the midst of the fire. . . .[37]

For Jews—and Christians—the image of the fiery revelation on Mount Sinai helped to endow the Ten Commandments with a divine stature that compelled people of later generations to preserve and observe their precepts.

As a mountain of memory, Sinai acquired a transcendent reality that enabled it to overcome the limitations of space and time so that the thunder of its message could continue to reverberate in the hearts and minds of Jews, wherever they might find themselves. The very uncertainty of its location served to enhance the aura of mystery that enshrouded the peak and made it such a powerful force in the Jewish tradition. Identified with no particular place or feature of the material world, Mount Sinai could assume a spiritual majesty and grandeur limited only by the human imagination. Above all, the lack of fixed location placed it in a realm beyond reach and prevented the mountain from succumbing to the evils of idolatry. Since no one could say for sure where it lay, no physical mountain could ever become a substitute for the spiritual peak of revelation that it embodied. Just as no one knew the site of Moses' grave, where he met death, so no one knew the location of the true Mount Sinai, where he met God. The mystery that surrounded the two kept a cult of Jewish worship from developing around either one.

Early Christians, however, sought to find the blessings of God's revelation at the actual site of Mount Sinai. Drawn by the awesome landscape and the spiritual power they felt there, wandering monks and hermits began in the third century A.D. to congregate around Jebel Musa, the Mountain of Moses, a remote and rocky peak hidden among the highest and most spectacular mountains of the Sinai Peninsula. A perfect setting for inducing the visionary experiences of solitary contemplation, Jebel Musa rises to a height of 7455 feet above a haunting wilderness of shadowed ravines and twisting ridges. Rounded cliffs of reddish brown granite, sculpted and scoured by wind and water, blasted and burned by sand and sun, give the mountain the appearance of having gone through the hardening fires of a cosmic furnace. The eerie, primordial quality of the barren landscape, swept clean of all contaminating influences, evokes a sense of the pure and timeless place where Moses conversed with God. The impressive appearance of the mountain led the devout to identify it with the legendary peak of scriptural tradition and elevate it to divine status. And who was to say that it was not, in fact, the true Mount Sinai of the Bible?

The presence of holy men communing with God on the sacred mountain attracted monks and pilgrims, who came in increasing numbers to partake of the spiritual atmosphere surrounding Mount Sinai. The writings of one such pilgrim reveal the kind of inner force that impelled him and others to make the difficult journey to the desert peak: "A powerful longing towards Sinai seized me, and neither with my bodily eyes nor with those of the spirit could I find joy in anything, so strongly was I attracted to that place of solitude."[28] By the beginning of the fourth century, hermits living in simple caves and huts scattered among the rocks and crags had established a monastic community and identified features of the local landscape connected with the story of Moses, such as the site of the burning bush and the place where the Golden Calf had stood.

Bloody massacres by roving bands of desert marauders interfered with the contemplative life of Mount Sinai, although the attacks did provide ample oppor-

The Monastery of Saint Catherine sits on the reputed site of the Burning Bush at the foot of Mount Sinai. (Andrew L. Evans)

tunity for inspirational martyrdoms. Fed up with these unsettling distractions, the monks finally appealed to the Byzantine emperor, Justinian I, for protection. In the middle of the sixth century, Justinian had a fortified monastery built on the site of the burning bush at the base of the mountain and stationed a military garrison there to allay the anxieties of the harried monks. Later on, around the eleventh century, after the Arabs had gained control of the region, a mosque constructed beside the church within the compound appeased the Muslims and guaranteed the survival of the Christian monastery. The monks also used a document attributed to Muhammad to persuade the Muslim rulers of Sinai that the Prophet had granted them the divine protection of Islam itself.

Originally named the Church of the Transfiguration, in memory of the transfiguration of Jesus on a mountain in the presence of Moses and Elijah, who had themselves been blessed on Mount Sinai, Justinian's monastery acquired in the eleventh century another name, the one by which most people know it today, the Monastery of Saint Catherine. A woman of noble family who converted to Christianity in the

fourth century, Saint Catherine was martyred for her beliefs in the Egyptian city of Alexandria. According to tradition, angels miraculously carried her body to the top of Jebel Katerina or Mount Catherine, the highest peak in the Sinai Peninsula. Three centuries later monks found her remains there and brought them down to the monastery, situated within sight of her mountain. When some of these relics went to France in the eleventh century, her cult spread rapidly through Europe and made her final resting place at the foot of Mount Sinai famous as the Monastery of Saint Catherine.

Along with the fame of its saint, the influence of the desert monastery swept far beyond the isolated mountains of Sinai. It became the center of a major monastic order with branches located as far away as Greece, Rumania, Russia, and even India. Numerous monasteries in Eastern Europe adopted the contemplative tradition of Mount Sinai in which monks withdrew to the solitude of mountain hermitages to cleanse themselves of sin and seek perfection in spiritual union with Jesus through the cultivation of ever-deepening love and humility. The writings of major

Dawn from the summit of Jebel Musa, regarded by many as the biblical Mount Sinai. (Bruce Klepinger)

figures connected with Saint Catherine's deeply in-fluenced meditative thought and practice in the East-ern Orthodox Church. John Climacus, an early abbot of the monastery, authored *The Ladder to Paradise,* a famous guide for monks describing the steps of the spiritual path leading from renunciation of the world to the attainment of peace in the perfect love of God. The stone staircase of three thousand some steps ris-ing up from St. Catherine's to the summit of Mount Sinai may have provided a model for John's image of a divine ladder reaching to heaven. In any case, Greg-ory of Sinai, who did much to inculcate these ideals in Eastern Europe in the fourteenth century, made it a religious practice to climb every day to the top of the mountain where Moses had come into the presence of God.[39]

The monastery built by Justinian in the sixth cen-tury still stands today as one of the oldest surviving Christian monasteries in the world. The massive walls of granite blocks that he constructed enclose a rec-tangular compound filled with a church, mosque, numerous chapels, a library, archives, courtyards, and living quarters for monks. The most important struc-ture, a chapel, sits on the site of the burning bush,

whose miracle Christian tradition compares with that of the Virgin Mary, who in conceiving Jesus was also infused with the blazing power of God. A magnificent mosaic imbedded in the apse of the church just after the death of Justinian depicts in a golden glow the transfiguration of Jesus with Moses and Elijah stand-ing to either side. The monastery itself serves as a repository of one of the greatest collections of Byzan-tine art and literature to be found anywhere in the world—a treasure house of priceless icons, ancient manuscripts, and beautiful mosaics.

The visitor who looks up from contemplating these works of inspiring art beholds above them the spiritual treasures of the heights—the monumental peaks of Sinai, arranged like colossal altars beneath the sky. Long ago monks laboriously carved the stairway of stone steps that winds up through steep ravines and crags to emerge in a beautiful little valley hidden beneath the summit of Jebel Musa itself. An arch of stone near the top of the stairs marks the place where a sixth-century monk took confessions and turned back those whose sins, he felt, would cause their deaths if they should venture onto the holy ground above. Just beyond this arch, to the pilgrim's surprise, he comes upon an enormous cypress with thick green boughs, strangely out of place in the rocky wilderness surrounding it. Close by the tree, not far from a spring, stands a simple hut of stone with whitewashed walls, marking the spot where Elijah heard the still, small voice of God after the thunderous roar of wind and storm. An atmosphere of profound peace settles over the pilgrim, removing him from the cares of the outside world and preparing him for the summit of the sacred mountain.

A steep climb up the final slopes of naked rock comes to an abrupt end at a small chapel set precariously on the very top of Mount Sinai. An immense vista of undulating ridges and distant deserts opens around the awestruck pilgrim. Even those who have no spiritual attachment to the place feel themselves in the presence of something that far exceeds their limited conceptions of the world they think they know. In such a place, stunned to silence by a view of incredible grandeur, a nonbeliever might even hear, for one brief moment, the whisper of a still, small voice. A modern Englishwoman climbed the steps of the sacred mountain, expecting to feel nothing, and experienced this on the summit of Mount Sinai:

I was standing at the top of a giant mountain altar. It rose gradually, majestically, over many miles, flowing upward in straight line from the multiple crests of distant low ridges to the framed triple peak, up and higher to the double peak, and then up to the apex of the altar—to the single peak of Jebel Musa and myself, standing upright, atop this single peak. Suffused with the sheer physical logic of this being the holy mountain, the metaphysical logic of mysticism, I felt that never had I stood higher than this, that there could be no higher place in the world.[40]

Mount Zion

Although Christians have lavished devotion on the physical site of Mount Sinai, the mountain that has occupied the attention of the Jews as an actual place of pilgrimage and veneration throughout most of their history has been Mount Zion. Originally a hill in the Kidron Valley next to Jerusalem,[41] Mount Zion became identified in the Bible with the city of David and the temple of Solomon. The epithets "Mount Zion" and "the Holy Mountain" appear more frequently than any others in biblical passages referring to Jerusalem. The authors of the Old Testament viewed the holy city of Judaism preeminently as the sacred mountain of God, the high place where he had chosen to make his temple and dwell among them. In going up to Jerusalem, pilgrims could go up to the Lord as Moses had on Mount Sinai. And indeed, a number of psalms sung on pilgrimages to the holy city bear the title "A Song of Ascents."

The hard and polished quality of their ancient limestone makes the rounded hills of Jerusalem shimmer with light as though some higher, more luminous reality were shining through them. Standing beneath the brilliant sky, gazing over the sacred city, one can easily see why rabbinical writers regarded Mount Zion as the earthly reflection of a heavenly counterpart suffused with the glory of God enthroned on high. One can also understand how biblical prophets such as Isaiah could make of a low and unimpressive hill the greatest mountain on earth, overshadowing all others in the majesty of its spiritual height and grandeur.

Although they stress its awesome nature as the holy mountain of God, the more poetic passages of the Bible also dwell on the beauties of Zion. The 48th Psalm, which Jews traditionally recite every Monday morning, begins with the words:

Great is the Lord and highly to be praised,
In the city of our God, His holy mountain,
Fair in situation, the joy of the whole earth;
Even mount Zion, the uttermost parts of the north,
The city of the great King.[42]

The 50th Psalm makes the mountain a sublime source of divine power and glory:

Out of Zion, the perfection of beauty,
God hath shined forth.
Our God cometh, and doth not keep silence;
A fire devoureth before Him,
And round about Him it stormeth mightily.[43]

The Psalms have enveloped Zion in the image of fire and storm associated with Sinai, but they have added to it an atmosphere of softness and beauty totally lacking in biblical descriptions of the earlier mountain, which appears only as a harsh and terrifying peak of awesome power and revelation.

In keeping with its softer and friendlier appearance, Zion is a humanized mountain, like T'ai Shan, the imperial peak of China. Where no one but Moses could climb to the summit of Sinai and live, his descendants found an entire city already established on the crest of Zion, which they took and made the capital of their nation. Today, not only shrines and temples—as in the case of T'ai Shan—but houses, streets, and markets cover the hills of the sacred mountain, which reverberates with the sounds of human activity. It is, in fact, one of the most densely populated places in the world. In addition, just as T'ai Shan was the sacred peak of the Chinese emperors, the place where they offered sacrifices to heaven, so Zion was the holy mountain of the kings of Israel, the site of the temple where they made burnt offerings to God.

King Solomon constructed the original temple on the traditional site of the summit of Mount Moriah, the place where Abraham bound Isaac and where God appeared to David. Solomon brought up to the top of the mountain, which the Bible regards as the highest point of Zion, the ark containing the stone tablets that Moses had received on Mount Sinai and that David had brought to Jerusalem. In this way, the spiritual essence of Sinai, embodied in the Torah, was finally transferred to Mount Zion, the site of the temple where God would henceforth reveal his presence and make known his will to the Children of Israel. The Temple Mount became, and has remained, the very center of the Jewish world. Today, the first place that Jews visit on coming to the holy city of Jerusalem is the Western Wall, the remnant of the Second Temple, destroyed in A.D. 70. There the observants, wrapped in prayer shawls, stand before the wall of ancient stones, rocking back and forth, lamenting the past and praying for the future.

In the famous prophecy of Isaiah, one of the most influential and often quoted passages in the Bible, Mount Zion plays a prominent role as the divine center from which a new age will emerge, a golden age of peace and righteousness in which all will acknowledge the supremacy of God:

An elderly Chasidic Jew reading the Bible at the Western Wall, the holy site of the ancient temple in Jerusalem. (Robert Apte)

And it shall come to pass in the end of days,
That the mountain of the Lord's house shall be
*   established as the top of the mountains,*
And shall be exalted above the hills;
And all nations shall flow unto it.
And many peoples shall go and say:
'Come ye, and let us go up to the mountain of the Lord,
To the house of the God of Jacob;
And He will teach us of His ways,
And we will walk in His paths.'
For out of Zion shall go forth the law,
And the word of the Lord from Jerusalem.
And He shall judge between the nations,
And shall decide for many peoples;
And they shall beat their swords into plowshares,
And their spears into pruning-hooks;
Nation shall not lift up sword against nation,
Neither shall they learn war any more.[44]

In the future Zion will replace Mount Sinai as the mountain of the law and the place of the final covenant between man and God.

The prophecy of Isaiah portends the central role that Zion and Jerusalem play in the messianic hopes and expectations of Christianity. As the site where the Messiah will appear, Zion becomes of necessity the place where Jesus must fulfill his destiny as the savior of mankind. There, on the summit of the sacred mountain of God, his life and teachings reach their dramatic climax in his crucifixion and resurrection. Pilgrims from all over the world come to take up his burden and follow the Via Dolorosa as it proceeds through the stations of the cross, winding up through the narrow streets of Jerusalem to the Church of the Holy Sepulcher, where Jesus was laid to rest and rose from the dead to proclaim the salvation of the world. For the authors of the New Testament and later theologians of the Christian tradition, Mount Zion represented the New Jerusalem, where the kingdom of God was, and will be, made manifest on earth. Zion also came to symbolize the spiritual home of all Christians.

Mount Zion as Jerusalem ranks as the third most important pilgrimage site in Islam after Mecca and Medina. The Mosque of Omar, also known as the Dome of the Rock, stands on the site of the Temple of Solomon, right over the rock that marks the summit of Mount Moriah, the place where Abraham offered his son in sacrifice and where Muhammad took off on his night journey to heaven. According to Muslim tradition, on the Day of Judgment, an angel will appear on this rock to sound the trumpet announcing the end of the world. Set on an octagonal structure of blue-and-white marble inlaid with the

The holy city of Jerusalem, identified with Mount Zion in the Bible, appears as a hill seen from the Mount of Olives with the Dome of the Rock on the right and an old Jewish cemetery in the foreground. (Robert Apte)

graceful characters of Arabic script, the gold dome of the mosque gleams in the sun, providing a center to orient the eye as one gazes from a distance on the sacred hill and city of Jerusalem. Within the mosque's cool and softly lit interior, a marble balustrade surrounds the worn and wrinkled slab of ancient rock that bears witness to the hopes and sorrows of so many people from so many faiths.

The overwhelming importance that Judaism, Christianity, and Islam have placed on Jerusalem and Mount Zion has produced a long history of bloody conflict over possession of the sacred mountain. Jewish rebellion against Roman rule led to the destruction of the Second Temple in A.D. 70 (the Babylonians had destroyed the First Temple in 586 B.C.). After the Roman emperor Constantine converted to Christianity in the fourth century A.D., Christians gained control of Jerusalem and built the Church of the Holy Sepulcher. In the seventh century Persian and Arab forces of the new religion of Islam wrested the city from the Byzantine Empire—which had succeeded the Roman Empire in the east—and constructed the Dome of the Rock over the site of the Jewish temple in 691. In a holy war against the Muslims, the Crusaders took the sacred mountain of Christianity in 1099 and established the Kingdom of Jerusalem on its heights. The forces of Islam recaptured the city in 1187, and the Ottoman Turks, who took over in 1517, retained Muslim control until 1917, when Palestine fell under the rule of the British Empire. In 1948, following the deaths of more than six million Jews in the Holocaust, the Zionist movement, started in the nineteenth century and named for Mount Zion, succeeded in reestablishing a Jewish homeland with its capital in Jerusalem—a capital that most other nations refused to recognize. The city itself remained divided between Jewish and Muslim control until 1967 when Israel took the eastern half from Jordan in the Six-Day War. Conflicting religious claims to the sacred site of Jerusalem lie at the heart of the most difficult issues to be resolved on the way to establishing a lasting peace in the Middle East.

From the point of view of a child, the history of Jerusalem must look very much like a game of king of the mountain. Followers of the three religions have fought among themselves, pushing and shoving to get to the top and gain control of Mount Zion. And all the while, they seem to have forgotten who the true king of the mountain is—the One who according to each of their traditions rules them all. The comparison with a child's game points out the ultimate absurdity of all attempts to possess the sacred. Those who truly experience it find themselves, on the contrary, overwhelmed and possessed by the power and grandeur of its awesome presence. A return to this sense of the sacred, to an appreciation of the true significance of Jerusalem, would go a long way toward resolving the conflicts that continue to swirl like storm clouds around the heights of Mount Zion.

Sinai and Zion, the two mountains of God in the Bible, represent the opposite but complementary poles that define the full range and diversity of the Jewish tradition. Sinai is the awesome peak of the covenant and the law, the place of the prophet and his revelation. Zion, on the other hand, is the beautiful site of the temple, the place of the priest and his sacrifice. The conflict between prophets and priests, righteousness and ritual, runs through the later books of the Old Testament and helped to give rise to Christianity and its subsequent schisms. Unlike Sinai, which has no political associations, Zion became the capital city and mountain of kings, beginning with David and Solomon. In the 2nd Psalm God declares, "Truly it is I that have established my king upon Zion, my holy mountain."[45] The tradition of sacral kingship associated with Mount Zion radically altered the nature of Judaism and prepared the way for the messianic thought and prophecy of Isaiah. At the same time this tradition placed a heavy responsibility on the kings of Israel to uphold the commandments of Moses and be true to the legacy of Sinai.

Mount Sinai lay in a remote desert, far from the world of everyday life. The events that took place on its summit were veiled in fire and cloud, hidden from the sight of all but Moses, the greatest of prophets. Zion, on the other hand, rises in the middle of a city. The sacrifices performed in the temple on its heights were open and visible to all. In the history of Judaism, Sinai, the rugged peak of the wilderness, gives way to Zion, the cultivated mountain of civilization. Although the values they represent may appear to be in conflict, the message of each comes from the same source and reveals the presence of the same God, the God who demands justice and righteousness of his people. The voice that sounded in the open space of the desert now echoes in the narrow streets of the city. Zion incorporates and fulfills in human society the meaning of the lonely encounter on Sinai.

Sinai is the mountain of the beginning, Zion the mountain of the end. With the covenant on Sinai, Judaism as we know it takes form: a holy nation, a nation of priests, is born. A people of wandering desert origin, a disparate band of liberated slaves, comes together to forge a new tradition based on the laws received by Moses on the heights of the sacred peak. In the future, in the last of days, according to the prophecy of Isaiah, the covenant made on Sinai will reach its fruition in Zion, when all the nations of the world will come up to Jerusalem to receive and accept the word of God. There, on the heights of the holy mountain, exalted above all other mountains, the Jewish tradition will witness the end for which it began: the establishment of the kingdom of God on earth, a golden age of peace and righteousness in which "the lion shall lie down with the lamb" and "nation shall not lift up sword against nation." Together Sinai and Zion span the history, past and future, of Judaism itself.[46]

The tension symbolized in the opposition between the two mountains led the disciples of Jesus to break with their own tradition and form the new religion of Christianity. Like the prophets who inveighed against the immorality of priests and kings, Jesus rebuked the religious establishment for its hypocrisy, exemplified in the practice of allowing money changers to ply their trade in the temple of Zion. His followers saw his life and teaching as the sign of a new dispensation, made by God to replace and fulfill the old law given to Moses on Mount Sinai. As the mountain of the future, Zion became for early Christians a symbol of the messianic kingdom of God proclaimed in the prophecy of Isaiah and confirmed in the person of Jesus, a divine descendant of the House of David.

Although the salvation of Zion seemed to have replaced the law of Sinai, the tension between the two remained alive, producing the schisms that eventually splintered the Christian Church. In the eleventh century the Eastern Orthodox branch based in Constantinople split with the Latin church of Rome, rejecting the authority of the Pope as the high priest of Christianity. The Eastern Orthodox church took a particular interest in the experience of Moses on Mount Sinai, which it saw as a prefiguring of the transfiguration of Jesus on Mount Tabor. Roman Catholics, on the other hand, made the temple of Jerusalem on Mount Zion a model for the Vatican in Rome. When Martin Luther started the Protestant movement in the sixteenth century as a protest against the priestly excesses of Rome, he turned to the Old Testament and revived Western interest in the revelations of prophets such as Moses. Like Jews who based their fundamental beliefs and practices on the Torah received on Mount Sinai, Protestants put primary emphasis on Scripture as the word of God.

The conflict over religious authority that Sinai and Zion symbolize also produced schisms in Islam. Shortly after Muhammad's death in 632, a dispute over his successor split the new religion into the two sects that divide the Islamic world today. The Sunnis, the larger sect, decided to follow the Caliphs, a line of leaders elected from the Prophet's tribe. The Shiites, however, insisted that the leadership of Islam should pass through direct descendants of Ali, the cousin of Muhammad and husband of his sister, Fatima. Unlike the Sunnis, they held that their first twelve leaders, called Imams, were, like the Pope, infallible and guided by God. Shiites believe that the last Imam is the Mahdi, the Muslim version of the Messiah. In this belief they express a messianic ideal associated in Judaism and Christianity with visions of Mount Zion.

The tensions revealed through the symbolism of Sinai and Zion run through religions and cultures around the world, both traditional and secular. In modern democracies they express themselves most pointedly in the dilemma of balancing the rights of the individual with the good of society. Although such tensions produce conflict and division, they also provide the impetus for growth and renewal. Out of the interaction of the priest and the prophet, tradition and revelation, city and wilderness, individual and society, come the changes and innovations required to realize the highest aspirations of the heart and mind. The tensions symbolized in the opposition between the two mountains force us to lift our gaze and seek something higher from which the good of each may come. In the reconciliation of Sinai with Zion lies a vision of the future with the power to inspire us in our efforts to resolve the conflicts that rack the world today.

In the end the sacred mountains of the Middle East hold out the promise of paradise on earth. Their shining heights draw the eye up to the heavens and a vision of the golden age to come. For the prophets and poets of the Bible, Sinai becomes Zion, which becomes, in turn, the entire world, transformed and revealed in its true nature as the dwelling place of God. Gazing up to the sacred summit of Zion, bright beneath the sunlit sky, the psalmist prays:

O send out Thy light and Thy truth; let them lead me;
Let them bring me unto Thy holy mountain, and to
  Thy dwelling-places.[47]

EUROPE

PARADIGMS OF PERFECTION

FROM THE BRIGHT PEAKS of the Mediterranean to the dark fjords of the Arctic, mountain ranges weave intricate patterns across the landscape of Europe, supplying the texture and definition responsible for much of the continent's beauty and appeal. Cliffs and ridges of clean rock dropping into the Aegean provide spectacular settings for the temples and monasteries that grace the mountains of Greece. North of the Mediterranean the snow peaks of the Alps swirl in a great arc to divide the cloudy reaches of northern Europe from the sunlit regions of the south. The forested mountains of eastern and central Europe, dark and green, conceal valleys with castles and shrines that evoke the haunted landscapes of myth and legend. To the west the high ridges and cirques of the Pyrenees have provided a natural sanctuary for witches, rebels, bandits, and dissident sects fleeing the oppression of the outside world. The spirits of legendary figures of the past still stalk the crags and moors that cover the hills of the British Isles. Stories of trolls and other supernatural creatures haunt the rocky mountains of Scandinavia, adding a pleasant shiver of childhood fear to the atmosphere of mystery that hovers like mist over the lonely landscape of the Arctic north.[1]

Much of our appreciation of mountains derives from images of European peaks, such as Mont Blanc and the Matterhorn, which provide us with our standard of mountain beauty and perfection. For all their appeal and grace, however, none of these mountains is comparable in religious stature to Zion in the Middle East or Kailas in Asia. Mount Olympus, the only

European mountain to rival Zion or Kailas in the popular imagination, long ago lost the kind of living power and reality the latter still possess for the millions who revere them. Yet, even though Greeks no longer venerate its lofty summit as the home of the ancient gods, Olympus does provide the basic model for modern conceptions of mountains as abodes of deities in traditional cultures elsewhere in the world. And if we look beneath their superficial appeal, we find a reverence for many other peaks in Europe that imbues them with a sacred quality, even among people who claim to admire them for their scenic beauty alone.

Because of the vast cultural and geographical diversity of the continent, I have chosen to focus on three representative mountainous regions—Greece, northern Europe, and the Alps. The oldest written records of religious beliefs and practices devoted to European peaks survive in the literature of ancient Greece. Images of the classical gods on Olympus and Parnassus continue to shape our views of sacred mountains. The way these deities succumbed to Christianity on Mount Athos exemplifies what transpired on many other peaks in Europe. Fragments of Germanic and Celtic mythology preserved in Scandinavia and the British Isles give us valuable clues to the nature of pre-Christian views of higher peaks to the south and west. Changes in European attitudes toward the Alps have given rise to our modern appreciation of mountain ranges throughout the world. And the sense of the sacred hidden in these attitudes has played a major role in the development of mountaineering as a sport.

The pyramid of the Matterhorn looms over the silhouettes of trees on a ridge near Zermatt.
(John Elk III)

GREECE

Ranges of folded limestone with valleys opening toward the sea characterize the classic landscape of Greece. From Athos and Olympus in the north to Parnassus and Ida in the center and the south, views of ridges and peaks blend with vistas of the blue Aegean to create the boundaries of a well-ordered world. A sparseness of vegetation combines with a brightness of light to give the Greek mountains a sharpness and clarity of definition that leave a powerful impression. Neither too high to be forbidding nor too low to be ignored, they possess a beauty and harmony of proportion that invite visitors to enter them as they would a temple like the Parthenon in Athens. Sensitive to these effects, the ancient Greeks placed many of their temples on high places oriented toward views of distant mountains.

The earliest civilization in the region of Greece, the Aegean or Minoan, flourished on the mountainous island of Crete between 3000 and 1500 B.C. The religion of this archaic culture centered around the cult of a female deity worshipped on mountains. The peaks sacred to this deity, a mother goddess associated with the fertility of the earth, tended to have the rounded shape of breasts or the cleft form of the female opening. The Cretan palaces of Knossos and Phaistos were built facing two of this goddess's most important mountains: Mount Jouctas and Mount Ida, both crowned with double summits shaped like horns.

Starting around 2000 B.C., successive waves of invaders from the north gradually took over the Greek mainland. Out of the interaction of their culture with that of the Minoan came the classical civilization of ancient Greece. The Indo-European religion the invaders brought with them from the steppes of Eurasia was dominated by masculine deities ruled by a god of thunder and lightening. When these warlike deities took over the religious sovereignty of Greece around the end of the second millennium B.C., the Minoan goddess became Rhea, the mother of Zeus, king of the gods and ruler of Olympus. Fleeing her husband Cronus, who was eating their sons to forestall a prophecy that one of her children would supplant him, Rhea went to Crete, where she gave birth to Zeus in the secrecy of a cave on Mount Aegeum. Gaea, the Greek deity of the earth and another transformation of the mother goddess of Minoan civilization, carried the baby to nearby Mount Ida for safekeeping. There, in the care of nymphs and warriors, he grew up, playing on the sacred mountain of Crete. When he reached manhood — or in this case adult godhood — Zeus returned to overthrow his father and become king of the gods.[2]

Like the ancient mother goddess of Crete, most of the Olympians, the twelve great deities of classical mythology, were closely associated with mountains. As the god of rain and thunder, the ruler of the sky whose place belonged in the clouds, Zeus was worshipped on numerous peaks, from Olympus in the north to Ida in the south. One of his sanctuaries, a mound of earth on the summit of Mount Lycaeus in Arcadia, was said to have been the scene of human sacrifice. Two columns engraved with eagles stood before the shrine, sharply outlined against the blue and windy sky. Artemis, goddess of forests and the hunt, loved to roam the hills and valleys of Arcadia in the company of nymphs. One of her epithets was Lady of the Wild Mountains. Apollo, the youthful god of light and reason, had his sanctuary at Delphi on the slopes of Mount Parnassus, where the Muses also dwelled. The favorite haunt of these goddesses of poetic inspiration, however, was Mount Helicon, a beautiful mountain of forested slopes covered with fragrant plants and abounding with springs of cold, clear water.

Hephaestus, the gnarled god of fire and forge, was born on the summit of Mount Olympus. Ashamed of his lame and twisted appearance, his mother Hera, the wife of Zeus, cast him off the mountain into the sea. Hephaestus eventually moved to Sicily, where he took up residence inside Mount Etna, the highest volcano in Europe. There he placed his heavy anvil on the head of Typhon, a monster whom Zeus had crushed and imprisoned within the fiery mountain. The smoke and steam issuing from the crater of Etna indicated to Greek sailors on passing ships that Hephaestus was hard at work, giving Typhon a truly monstrous headache. The Romans later assimilated Hephaestus into their own divine blacksmith and god of fire, Vulcan, from whose name has come the English word *volcano*.

One of the most famous and influential of all the Greek deities associated with mountains was the Titan Prometheus. As punishment for having given mankind the sacred gift of fire, Zeus had him chained to the highest, most barren peak of the Caucasus, far off on the borders of Europe, where Russia today abuts against Turkey. To make Prometheus writhe in even greater torment, the ruler of the gods dispatched a vulture to peck and gnaw at his liver each day. When Jason went off to the limits of the known world in search of the Golden Fleece, he spied the Titan freezing in agony on the grim heights of the Caucasus Mountains. The image of Prometheus suffering for the sake of mankind has haunted the imagination of innumerable authors from the Greek playwright Aeschylus, who wrote *Prometheus Bound*,

Zeus and Hera preside over the court of the Olympian gods on the summit of Mount Olympus. Painting by Andrea Appiani, Brera Pinacoteca. (Scala/Art Resource, N.Y.)

to the English poet Percy Bysshe Shelley, who composed *Prometheus Unbound.* In this story of classical mythology, the mountain as a place of martyrdom of a savior figure plays a remarkably similar role to that of the Cross in the Christian religion: Prometheus chained to the rock of the former, Jesus nailed to the wood of the latter.

Mount Olympus

Originally the dwelling place of Zeus, the lofty peak on which the god of storms stood to hurl his thunderbolt and gather his clouds, Mount Olympus became the principal abode and palace of the twelve Olympians, the highest deities of the Greek pantheon. There on its summit, suspended above the world of mortals in a cloudless realm of bliss, Zeus ruled over the affairs of men and gods with his brothers, Poseidon and Hades; his sister, Hestia; his wife, Hera; and his children, Ares, Athena, Apollo, Aphrodite, Hermes, Artemis, and Hephaestus. Lines from the *Iliad,* one of the two great epics composed by Homer around 700 B.C., stating that Zeus, Poseidon, and Hades—the deities who ruled over the separate realms of sky, sea, and underworld—had Mount Olympus in common suggest that for the ancient Greeks the mountain transcended any simple conceptions of a heaven on high, removed from the cares of the world below. Indeed, elsewhere in the epic Zeus boasts that he

could pull up the earth and sea on a golden rope and hang them from a pinnacle of the sacred peak.[3]

In addition to being the divine abode of the gods, Olympus served also as their impregnable fortress. After Zeus wrested sovereignty from his father, Cronus, the Titans rose against his rule and attacked him on his sacred mountain. Ensconced on the heights of Olympus, Zeus hurled thunderbolts down at his attackers, scorching the earth and boiling the sea before defeating the Titans and casting them into the depths of the underworld. No sooner had he done so than the Giants, even more formidable foes, emerged from the ground and tried to storm the sacred peak by piling the nearby mountains of Ossa and Pelion on top of each other. Since an oracle had proclaimed that only a mortal could kill these sons of Gaea, goddess of the earth, Zeus and the gods of Olympus had to call on Heracles to protect their mountain fortress by finishing off their enemies for them.

Although much of Greek mythology takes place in the shadow of Mount Olympus, we find surprisingly little description of the mountain itself in classical literature. Numerous passages in the *Iliad* and *Odyssey* refer in passing to the peak as the setting for the comings and goings of the gods, the exalted place where the immortals on high meet to decide the fate of heroes on earth. References to Zeus, for example, generally portray him on the summit of Olympus. But neither the epics nor other works of Greek literature describe the mountain in detail. Instead, they give us only brief glimpses from which to form our own picture of the peak.

Here, for example, is one of the more detailed descriptions of Mount Olympus in the *Iliad*, a passage describing the ascent of Thetis, the divine mother of the hero Achilles, to make an appeal to Zeus on the summit of the sacred mountain:

> *. . . nor did Thetis forget the entreaties*
> *of her son, but she emerged from the sea's waves early*
> *in the morning and went up to the tall sky and Olympus.*
> *She found Cronus' broad-browed son apart from the others*
> *sitting upon the highest peak of rugged Olympus.*[4]

A few words are enough to convey the vivid impression of a peak of great height and ruggedness. We see Zeus perched like an eagle on its topmost crag at the very top of the sky, overlooking the ocean far below. Elsewhere the *Iliad* describes Mount Olympus as tall and sheer. The poet Hesiod in his *Theogony*, a major source of Greek mythology, reinforces the impression

The snow-streaked precipices of Mount Olympus, abode of Zeus and the major gods of ancient Greece. (Gene White)

of height by making repeated references to "snowy Olympus" and "the highest snow summit of Olympus."[5]

The *Iliad* also presents an image of a broad massif of many peaks and ridges. It tells us how the goddess Iris made her way to "the utmost gates of many-folded Olympus." The phrase "many-folded Olympus," which appears elsewhere in Homer and Hesiod, gives us a poetic image of a mountain, or mountain range, creased with ridges and ravines resembling the folds in a curtain hung from the heights of the sky. The "utmost gates" mentioned in the passage refer to the clouds and darkness that enshroud Olympus and bar the way to the realm of the gods on the summit above. These clouds and darkness form two great doors that the Horae, goddesses of time entrusted with guarding the sacred mountain, open and close for the immortals as they come and go.[6]

The epics give us a more detailed description of the abode of the gods on Olympus' summit. There, above the gate of clouds, in a realm of light and bliss, the immortals dwell in perfect comfort, untouched by wind and weather. The *Odyssey* describes the rarefied atmosphere of their abode in heavenly terms:

> *Never a tremor of wind, or a splash of rain,*
> *no errant snowflake comes to stain that heaven,*
> *so calm, so vaporless, the world of light.*
> *Here, where the gay gods live their days of pleasure. . . .*[7]

Seated on golden thrones beside golden tables, the gods pass their time feasting on ambrosia and nectar and savoring the fragrance of burnt offerings sent up from altars in the world below. The Muses and Graces entertain them with beautiful songs and dances, accompanied by the sweet music of Apollo's lyre. From time to time, Zeus calls them to assemblies to resolve disputes and intervene in the affairs of mortals on earth. These interruptions in the harmonious life of the gods on Olympus occupy many of the passages mentioning the mountain in the *Odyssey* and *Iliad*.

Like Mount Kailas in Indian mythology, Olympus seems to have originally existed as an idealized mountain only later projected onto an actual feature of the physical landscape. The epics portray the peak as a heaven on high; they give us very little specific information with which to determine its geographical location. As a result, a number of mountains associated with Zeus, from Greece to as far away as Turkey and Cyprus, bear the name of Olympus, a word meaning "mountain" in the pre-Greek languages of the Mediterranean. Tradition, however, has settled on one as the most likely abode of the gods—Mount Olympus in Thessaly, 160 miles north of Athens, the highest and most impressive mountain in Greece. A passage in the *Iliad* relating how Hera leaves the dwelling of Zeus and immediately passes over Pieria, a landmark that lies near this particular mountain, reinforces its identification with the supreme peak of Greek mythology.[8]

Like a great wave poised over the coast, the mountain rises in one smooth sweep directly up from the sea to the heights of the sky, reaching an altitude of 9570 feet. Its cloud-fringed silhouette, visible from far out in the Aegean, must have made a great impression on seafaring Greeks, especially those on their way to the wars in Troy. It would not have taken much to see the gods coming and going in the golden beams of light radiating from Olympus' highest peak. Viewed from closer up, the solid bulk of the mountain, composed of weathered limestone resting on metamorphic rock, breaks up into a cluster of jagged summits separated by frightening precipices, grim and gray with streaks of polished snow. Clouds frequently gather about them, swirling through their pinnacles and cutting them off from the world below, making Olympus the perfect throne for Zeus, the awesome god of thunder and lightning. Beneath the stark and forbidding realm of its unearthly heights lies a gentler domain where mortals can come to gaze up toward the place of the gods on high. Here, on the lower slopes of the mountain, laurel and pine intermingle with crags of clean white rock to create scenes reminiscent of Chinese landscape paintings, delicately lit with touches of yellow and green.

Although the ancient Greeks probably venerated Olympus as the abode of Zeus, there are no shrines or temples to show that they ever reached its highest summit, which requires some rock climbing technique. Over the centuries, however, people did ascend to the upper reaches of the mountain. In the eleventh century A.D. Saint Dionysius of Hallicarnassus built a shrine on top of Hagios Ilias, one of the lower peaks of Olympus. The first recorded ascent of the mountain by a mortal—rather than a god—took place in 1913, when two Swiss mountaineers climbed Mitka, its highest summit. Today large numbers of climbers and tourists use a luxurious hut built by the Greek Alpine Club to make the climb with relative ease and comfort.

My own ascent of Olympus in 1973 involved a couple of incidents that seemed to come out of Greek mythology. In the course of a journey up the coast of Greece, I met a blind Englishman who wanted to climb Mount Olympus. He was traveling alone, sightseeing, which the Greeks found quite puzzling. "How can you sightsee if you can't see?" they would ask him. He would respond that he liked to touch and feel the stones of ancient ruins. Since I was planning to climb Mount Olympus, he asked if I would take him up

the mountain. Having no rope, I was hesitant to do so until we ran into a couple of Swiss with the necessary equipment. They agreed to accompany us, and we set off to climb Mount Olympus together.

The last part of the ascent involved a scramble up easy cliffs. With one of us ahead and the others behind the blind man, we guided his hands and feet from hold to hold. Secured by the rope, he moved with confidence, delighting in the cold touch of rock and wind. As we approached the summit, I thought of Teiresias, the blind seer of Greek drama and myth. Did our companion, like the ancient sage, see more in darkness than we in light? Was he closer to a blinding vision of the gods on their sacred peak? We reached the top without incident, and never have I seen anyone experience greater joy on the summit of a mountain than he that morning, standing there blind in a blaze of sun and sky.

After helping him come down, I left the Englishman with the Swiss and went off on my own to climb a slightly lower but more spectacular peak of Olympus, nicknamed the Throne of Zeus. By this time clouds had gathered around the mountain, and as I picked my way up a narrowing ridge of rock, I entered a blurred and shifting world of mist. Great precipices opened to either side, falling away into formless space. Not far from the summit I came to a tricky spot—a bulging tower that overhung drops of a thousand feet on one side, five hundred on the other. I started to wriggle past it, but without the reassuring presence of a rope or companion my legs began to tremble and I had to withdraw. I sat down, considering what to do next—whether to try again or turn around and go back. At that very moment a flight of black birds came sweeping out of the mist to pass right over me and vanish into the gray void surrounding the Throne of Zeus.

A bird omen, telling me what to do. In the *Iliad* Zeus takes the form of a black eagle and flies down from Olympus to communicate with mortals. I remembered that for the ancient Greeks the sight of birds flying from one direction was a good sign, from the other a bad. The only trouble was that I could not remember which was which. After pondering this problem for a number of minutes, I stood up, said to myself, "What the hell," and made my move. The birds must have been flying the right way for I easily crossed the difficult passage and reached the summit, there to commune with the spirit of Zeus in the clouds. Some years later, I asked a friend of mine, a professor of classics at Harvard University, about bird omens, but he could not remember which direction was which either, and we concluded that I must have been lucky—or favored by the gods.

Mount Parnassus

South of Olympus, less than eighty miles from Athens, rises the other famous mountain of ancient Greece, Mount Parnassus, sacred to Apollo and the Muses. A long massif capped by a smooth, burnished ridge with two summits—the highest 8061 feet in elevation—Parnassus stands serene and tall above the Gulf of Corinth, overlooking the northern Peloponnesus. The sight of the mountain's broad outline spread across the gleaming sky has inspired poets throughout the ages, making it a favored abode of the goddesses of poetic inspiration. The ancient Greeks also revered the numerous groves, ravines, and caves along its flanks as the haunts of gods and spirits who influenced the events of their daily lives. The sanctuary of Delphi on the lower slopes of Parnassus was the seat of the most celebrated oracle in all of Greece, a priestess on whose cryptic words in trance depended decisions affecting affairs of state and the fate of nations.

Greek mythology tells us that Apollo established Delphi as one of his principal places of worship. Shortly after his birth he pursued the dragon Python to her lair in a gorge on the side of Parnassus. There, in the sacred grove of Pytho, he killed her and built an altar to mark the spot. In search of attendants for the sanctuary he wished to establish, he took the form of a dolphin and leapt out of the sea to commandeer a passing ship. The sailors and their descendants he made the priests of his temple at Pytho. Since they had first seen Apollo in the shape of a dolphin, he told them to call him "Delphinian," from which came the name of the sanctuary itself—Delphi.

Through his oracle, a priestess named the Pythia, Apollo would make pronouncements concerning the destiny of men and women. According to legendary accounts, Pythia would take her seat on a golden tripod set up over a fissure in Apollo's sanctuary at Delphi. Partly intoxicated by fumes issuing from the depths of the earth, the oracle would fall into a trance and become possessed by the god. Attendant priests would then interpret the strange and convoluted prophecies that would stream from her mouth. In actual fact her possession by the deity probably took place in a calmer fashion without fumes or frenzy. When the Persian emperor Xerxes attacked Athens in 480 B.C., the people of the city sent emissaries to Delphi to ask the current oracle for her advice on what to do. In her enigmatic manner she advised them to seek refuge in wooden walls. Themistocles, the leader of the Athenians, took her words to mean that they should leave the city and take to the sea in wooden ships. The victory of the Greeks in the famous naval battle of Salamis, which resulted from his interpretation of the oracle's prophecy, changed the course of history—and determined the subsequent character of Western civilization. After this disastrous defeat, the Persians, who might have otherwise imposed their culture on Greece, never returned to the shores of Europe.[9]

The Muses, goddesses of song who accompanied Apollo, also frequented Parnassus and served as guardians of his oracle, revealing the close connection between poetry and prophecy in the literature of ancient Greece. The spring of Castalia, which was sacred to the Muses, issued from a cleft in the cliffs above Delphi. It's clear and shining waters inspired poetry in poets and prophecy in the priestesses of Apollo. The Greeks also used the sacred spring in purification rites intended to make supplicants fit to receive the words of the god at Delphi. The ancients regarded a carved rock at the place of prophecy as the *omphalos*, the very naval and center of the earth. Today only the ruins of the sanctuary remain, strewn like bones of marble upon the ground. Yet something of the life and spirit of the place endures—in the wild call of eagles circling over the cliffs of Parnassus, in

The Temple of Athena at Delphi on the flower-covered slopes of Mount Parnassus. (©Gary Braasch)

The Monastery of Saint Paul (Ayiou Pavlou) nestles in a gorge secluded beneath the white cliffs of Mount Athos. (James Chotas)

the deep and fluid light of the setting sun, in the silence of the stars shining through the gorge that once marked the spiritual center of the ancient world.

A Greek myth about Parnassus bears a remarkable similarity to the biblical story of Noah and Mount Ararat. Outraged at the theft of fire by Prometheus, who had given it to humanity, Zeus decided to wipe out the human race by flooding the earth. Prometheus warned his mortal son, Deucalion, about the impending disaster, and told him to build an ark for himself and his wife. Zeus gathered his wind and clouds and caused rain to pour from the sky for nine days, inundating the world. On the tenth day the torrent abated and the ark came to rest on the summit of Mount Parnassus. Deucalion and his wife disembarked and immediately offered a sacrifice to Zeus to protect themselves from his wrath. Pleased by their action and satisfied with their contrition, the king of the gods offered to grant them a wish. They asked him for the restoration of the human race. In another version of the myth, they descended to Delphi on the lower slopes of Parnassus and made the same request of Themis, an early wife of Zeus, who told them to throw the bones of their first ancestor behind them. Realiz-

ing that she was referring to the rocks that formed the skeleton of their original mother, the earth from whom their forefathers had sprung, they picked up some stones and tossed them over their shoulders. The stones turned into men and women, making Deucalion the Greek equivalent of Noah, the man who renewed the human race.

Mount Athos

The floods of time have swept away the gods of ancient Greece. Converted to Christianity more than fifteen hundred years ago, the Greeks no longer regard Olympus and Parnassus as places of sacred power and inspiration. They look instead to Mount Athos, the holy mountain of the Eastern Orthodox church. A mountainous sanctuary inhabited by monks and hermits, Athos forms the easternmost of three narrow peninsulas that reach out like rocky claws to rake the blue waters of the northern Aegean. A wooded ridge thirty miles long and two to five miles wide, the peninsula rises sheer from the sea to culminate near its tip in Mount Athos itself, a beautifully shaped peak of delicate white marble, poised 6670 feet above the waves that crash and swirl around its base. A narrow isthmus links the peninsula to the mainland of northern Greece in Macedonia; here in the fifth century B.C. the Persian emperor Xerxes dug a canal so that his fleet could avoid the perilous passage around the rocky end of Athos.

Both the peak of Mount Athos and the peninsula that bears its name abound with wild streams that water a rugged landscape of hanging forests and flowering meadows. Along the flanks of Athos, sharply defining its edges, bluish gray cliffs drop straight into the sea, forming a coastline so sheer and smooth that boats have difficulty finding a safe harbor to land. Frequent storms sweeping in from the Aegean make the peninsula even more remote and inaccessible, a perfect sanctuary for anchorites seeking a place of quiet contemplation far from the distracting influence of the outside world. Perched like white gulls on rocks, monasteries and hermitages dot the cliffs and crags of Mount Athos, overlooking vast blue views of sea and sky. There, suspended between heaven and earth, monks and hermits live in the humbling awareness of their total dependence on the grace of God.

The modern Greek name for Athos, Aghion Oros, means the "Holy Mountain." Sometime after Christianity took over Greece in the fourth century, the peak and peninsula that had been sacred to ancient gods such as Zeus and Apollo became the special preserve of the Virgin Mary, who assumed the role of its divine patroness and guardian. According to Greek legend, she joined the apostles in their mission to spread the gospel. She boarded a ship for Cyprus, but the vessel was blown off course. When the long ridge and marble peak of Mount Athos came into view, she said, "This mountain is holy ground. Let it now be my portion. Here let me remain."

The ship put into shore beside a temple and oracle dedicated to Apollo. When Mary set foot on land, the statues of the pagan gods shattered and fell to the ground. A great stone image of Apollo himself spoke out, declaring itself a false and empty idol. It appealed to the people of Athos to come down to the harbor and pay homage to the mother of the great god Christ. She thereupon baptized and converted the entire population to Christianity. When she blessed the mountain and all who lived there, a voice from heaven responded and declared Mount Athos to be her special place.[10]

The history of Athos as a place of hermitage begins with Peter the Athonite, a monk and saint who lived in the ninth century. One night, as he was sailing back from his ordination in Rome, Mary appeared to him in a dream to tell him that he would spend the rest of his life on her holy mountain. A few days later his ship passed by Athos and came to a stop, despite the presence of a strong wind. Peter took this as a sign that he should disembark and stay on the mountain, which he did, much to the distress of the sailors who left him on the shore. Climbing up the wild and uninhabited slopes of Athos, Peter came to a cave infested by demons and made it his hermitage. Satan and his host mounted a number of attacks on the hermit, but he fended them off, and after seven years of trials, he attained a state of complete humility and spiritual perfection. For fifty years he lived alone in the cave, surrounded by demons and wild beasts. His story probably reflects Christianity's final conquest of the ancient gods.[11]

Following his example, others came to dwell in solitary hermitages on Mount Athos. By the end of the ninth century, many of them had gathered together in *lavras*, informal communities of hermits clustered around spiritual leaders, such as Saint Euthymios, the founder of the first *lavra*. Over time a number of these *lavras* crystallized into major monasteries, complete with elaborate buildings and monastic structures. Saint Athanasios founded the first and most famous monastery on Athos, the Great Lavra, in the middle of the tenth century. In recompense for breaking a vow he had made to take up the monastic life, Nikephoros Phokas, a mystic and general who became emperor of Byzantium, gave Athanasios the money and authority to build the Great Lavra. He also overwhelmed the saint with gifts of all kinds, including two of the greatest treasures on Athos: a jeweled Bible and the reliquary of the true Cross.

Over the centuries a number of Byzantine emperors lavished wealth and patronage on the monasteries of Mount Athos. They also granted them titles and independence from outside control. A decree by Emperor Alexios Comnenos reflects the high esteem in which he and his compatriots held Mount Athos: in 1094 he exempted its monasteries from taxes in the hope that "the most royal and divine Mountain should stand above other mountains of the universe, as Constantinople stands above other cities."[12] As a result of such imperial patronage, the great monasteries of Athos, like the Monastery of Saint Catherine on Mount Sinai, became repositories of priceless works of Byzantine art and literature. They also assumed the form of fortresses designed to guard their treasures and protect their monks from the attacks of maraud-

A sacred well at Docheiariou Monastery stands in front of murals depicting important saints and patrons of Mount Athos.
(James Chotas)

ing pirates. Some even had overhanging walls and stone turrets equipped with cannon. During various periods of turmoil in the history of Athos, a number of these monastic fortresses were destroyed. Of the three hundred or so monasteries that at one time adorned the slopes of the holy mountain, twenty established during the late Byzantine Empire still survive, including the Great Lavra itself.

Around the beginning of the twentieth century, the population of Athos reached its peak with over seven thousand monks, many of them Russian Orthodox. Since the decline of Christianity in Russia as a result of the Communist Revolution, the number has fallen to about three thousand today, most of them Greek. No women are allowed on Athos, a truly monastic mountain: its patroness, the Virgin Mary, is a jealous mistress who brooks no rivals. The three forms of monasticism that developed over the ages still coexist side by side. Monks continue to live in solitary hermitages, often perched in inaccessible eyries; in *lavras* consisting of loose collections of individual cells; and in the large established monasteries, such as the Great Lavra. Daily life includes physical labor as well as spiritual contemplation. Gardening forms an important part of their religious practice, anchoring them to the more concrete realities of existence. The following passage written by a monk reveals the spiritual delight and satisfaction he and his fellows derive from working with the soil of the holy mountain:

And I had a small axe and I cleared pine-trees, olive, holm-oaks, and I chopped them up. And sometimes I planted olives, sometimes pears, or apples, or almond-trees, or vegetables, leeks and garlic, and I rejoiced in the soil as the worldly man in money. I found myself in a garden of graces, in a true paradise of delight. . . . The place was full of fragrance, the trees gave out their odors, birds flew around about, singing while one chanted, and the ground was covered with various flowers and lilies, delighting the eye and ear and filling one with gladness. . . . Hearing, sight, touch, smell, all offered thanks to God.[13]

Such work serves the higher purpose of life on Mount Athos—to resolve the conflict between the worlds of matter and spirit and to recover the spiritual vision and state of man before the Fall. The path of contemplation leading to this goal ascends through three stages in which the mind gradually turns from the sensual pleasures of the body to the intelligible delights of the soul. In the first stage the monk purifies himself through the practice of austerities and cultivates virtue through the recitation of prayers. He also learns to watch his mind and refrain from responding to his baser inclinations. The sense of space embodied in the blue immensities of sea and sky around him infuses his contemplation, preparing him for the second

stage in which he comes to perceive and know the glorious mysteries of God. In the third and final stage, he goes beyond knowledge to attain union with divinity itself, becoming himself a godlike person, higher even than angels, who have not had to overcome and transform their lower selves. Having recovered the state of perfection that Adam possessed before the Fall, he assumes the divine nature of light, the pure and perfect light that glows in the air and glitters on the waves around Mount Athos. He becomes, in the words of Saint Symeon,

. . . one who is pure and free of the world
and converses continually with God alone;
He sees Him and is seen, loves Him and is loved,
and becomes light, brilliant beyond words.[14]

NORTHERN EUROPE

From the grassy hills and peaks of the British Isles to the ice-capped ranges of Scandinavia, the mountains of northern Europe form the scattered remnants of an archaic plateau that rose up long before the rest of the continent came into existence. Once as high and jagged as the Alps, these mountains have been worn down by wave after wave of glacial ice. Their ancient forms, wrinkled and gouged by ages of erosion, provide an ideal setting for myths and legends that tell of gods and spirits who have haunted the earth since the beginning of time. Walking across their moors and through their forests, scrambling over their crags and up their peaks, one feels the presence of an older, more mysterious reality waiting to reveal itself when the mists blow in and blot out the world of familiar experience.

The people who populated the mountains of Scandinavia with the gods of Norse mythology—the only Scandinavian deities of whom we know much of anything—came from the region of Europe north of the Alps. Beginning in the second millennium B.C., Teutonic tribes speaking Germanic languages gradually moved up to the area occupied today by the countries of Denmark, Norway, and Sweden. They brought with them an Indo-European religion that dominated a vast region that included the territory of modern-day Germany. As Christianity advanced north from Rome in the first millennium A.D., destroying pagan beliefs and practices, this Norse religion withdrew to ever remoter regions of the far north. In the ninth century Vikings carried the religion to Iceland, where the only substantial records of the pre-Christian Germanic and Scandinavian deities have survived, preserved in texts written down in the twelfth and thirteenth centuries.

Although Scandinavia belongs to the same Indo-European family of cultures as Greece and India, its mythology — or what we know of it — makes no mention of a sacred mountain comparable in stature to Mount Olympus or Mount Meru. In place of such a mountain, we find a mythical tree named Yggdrasill that acts as a cosmic axis linking together the various levels of existence. From the twisted depths of its gnarled roots to the open heights of its overarching branches, Yggdrasill reaches up to spread out and encompass the four major divisions of the cosmos: the realm of the dead, the land of the giants, the world of human beings, and Asgard, the abode of the gods. Given the deep and mysterious forests that dominated their environment, the people of ancient Germany and Scandinavia naturally looked to the trees around them for a model on which to pattern their conceptions of a cosmic axis, standing straight and tall at the center of the universe in which they lived.

Although Asgard does not perch on top of a lofty peak like the court of Zeus on Olympus, two of its most important deities reside on high places that appear to be mountains of a divine nature. Odin, the chief of the gods, master of war and wisdom, has an elevated seat from which he can survey all that takes place below him. The *Eddas*, medieval works of Scandinavian mythology preserved in Iceland, describe this seat as though it were placed on top of a mountain:

There is a place there [in Asgard] called Hlidskjalf, and when Odin sat there on his high seat he saw over the whole world and what everyone was doing, and he understood everything he saw.[15]

The image recalls that of Zeus in Homer, seated on a crag of Olympus. The name of Odin's high place, Hlidskjalf, means "Hill or Rock with an Opening," reinforcing the impression of a rocky peak overlooking a view of the world of mortals spread out beneath the abode of the gods.[16]

The other deity, Heimdall, the "great and holy" guardian of Asgard, dwells on Himinbjörg, the "Mountain of Heaven." His residence, whose name also means the "Cliffs of Heaven," stands at the upper end of Bifröst, the rainbow bridge that reaches up in a shining arc to link the world of men to the abode of the gods. The *Eddas* tell us that Heimdall, known as the White God, can see hundreds of leagues by day or night and hear all sounds, even the whisper of grass as it grows on the earth below. Seated on his sacred mountain, he guards the bridge against the giants who threaten to storm the heights of Asgard. When they finally do, at the end of the world, he will stand up on Himinbjörg and blow his horn to announce the dreaded beginning of Ragnarök, the "Twilight of the Gods."[17]

Valhalla, or Valhöll, the name of Odin's palace in heaven, where warriors killed in battle go, probably comes from an older word meaning the "Rock of the Slain." Such a meaning would have had its origins in an ancient belief, widespread in Scandinavia, that the dead entered rocks and mountains to dwell within them, continuing to feast and fight as they did in life.[18] The Viking sagas give us a vivid impression of the veneration accorded one such mountain in Iceland by a chieftain named Thorolf:

On this headland is a mountain held so sacred by Thorolf that no one was allowed even to look at it without first having washed himself, and no living creatures on this mountain, neither men nor beasts, were to be harmed unless they left it of their own accord. Thorolf called this mountain Helgafell, the Holy Mountain, and believed that he and his kinsman would go into it when they died.[19]

When an important landowner dies, his shepherd has a vision in which he sees the side of Helgafell open up to receive his master, who is welcomed by a host of merry Vikings noisily drinking and feasting by great fires burning brightly within the mountain.[20] As this passage indicates, the old Norsemen viewed the underground abodes of the dead inside mountains as heavens rather than hells.

Mountains also had a somber, demonic aspect. According to Scandinavian mythology, the realm of the frost giants, who threatened the gods and would someday bring about the end of the world, lay somewhere to the north in a place of ice and snow far from human habitation. In the nineteenth century modern Norwegians, looking back to their Viking heritage, took the name of this place, Jotunheim, or the "Land of the Giants," and applied it to the highest mountains in Scandinavia, a wild range of peaks and glaciers rising to 8097 feet above the fjords of southern Norway. Henrik Ibsen drew inspiration from these rugged mountains of misty rock and ice to create the eerie, troll-infested setting for his well-known play *Peer Gynt*. Today an extensive network of trails and huts allows cross-country skiers to skim with ease across the high and dangerous snowfields once haunted by the frost giants of ancient myth.[21]

With the introduction of Christianity and the passing of the Norse gods, mountains became the embodiment of the other world — the strange and frightening realm of giants, trolls, dwarfs, fairies, elves, and other supernatural beings. Among the demonic creatures believed to inhabit the mountainous regions of northern Europe, the ones known best outside of Scandinavia are the trolls. Closely related to the giants of Norse mythology, perhaps derived from them, they take a number of different forms, some more menacing than others. In the folklore of Norway and Iceland, they usually have huge bodies and grotesque faces with bloated features. They are strong, but stupid. People can trick them into staying outside past daybreak so that they turn to stone in the light of the sun: folk legends identify striking rocks and pinnacles of the natural landscape as petrified trolls. Swedish versions of these malevolent creatures, on the other hand, look like ordinary men and women, but with a crucial difference: they resemble no one known in the local community—and they do not belong to the Christian faith. They are the dangerous outsiders who threaten the established morality and order of things.

The Swedish word for kidnapping, *bergtagning*, means "taking into the mountain." It has its origins in a belief that trolls either take or lure people away into their mountain palaces, where they enslave them as servants—if they do not eat them outright. A typical story tells of a girl who was kidnapped in this way. The trolls put her under a spell and forced her to perform the most loathsome tasks. They also gave her a cap that made her invisible and ordered her to steal things from her village. One day the hat fell off her head, and the villagers were able to see her and return her to her parents. She told them about the trolls and the riches concealed within their mountain—a common theme. In other stories the victims are often

A boy chops off the heads of a troll who has kidnapped a princess. Illustration by Erik Werenskiold.

rescued by the ringing of church bells, a holy sound that no troll can tolerate.[22]

In Scandinavian folklore, the Christian Devil joins the remnants of the pagan past to take over the mountains and make them a place of fear and evil. During the Middle Ages dancing was considered a sinful activity, especially on the Sabbath. A number of stories tell of a well-dressed stranger who joins a group of people engaged in this activity, usually on top of a mountain. When he pulls out a fiddle and begins to play, they lose control of their legs. Unable to stop, they dance faster and faster. Soon their shoes wear out, then their legs, and finally their bodies. Eventually only their skulls remain, still dancing, jumping about like bouncing balls. A cloven hoof or a tail reveals the true identity of the fiddler, who can sometimes be stopped by the timely arrival of a parson. Such stories probably inspired the haunting last scene of Ingmar Bergman's movie *The Seventh Seal*, in which Death leads a silhouetted line of enchanted dancers up a hillside to their doom.[23]

Because of their common heritage, the Germans to the south shared much of the same mythology and folklore. The most eerie and vivid belief in satanic spirits in the Scandinavian and Germanic world hovered around the misty heights of the Harz Mountains, a forested range in central Germany. The Specter of the Brocken, a ghostly image projected by the observer's own shadow on the clouds that often enshroud the highest peak, enhanced the mystique of these mountains as the sinister haunt of demonic beings, both human and inhuman. The Chinese, in striking contrast, viewed the same optical phenomenon on O-mei Shan as a divine manifestation of the Buddha. And indeed, the figure projected onto the mist looks more like a saint than a demon, surrounded as it is in a halo of light. Given a different attitude in Europe, however, the Harz Mountains became renowned as the site of the most famous gathering of witches and devils in Western folklore and literature, Walpurgis Night, immortalized most dramatically in Goethe's *Faust*. The following passage from the play describes the uncanny atmosphere of the scene as Faust, in the company of the Devil, Mephistopheles, approaches the summit of the Brocken to take part in the orgiastic carnival of evil:

The winds are hushed, the stars are pale,
The mournful moon puts on her veil.
In wild career the witches' choir
Scatters a thousand sparks of fire.[24]

When Christianity took over the Roman Empire in the first half of the first millennium, it reduced the divinities of springs and other features of the natural landscape to demons and evil spirits antagonistic to the new religion. Christian missionaries deliberately cleared away forests and cut down sacred groves where pagan rituals traditionally took place as a means of ending such practices. The forested slopes of the Harz Mountains were probably a major site for druidic rites that became associated with witchcraft in the beliefs of later Europeans. In any case, Christians inspired by the writings of early theologians such as Augustine tended to view the wilderness—and the mountains that formed a particularly wild and uncontrollable part of it—as the corrupt domain of the evil powers of nature that the church had to suppress in order to establish the kingdom of heaven on earth.

Under the influence of medieval Christianity and its negative views of the natural world, Mount Hekla in Iceland became the entrance to hell, the abode of Satan himself. One of the most active volcanoes in the world, Hekla lends itself easily to such a view. Since its first recorded eruption in A.D. 1104, the mountain has erupted one or more times a century. The sight and sound of molten lava writhing out of its red-hot craters convinced many observers that they were gazing into the fiery depths of hell. A sixteenth-century physician from the mainland of Europe described the volcano in particularly vivid and horrifying terms:

Out of the bottomless abyss of Mount Hekla, or rather out of Hell itself, rise miserable cries and loud wailings, so that these lamentations may be heard for many miles around. Coal-black ravens and vultures hover around this mountain and have their nests there. There is to be found the Gate of Hell, for people know from long experience that whenever great battles are fought or there is bloody carnage somewhere on the globe, then there can be heard in the mountain fearful howlings, weeping and gnashing of teeth.[25]

Monks throughout Europe used descriptions like this as tangible proof of the existence of hell in sermons intended to discourage sinners from continuing in their evil ways.

Even the local Icelanders, who viewed the mountain in a more matter-of-fact way, took the birds flying around the volcano to be the souls of the damned, crying in endless torment. A Frenchman writing in the seventeenth century added an interesting touch to the beliefs concerning these souls: he reported that the Devil occasionally hauled them out of the fires of Hekla to cool them off on the pack ice floating in the nearby ocean.[26] Such views lingered like sulfurous fumes about the volcano, contaminating the atmosphere surrounding it, until well into the nineteenth century. Even today, when cursing someone, Swedes will sometimes say, "Go to Hekla!" And, in apparent confirmation of older beliefs, the mountain has continued erupting with the force and fury of the Devil,

most recently in the spectacular eruption of 1970.

Like the Germanic tribes who took over Scandinavia, the ancient Celts of Gaul, a group of Indo-European peoples who controlled a region that includes modern-day France and Switzerland, regarded mountains as places sacred to the gods. Around 500 B.C. they carried their deities across the English Channel to install a number of them on the hills and peaks of the British Isles. Whereas the memory of these gods died on the continent, succumbing to Christian missionaries in the first millennium A.D., they survived in remote strongholds of Celtic culture such as Ireland—and in the humanized form of heroes and heroines of Arthurian legend. Unlike their colleagues on the mainland, Irish monks took an interest in their cultural heritage and wrote down extensive descriptions of the pagan beliefs and practices they destroyed.

According to the epic cycles of Irish mythology preserved by these monks, the Tuatha Dé Danann, a tribe of Celtic deities descended from the goddess Danu, came riding in mist to conquer the land and become the gods of ancient Ireland. Their divine rule continued until the religious power of Christianity overwhelmed them in the first half of the first millennium A.D. After their defeat the Tuatha Dé Danann retreated into underground worlds and palaces concealed within mound-shaped hills called *sidh*. Hidden from the sight and knowledge of the outside world, the gods underwent a transformation in which they became fairylike beings of later Irish folklore. Known as the *sidhe*, the people of the enchanted hills, they roamed the countryside with the wind, casting spells on whomever they met and bearing them away to their magic realms within the earth.[27]

In tales told of the *sidhe*, the fortunate mortals they meet in the mist they lead away to paradises hidden on or within the hills in which they dwell. There in the realm of the fairy gods lies all a man or woman could want: every delight of body and soul, eternal youth and soothing peace, and not the slightest trace of death or grief. In some stories the human hero seeks, and occasionally finds, within a subterranean paradise of the *sidhe* a magic cauldron of plenty, which also has the power to restore the dead to life. This huge bowl of Celtic myth mixed with Christian ideas of a sacred chalice to become in time the Holy Grail of Arthurian legend, sought by the knights of the Round Table.[28] In the epic poem *Parzival* by the German poet Wolfram von Eschenbach, the purest of these seekers, Sir Perceval, happens on the Grail in a magic castle located, significantly, on a hill named Munsalvaesche, the Wild Mountain—which Richard Wagner in his well-known opera *Parsifal* renamed Montsalvat, the Mount of Salvation. The company of mysterious knights who guard this castle and its

treasure probably have their origins in Celtic gods who once dwelled and were worshipped on hills throughout the British Isles.

In legendary accounts of the life of King Arthur, when it comes time for him to die, ethereal maidens come to take him away in a boat to the Isle of Avalon, an idyllic land of perpetual spring and eternal youth. Later legends shifted the location of this earthly paradise from the distant reaches of the sea to the interior of hills that had been the abode of the *sidhe*, or their English equivalents. People came to believe that Arthur had not died, but lay asleep within some magic hill, waiting to awaken and return as the once and future king of Britain. Some identified his blissful resting place with the high and lonely peak of Snowdon, the highest mountain in Wales. According to Geoffrey of Monmouth, who composed a mythical history of Britain in the twelfth century, a pair of chained eagles watched over the grave of Arthur on the heights of this sacred peak. There, in a haunting landscape of windswept crags and misty tarns, one could easily imagine figures of the mythic past come back to life — awesome figures like Rhita Gawr, greatest of the legendary giants, and Brenin Llwyd, the Grey King of mysterious disappearances.[29]

Glastonbury Tor and Croagh Patrick

The place most commonly associated with the Isle of Avalon is the mound-shaped hill of Glastonbury Tor in southern England. Situated next to the ancient town of Glastonbury, it rises like an island over a green sea of surrounding meadowland that used to be covered with marshes. According to legend, some thirty years after the crucifixion of Christ, Saint Joseph of Arimathea came from Jerusalem with the sacred blood of Jesus in the chalice of the Holy Grail to establish the first church in Britain at the foot of this hill. Whatever the historical authenticity of the story, Glastonbury was one of the earliest Christian sites in England, probably founded in the first or second century A.D. Many years later, in 1191, guided by a prophecy and a vision, monks reportedly discovered the remains of King Arthur and Queen Guinevere, buried in a grave in the ancient abbey of the town. Since that time local tradition has identified Glastonbury with the resting place of Arthur and the Grail, the two most evocative icons of English history and romance.

The sacred hill of Glastonbury Tor with the ruined tower of Saint Michael's Church, according to legend the site of King Arthur's Avalon. (Roy Bonney)

A legend that may have originally inspired this identification tells us that Glastonbury Tor was once the castle of Gwynn, the ancient British god of the idyllic otherworld equivalent to the fairy realms hidden in the hills of Irish myth and folklore. Gwynn heard that a Welsh saint named Collen, who was meditating in a cell on the lower slopes of the hill, had made disparaging remarks about him. Summoned by the deity to explain himself, Collen climbed to the top of Glastonbury Tor. There he found a magnificent castle filled with lovely people and beautiful music. Gwynn politely offered him some food, but Saint Collen refused to take any, gazing with scorn on the merriment around him. Declaring the Celtic god a demon, he cast holy water over the crowd, and the castle vanished, leaving only a barren summit, swept by the wind. As a monument to the triumph of the Christian faith, a ruined chapel now stands on the hill once revered as a paradise of the ancient gods. It bears the name of Saint Michael, the archangel who in the Bible slays Satan in the form of a dragon and leads the hosts of heaven against the forces of evil.[30]

The aura of Celtic myth and Arthurian legend that envelopes the sacred hill has made it a major center for people with a variety of esoteric interests. They see in Glastonbury a mysterious source of primordial power and knowledge. Many come to draw spiritual sustenance from the reddish waters that flow from Chalice Well, a spring believed to issue from the buried Grail. A recent hypothesis that has aroused a great deal of interest speculates that a series of grassy terraces ringing Glastonbury Tor represents an ancient labyrinth imposed on the hill for ritual purposes. People who have weaved back and forth across the hillside, following the pathway of this labyrinth to its end on the summit, have reported profound experiences of power and awakening, as though they were tracing a spiritual path toward enlightenment.[31]

As the placement of a chapel dedicated to Saint Michael on top of Glastonbury Tor suggests, the introduction of Christianity did not always lead to the demonization of hills sacred to older deities. Rather than become the haunt of a devil, a prominent hill could become the shrine of a saint who overcame pagan gods perceived as demons. Such appears to have happened in the case of Croagh Patrick, the major holy mountain of modern Ireland. An impressive pyramid of quartzite rock, it rises in an inspiring form of grand simplicity to a height of 2510 feet over the waters of Clew Bay in County Mayo. Nearly fifty thousand pilgrims a year climb the mountain to pay reverence to the spirit of Saint Patrick, enshrined in a chapel on its summit.

According to legend, Patrick, the patron saint of Ireland, climbed to the top of this mountain in the fifth century and rang a bell to banish snakes and other noxious creatures from the country as an important part of his mission to convert the Irish to Christianity. Most accounts claim that like Jesus in the test that culminated on the Mount of Temptation, Patrick spent the forty days and nights of Lent on the summit of Croagh Patrick. Since the Christian tradition associates the serpent with Satan in the story of Adam and Eve, it seems likely that the legend of banishing snakes on top of the peak is telling us that Saint Patrick exorcised a Celtic god viewed as a demon or as the Devil himself. In the process Patrick replaced the pagan deity as the source of the mountain's sanctity. This impression is reinforced by accounts that some of Patrick's Irish contemporaries referred to him as one of the *fir sidhe*, the Celtic gods of the hills.

In any case, whatever actually happened on its summit, Croagh Patrick managed to retain its status as a mountain of sacred power—even to this day. Each year, on the last Sunday of July, thousands of Catholic pilgrims gather to climb the peak in a great procession. In an effort to expiate their sins, many make the ascent barefoot, creeping in pain over the sharp rocks that make up the broad white path leading to the top. Their steps take them along a route marked by the stations of the cross and numerous sacred statues. When they reach the chapel on the summit, they make fifteen ritual circuits around it, expressing their deep devotion to the saint and the Holy Spirit he brought to Ireland. Then, standing on the top of the sacred mountain, they listen to a priest recite the Mass.[32]

THE ALPS

Raised up in successive folds by the pressing together of tectonic plates, the Alps curve in an arc from southern France in the west to northern Yugoslavia in the east. Along the way they pass through or skirt the countries of Switzerland, Italy, Lichtenstein, Germany, and Austria. The peaks themselves range in type from the granitic massif of Mont Blanc to the limestone towers of the Dolomites. To complement the complexity of their geological structure, the mountains contain within their deeply cut valleys a multitude of different peoples and cultures. Reflecting this cultural diversity, Switzerland, the country most closely identified with the Alps, has four national languages—French, German, Italian, and Romansh.

Although Olympus and Athos stand out as the major sacred mountains of Europe, the Alps define, for the Western world, the paradigm of mountain perfection. Three of these mountains—Mont Blanc, the Matterhorn, and the Eiger—come immediately to mind as embodiments of three essential attributes of what many would consider the perfect peak. The clean white summit of Mont Blanc, the highest in the Alps, embodies our image of the ultimate mountain as a vision of transcendent purity, floating like a cloud in the deep blue sky. The finely pointed peak of the Matterhorn represents the inspiring ideal of a perfect pyramid of rock sharpened with glistening streaks of ice and snow. The grim north face of the Eiger, whose name means the "Ogre," sets the standard for extreme and terrifying ascents that take mountaineers to the edge of death—and beyond.

Today the Alps are esteemed as places of beauty and inspiration. Many regard them as a kind of paradise on earth, a heavenly refuge from the ugly realities of the modern world. Millions of tourists come up to their valleys each year to ski, climb, hike, and generally escape from the anxiety and boredom of their daily lives. It may come as a surprise to learn that Europeans have not always regarded the Alps in such a positive light, that, in fact, only three hundred years ago they shunned them as horrifying mountains to avoid or, if that should prove impossible, to cross in the greatest possible haste. John Evelyn, an Englishman who traveled over the Simplon Pass on his way to Italy in 1646, described the Alps as "strange, horrid, and fearful crags and tracts, abounding in pine trees, and only inhabited with bears, wolves, and wild goats."[33] Whereas the modern traveler finds peaceful valleys surrounded by beautiful peaks, Evelyn, in a response typical of his times, found a fearsome haunt of terrifying scenery and ferocious beasts.

Others had even stronger reactions to the Alps, finding them aesthetically offensive and even morally repugnant. Thomas Burnet, an Englishman who visited the Alps shortly after Evelyn, wrote in his influential work, *The Sacred Theory of the Earth*:

'Tis prodigious to see and to consider of what Extent these Heaps of Stones and Rubbish are! . . . in what Confusion do they lie? They have neither Form nor Beauty, nor Shape, nor Order. . . . There is nothing in Nature more shapeless and ill-figured than an old Rock or a Mountain. . . . I fancy, if we had seen the Mountains, when they were new born and raw . . . the Fractions and Confusions of them would have appeared very ghastly and frightful."[34]

So much for the Alps as sacred mountains—at least in Burnet's sacred theory of the earth. The sentiments expressed in this passage derive from a view, widely held in Europe at the time, that regarded mountains as irregular blemishes disfiguring the smooth and perfect surface of the land. English poets of the period commonly referred to them as warts, pimples, and blisters on the face of the earth. The grandeur and wildness of mountains that makes them so attractive to us seemed to them to violate the sense of proportion and symmetry, balance and harmony, required of an ideal landscape pleasing to God.

Yet just as the sacred both repels and attracts, mountains also held a paradoxical fascination for Burnet, a fascination that evoked a sense of their divine, as well as demonic, character. In the very same work he writes:

. . . next to the Great Concave of the Heavens, and those boundless Regions where the Stars inhabit, there is nothing that I look upon with more Pleasure than the wide Sea and the Mountains of the Earth. There is something august and stately in the Air of these things, that inspires the Mind with great Thoughts and Passions; we do naturally, upon such Occasions, think of God and his Greatness.[35]

A different view was emerging, one that would lead to our modern appreciation of mountains as sublime, even divine, manifestations of nature. The mixture of fear and fascination characteristic of the experience of the sacred appears in the impressions of a traveler who visited the Alps only a few years later, in 1701:

At one sight of the walks, you have a near prospect of the Alps, which are broken into so many steps and precipices, that they fill the mind with an agreeable kind of horror.[36]

Here the author takes a shiver of delight in the horror that the mountains inspire.

Over the next hundred years a radical transformation took place in European perceptions of the Alps. The writings of Albrecht von Haller and Jean-Jacques Rousseau extolling the virtues of Swiss peasants and Alpine views combined with a growing scientific interest in nature to awaken a new appreciation of mountains as places of divine inspiration and spiritual renewal. Writers of the Romantic period came to regard the Alps as symbols of the infinite, worthy of depiction in literature and art.[37] In 1790 William Wordsworth crossed the Simplon Pass. The very same scenery that had aroused horror and disgust in John Evelyn 150 years previously inspired Wordsworth to compose some of the most memorable lines in English poetry:

. . . The immeasurable height
Of woods decaying, never to be decayed,
The stationary blasts of waterfalls,
And in the narrow rent at every turn
Winds thwarting winds, bewildered and forlorn,
The torrents shooting from the clear blue sky,
The rocks that muttered close upon our ears,
Black drizzling crags that spake by the wayside
As if a voice were in them, the sick sight
And giddy prospect of the raving stream,
The unfettered clouds and region of the Heavens,
Tumult and peace, the darkness and the light—

Were all like workings of one mind, the features
Of the same face, blossoms upon one tree,
Characters of the great Apocalypse,
The types and symbols of Eternity,
Of first, and last, and midst, and without end.[38]

In this passage the wildness and irregularity of the mountains have become sublime manifestations of a divine presence infusing the world and resolving its oppositions.

Most observers have interpreted this change of attitude as the emergence of an unprecedented new appreciation of mountains in Europe. According to this interpretation, the development of scientific understanding and the creation of an aesthetic of the sublime caused Europeans to cast off traditional fears and prejudices and venture up to the heights in search of knowledge and recreation. But if we look more deeply, what we actually see is not the simple appearance of an entirely new appreciation of mountains, but an interweaving of two attitudes toward the Alps in which one recedes into the background as the other comes to the fore. These attitudes both have their origins in the distant past and reflect two experiences of the sacred that we have encountered in other parts of the world: the sacred as the divine and the sacred as the demonic.

We have very little information about early perceptions of the Alps. Unlike the Greeks, the ancient Romans had no real appreciation for mountains. They regarded the Alps as an abomination and generally ignored them in their writings. The only positive role the range had was to serve as a barrier protecting Rome from the barbarians of the north. The Celts who controlled the Alps before the Roman conquest of northern Europe in the first century B.C. probably worshipped high peaks as the abode of a sky god, whom the Romans called Jupiter Poeninus, "Jupiter of the Mountains," from *penn*, the Celtic word for "mountain." The ruins of a temple to this deity stand on the summit of the Great Saint Bernard Pass, placed there in all likelihood for the protection of travelers. When Christianity took over the Roman Empire in the fourth century A.D., Christian missionaries destroyed all traces of pagan beliefs regarding the Alps. Records of Celtic and Germanic lore preserved in the British Isles and Scandinavia suggest that the Celts to the south regarded Alpine peaks as paradises of the dead as well as abodes of the gods. As an Indo-European people with a related language and mythology, the Celts probably shared with the ancient Greeks a reverence for the mountains among which they lived.

Written records of a positive perception of the Alps after the advent of Christianity extend back to at least the fourteenth century A.D., if not earlier. In 1335 the Italian poet Petrarch, responding to an inner urge, climbed Mount Ventoux, a minor but prominent peak 6430 feet high, and wrote a letter describing his experience. As he struggled up the steep and tiring slopes, he compared the physical ascent of the mountain to the spiritual progress of the soul, a comparison that gave him the energy to overcome his inertia and continue to the top. His awestruck impressions standing on the summit reveal a clear link in his mind between Mount Ventoux and the sacred mountains of Greece:

At first, I was so affected by the unaccustomed spirit of the air, and by the free prospect, that I stood as one stupefied. I looked back; clouds were beneath my feet. I began to understand Athos and Olympus, since I found that what I heard and read of them was true of a mountain of far less celebrity.[39]

However, another attitude toward mountains put an abrupt end to his enjoyment of the moment. Uplifted by his spiritual contemplation of the mountain scenery, Petrarch turned to a copy of Augustine's *Confessions* that he was carrying with him and opened the book, by chance, to the following passage: "There are men who go to admire the high places of mountains, the great waves of the sea, the wide currents of rivers, the circuit of the ocean, and the orbits of the store—and who neglect themselves." Dismayed and ashamed at his reaction to the mountain, he wrote, "I shut the book half angry with myself, that I, who was even now admiring terrestrial things, ought already to have learnt from the philosophers that nothing is truly great except the soul." He descended in glum silence and wrote later, "I looked back to the summit of the mountain, which seemed but a cubit high in comparison with the height of human contemplation, were it not too often merged in the corruptions of the earth."[40]

Although many have regarded Petrarch as unique for his time in his appreciation of mountains, other Europeans of the period shared his positive regard for the heights. In fact, some Germans were reportedly climbing peaks and expressing enthusiasm for them as early as the tenth and eleventh centuries.[41] Pilgrimages to Christian shrines on Alpine summits belie the simplistic notion that mountains were universally condemned by the Catholic Church. In 1358, in fulfillment of a vow he had made on escaping from the Moors, Bonifacio Rotario of Asti built a chapel dedicated to the Virgin Mary on top of Roche Melon, an outlying peak of the Alps 11,605 feet high. Shortly thereafter the people of the region initiated the festival of Our Lady of the Snows, in which a procession of pilgrims climbed the mountain each year on

August fifth, carrying a statue of Mary from the cathedral in Susa to her shrine on the summit of Roche Melon—a pilgrimage that has continued to this day.[42] In *The Divine Comedy*, written at the beginning of the fourteenth century, the Italian poet Dante thought enough of mountains to make the ascent of Mount Purgatory the allegorical route of spiritual redemption leading to the earthly paradise on its summit and communion with God in the heavens above.

Some two hundred years later, in the sixteenth century, the Swiss philosopher and scientist Conrad Gesner, in sharp contrast to Petrarch, felt no compunction at all over expressing his appreciation of mountains. He made it a practice to climb a few each year "partly for the sake of bodily exercise and the delight of the spirit." He regarded the Alps as fitting objects of spiritual contemplation, as his words in a letter to a friend so clearly show:

For how great the pleasure, how great, think you, are the joys of the spirit, touched as is fit it should be, in wondering at the mighty mass of mountains while gazing upon their immensity and, as it were, in lifting one's head among the clouds. In some way or other the mind is overturned by their dizzying heights and is caught up in contemplation of the Supreme Architect.[43]

Gesner was well aware of the other attitude toward the Alps, which regarded them not only as terrifying and ugly places to avoid, but also as the cursed haunts of evil spirits and demonic beings. In 1555 he climbed Mount Pilatus, a small peak near Lausanne, named for the spirit of Pontius Pilate believed to reside in a marshy pond just below its summit. According to the legend reported by Gesner, the body of Pilate, who had ordered the crucifixion of Jesus, was thrown into this pond after being rejected elsewhere. The people of Lausanne believed that his evil spirit would cause terrible storms if anyone threw anything in the water; the authorities, therefore, had forbidden anyone from climbing the mountain without an official along to make sure the ghost was not disturbed. The natives held this belief so strongly that in 1307 six clergymen had been sentenced to prison for attempting to climb Mount Pilatus without permission. Although Gesner debunked the legend as superstition, he had to take along an officially appointed guide. Thirty years after his ascent, in 1585, Johann Müller finally punctured the belief when, in the company of a large number of witnesses, he threw stones in the lake, defying Pilate to do his worst—and nothing happened.[44]

Beliefs in other demonic beings were not so easily dispelled. Many Europeans regarded mountains as the favorite lair of dragons, a view that lingered well into the eighteenth century. Stories about meeting them

on the summits of high peaks abounded, and almost everyone assumed as a matter of course that they existed, along with other flora and fauna of the Alps. Johann Jacob Scheuchzer of Zurich, a professor of physics who formulated the first scientifc theory of glaciers, also produced a matter-of-fact catalogue of Swiss dragons. The best specimens, he wrote, were to be found in the sparsely inhabited canton of Grisons: "That land is so mountainous and well provided with caves, that it would be odd not to find dragons there."[45] A belief that the devil resided near the summit of the Matterhorn and hurled rocks down into the valleys interfered with attempts to climb the mountain in the nineteenth century.

Europeans regarded the Alps and the Pyrenees as the special haunt of witches, believed to ride the wild winds and storms that swirled about their rocky peaks. The witch craze that sent many an innocent woman to a flaming death took possession of these mountains in the thirteenth century, two centuries before it descended to sweep like wildfire across the plains of Europe. The presence of heretical sects, such as the Albigensians and the Vaudois, who had sought refuge from persecution in remote mountainous valleys, reinforced a widespread belief that witches controlled the heights, where they practiced satanic arts abhorrent and dangerous to God-fearing Christians. In 1610 witches in the western Pyrenees were held responsible for brewing up storms that wrecked ships going

A terrifying encounter with an alpine dragon. Illustration from OURESIPHOITES HELVETICUS *by Johan Jakob Scheuchzer, 1723. (Courtesy of The Bancroft Library)*

in and out of the harbor at St. Jean de Luz on the Atlantic coast of southwestern France. Even as late as the early twentieth century, Basque peasants continued to believe in a mountain spirit who ruled over witch covens and stirred up tempests in the Pyrenees between Spain and France.[46]

In the folklore and beliefs of Alpine villagers, the spirits of the dead wandered the heights in torment and lay imprisoned in the icy depths of glaciers. Sinuous lines of lights flickering across meadows and mountains revealed the paths taken by ethereal processions of the damned. Various glaciers in the region of Mont Blanc acted as frigid purgatories in which the souls of those who had committed forgivable sins gradually tunneled their way out by scraping at the ice with a tiny pin. Taking pity on these lost souls, villagers would come up to dig on the surface and help them escape to paradise. One story about the Aletsch Glacier, the largest in the Alps, tells of a young student who peered over the shoulder of his religious teacher to see the blue depths of a crevasse crammed full of ghostly heads.[47]

Supernatural beings who appeared demonic sometimes took on a benevolent character, reflecting the ambivalent nature of the sacred. The villagers of Breuil on the Italian side of the Matterhorn used to tell of a wild man who dwelled in the chalets of the wind and taught their ancestors everything they needed to know to live in the mountains — such essentials as how to cure their cows of sickness and how to make cream and cheese. In some of their tales, this benefactor assumed the form of a giant named Gargantua. In ancient times, according to a legend recorded in the nineteenth century, the Matterhorn did not exist: a long horizontal ridge stood in its place, separating the valleys of Switzerland from those of Italy. One day the giant wanted to see what lay on the other side and stepped across the barrier, smashing through it with his boots. The ridge shattered, and the pyramid of rock left standing between his legs became the Matterhorn we see today.[48]

Although overshadowed by the sense of the sublime, the demonic view of the Alps survives to this day. Villagers in certain parts of Switzerland continue to believe that ghosts and spirits of the dead haunt the eerie world of the heights above them. For people with modern sensibilities who no longer hold such beliefs, accidents can still turn the mountains into hellish places of nightmare. One of the most memorable climbing films ever made was titled *The White Hell of Piz Palu*. Filmed in the Alps in 1929, it tells a story of death in the mountains and reveals the power over the imagination that the demonic view has managed to retain, even in the twentieth century. Today the Eiger, whose sinister north face has claimed

numerous lives, continues to elicit among onlookers who watch those who climb it the kind of ghoulish fear and fascination expressed in the meaning of its name — the Ogre.[49]

Mont Blanc

The highest peak in the Alps, Mont Blanc sits like a monarch on a throne made up of lesser mountains. Its summit, a snowy dome 15,781 feet high, overlooks a great massif of jagged pinnacles and sweeping glaciers. So many buttresses and subpeaks surround it that one has difficulty seeing it from its base. Mont Blanc was formed from an intrusion of igneous rock raised up from the bed of an ancient sea. Glaciers have scoured out the softer portions of the massif, leaving peaks such as the Aiguille du Midi, the Grandes Jorasses, the Dru, and the Grépon — granite needles arrayed like warriors with lances uplifted to guard their king.

The history of Mont Blanc and its ascents exemplifies the change in attitude that swept the Alps and other peaks of Europe. A story about the defeat of Satan on the Great Saint Bernard Pass illustrates the way in which Christianity converted pre-Christian mountain deities into demons and replaced them with saints. According to a version of the legend recorded in the sixteenth century, the Devil inhabited a colossal statue on top of the pass. Saint Bernard was engaged in a campaign to cleanse the region of evil spirits. When he and his followers approached the pass, seeking to open a route through it, Satan covered the mountains with deep snow and filled the sky with seething black clouds. Lightning flashed and thunder roared, making the air sizzle with fire. Unperturbed, Saint Bernard held forth his staff and advanced on the Devil. Despite Satan's efforts, the saint reached the top and dealt the statue a great blow, so that it crumbled and turned to dust. Then he exorcised the Devil from the pass and banished him to dwell within Mont Blanc until the end of time.[50] The statue in the legend derives from a memory of the temple to the Celtic sky god Jupiter Poeninus and the way in which Christians made him a manifestation of Satan, the angel who fell from heaven.

Perhaps because of this legend — or the daunting nature of the peak itself — until the eighteenth century the local people knew Mont Blanc as Mont Maillet or Mont Maudit, the "Accursed Mountain." Outsiders drawn by its height and beauty saw it in a new light and renamed it Mont Blanc, the "White Mountain." The German poet and scientist Johann Wolfgang Goethe visited the village of Chamonix at the foot of the peak in 1779. His impressions on seeing Mont Blanc at night as its summit came into view, hovering among the stars, reflect the great change in

Sunset illuminates the glaciers of Mont Blanc. (Brock A. Wagstaff)

attitude toward the Alps that took place in the eighteenth century:

The stars came out one by one, and we noticed above the summits of the mountains before us a light we could not explain. It was clear, without brilliance, like the Milky Way, but more dense, a bit like the Pleiades, only more extensive. For a long time the sight of it riveted our attention. Finally, as we shifted our position, like a pyramid illuminated by a mysterious, inner light—comparable to the phosphorescence of a glow worm—it appeared to soar above the summit of all the mountains; and we knew that it could only be the summit of Mont Blanc. The beauty of this view was quite extraordinary. Shining among the stars that surrounded it, glittering not as brightly, but with a vaster, more coherent mass, the mountain appeared to belong to a higher sphere.[52]

Although the sight of its summit evoked a sublime sense of the infinite, beyond the reach of human comprehension, the mountain itself remained finite, an inspiring but limited feature of the natural world. Goethe wrote of another view of the Alps, "We gave up at once all pretensions to the infinite, for even the finite that appeared before us eluded the grasp of our thought and sight."[52]

With the full development of Romanticism in the early nineteenth century, the Alps transcended the sublime to become themselves actual embodiments of the infinite. The English poet Percy Bysshe Shelley was so overwhelmed by the sight of Mont Blanc that he wrote a poem dedicated to it in which he addressed the mountain as a manifestation of divinity itself:

Far, far above, piercing the infinite sky,
Mont Blanc appears,—still, snowy, and serene—
. . . .

Thou hast a voice, great Mountain, to repeal
Large codes of fraud and woe; not understood
By all, but which the wise, and great, and good
Interpret, or make felt, or deeply feel.
. . . .

Mont Blanc yet gleams on high:—the power is there,
The still and solemn power of many sights,
And many sounds, and much of life and death.
. . . .

. . . The secret Strength of things,
Which governs thought, and to the infinite dome
Of Heaven is as a law, inhabits thee![53]

No longer finite, Mont Blanc has become for Shelley the infinite itself, fully endowed with the life and power of the source and ruler of all that is.

This sense of the sublime, verging on the divine, helped to launch the modern sport of alpinism, which began with the first ascent of Mont Blanc by Jacques Balmat and Dr. Michel Paccard in 1786, only seven years after Goethe visited Chamonix. Horace Bénédict de Saussure, the Swiss scientist who offered a prize for this achievement and who is regarded as the father of mountaineering, was driven to climb the mountain by the same feelings of awe and exaltation that moved Goethe and, later, Shelley to write about it. On his own attempt to make the first ascent of Mont Blanc in 1785, a night view of the surrounding mountains from high on the peak had a profound effect on him:

. . . the brilliance of the stars, which at this height had ceased to twinkle, cast over the mountain-tops a pale glow, extremely faint, but enough to show their power and distance. The restfulness and the utter silence reigning in the vast spaces spread out before my eyes, which imagination pictured vaster still, inspired in me a feeling akin to terror: I seemed to be the sole survivor of the universe, and that it was its corpse I saw stretched beneath my feet.[54]

After he fulfilled his dream by reaching the summit in 1787, he returned to Mont Blanc the following year and had a more joyous, but equally powerful, experience of the mountain, this one in moonlight:

These fields of snow and cliffs of ice, too dazzling to be looked at in the day, what a wondrous and enchanting spectacle they present under the soft beams of the torch of night! What a magnificent contrast the dark granite rocks afford, standing out in sharp, bold outlines against the gleaming snow! Was ever such a moment given for meditation? What pains and hardships are not paid in full by moments such as these! The soul of man is lifted up, a wider, nobler horizon is offered to his view; surrounded by such silent majesty he seems to hear the very voice of Nature, and to become her confidant, to whom she tells the most secret of her operations.[55]

From Mont Blanc practitioners of the new sport of mountaineering ventured to other peaks of the Alps and from them to other ranges of the world—the Andes, the Rocky Mountains, the Caucasus, the Himalayas. Many of these mountaineers sought the kind of spiritual experience that de Saussure described so eloquently in his account of his climb and scientific research on Mont Blanc. The mountain itself continued to provide inspiration for climbers of succeeding generations, offering ever more challenging routes up its many faces and ridges and its even more numerous peaks and needles. The words of Gaston Rébuffat,

one of the great French climbers of the twentieth century, reveal the kind of mystical fascination that Mont Blanc continues to have for modern climbers. He describes how as a boy he felt himself drawn to the mountain as a pilgrim to Mecca and the wonder he experienced on seeing it for the first time—and thereafter:

At last one day the narrow valley suddenly fanned out. The dream cherished for so many fervent evenings crystallized and took shape: there stood Mont Blanc, ideally beautiful. It was only later, after having seen it a thousand and one times, that I noticed its structure, so sober it is. There is equilibrium in the composition of the massif and measure in the grouping of the satellites around the main summit, somewhat set back, as it should be. The forms born of the union of snow and rock seem to soar, despite their mass. How simple and how just the name: Mont Blanc.[56]

Reverence for mountains in Europe continues to express itself in both traditional and nontraditional ways. Shrines on hills and peaks throughout the continent still attract Christian pilgrims in great numbers. They go to venerate saints who have, in many cases, replaced the pre-Christian gods who used to hallow the heights of sacred mountains such as Croagh Patrick and Mount Athos. Among the most popular of these saints are Saint Michael and the Virgin Mary. Shrines to Saint Michael, the archangel traditionally associated with the church's efforts to overcome the demonic forces of heathenism, stand spectacularly situated on prominent hills and crags, most notably the rocky promontories of Saint Michael's Mount in Cornwall and Mont Saint Michel on the northwest coast of France. More than a million pilgrims visit Mont Saint Michel each year, mixing with tourists who come to view the abbey on its miniature mountain above the sea.

The Virgin Mary draws her devotees to higher and more impressive peaks. In the festival of Our Lady of the Snows, a procession of pilgrims carries her statue up icy cliffs to a chapel near the summit of Roche Melon, at 11,605 feet the highest pilgrimage site in the Alps—and probably all of Europe. On more than one occasion, bad weather and dangerous conditions have forced the local authorities to cancel the ritual ascent. A much greater number of pilgrims journey up to a monastery perched in a gully high on the side of Montserrat, a 4054-foot mountain in northern Spain, near Barcelona. They go there to venerate a holy image of the Virgin Mary, said to have been miraculously discovered by shepherds in the Middle Ages. Because of the sandstone towers that line its serrated crest, Medieval legends associate Montserrat with the mysterious castle of the Grail quest. Richard Wagner used the mountain and its monastery as a model for Montsalvat, the name he gave this legendary castle in his opera *Parsifal*.

Shrines dedicated to the Virgin Mary also dominate the heights of hills and peaks in Eastern Europe. The most important and influential of these is the monastery of Jasna Góra, the "Shining Mountain." It sits on a low hill overlooking the town of Czestochowa in southern Poland. Within its central shrine it holds the most revered religious icon in the country: a portrait of the Virgin Mary known as Our Lady of Czestochowa. Each summer in August, a million Poles come on foot to the Shining Mountain to pay her homage, many of them walking for nine days from the capital city of Warsaw. Also called the Black Madonna on account of her dark, reddish brown complexion, this image of Mary is regarded as the Queen of Poland, a powerful symbol of Polish nationalism. In a gesture of religious devotion and national independence, members of Solidarity—most prominently Lech Walesa, leader of the independent labor movement that won control of Poland in 1989—wear badges showing the portrait of the Virgin in her sanctuary on the Shining Mountain of Jasna Góra.[57]

Many of the tourists, hikers, and climbers who throng to the heights of Europe also seek the inspiration of the sacred—but in their own, less traditional, ways. The beauty and challenge that draw them to peaks like Mont Blanc and the Eiger conceal a deeper, more spiritual source of attraction. A number of the tourists who come to gaze on views of the alps harbor a secret longing to glimpse a vision of transcendent power and mystery lacking in the mundane concerns of their everyday lives. Hikers who venture deeper into the mountains willingly put up with fatigue and discomfort to experience the world anew— as a primordial paradise untouched by the inventions of modern man. Many climbers who risk their lives on dangerous routes, such as the North Face of the Eiger, unwittingly seek a jolt of fear to awaken them from the slumber of spiritual complacency. Others climb mountains in search of the joy and freedom they find in transcending the earth and touching the sky.

Although they tend to feel embarrassed about discussing the subject, climbers occasionally acknowledge the religious underpinnings of their fascination with mountains. In 1918 H. E. M. Stutfield gave a talk to the British Alpine Club titled "Mountaineering As a Religion," in which he pointed out a number of striking similarities between the modern sport of mountaineering and the traditional practice of religion. His observations elicited an enthusiastic response from the Reverend F. T. Wethered, a climber and cleric, who wrote in a letter to *The Alpine Journal*, "The mountains have done the spiritual side of me more good religiously, as well as in my body physically, than anything else in the world. No one knows who and what God is until he has seen some real mountaineering and climbing in the alps."[58]

In an age that increasingly values material accomplishment above everything else, the quest for the sacred in the heights of Europe may seem a quaint anachronism. From its beginnings in the Alps in the eighteenth century, critics have questioned the value of mountaineering, asking what practical use it could possibly have. At first climbers justified the effort and risk of their sport by claiming to be doing it in the name of science. Later, in the nineteenth century, they forthrightly declared it a form of recreation. Sir Leslie Stephen, a noted spokesman for the sport, wrote an influential book about climbing in the Alps called *The Playground of Europe*. In keeping with this view of its purpose, mountaineering has found practical justification as part of the tourism industry that now plays a major role in the economies of countries such as Switzerland and Greece. The playground that Stephen extolled is rapidly turning into the Coney Island of Europe.[59] It has become increasingly difficult to find a place from which to see the Alps cleanly without the sight of a cable car or condominium to sully the view. Religious pilgrimage to mountain shrines in other parts of Europe has undergone a similar process of commercialization: guidebooks to sites such as Montserrat and Mont Saint Michel warn visitors not to be put off by the clamor of vendors hawking devotional items to busloads of pilgrims.

The purely recreational approach to mountains, making the peaks of Europe glorified amusement parks, diminishes our experience of them. As paradigms of perfection, redolent with histories of a sacred past, they have more to offer us than a good time. They hold out the promise of something higher, something we desperately need in the muddle of modern life: an inspiring vision of a more meaningful realm of existence in which we can find the freedom to be true to ourselves. Whether approached in the traditional way of a pilgrim or in the nontraditional manner of a climber, the mountains of Europe stand as reminders of another view of reality—a view that calls us to a truer, more spiritual awareness of what is best in ourselves and the world around us.

· 8 ·

AFRICA

FACING THE HEIGHTS

GLIMPSED THROUGH THE HAZE of the imagination hidden behind vast expanses of jungle, desert, and grassland, the mountains of Africa crystallize in rock—and traces of equatorial snow—the aura of mystery that imbues the continent with its special mystique. Just south of the Mediterranean, the Atlas Mountains rise in a great wall of folded ranges, barring the way to the empty wastes of the Sahara. Deep in the sands of the world's largest desert, undulating in waves of heat, stand the gaunt pinnacles of the Ahaggar and Tibesti massifs. Like most of the mountains of Africa, other than the Atlas and the Ruwenzori, the Ahaggar and Tibesti massifs were formed by volcanic activity. Remnants of ancient lava flows sprawl over other parts of the continent, capping the serrated ranges of South Africa and covering the highlands of Ethiopia. Deadly gases saturate strangely colored lakes that fill craters of volcanoes in West and East Africa: when these gases bubble up and spill over in invisible floods of carbon dioxide, thousands die. In both form and substance, the varied mountains of the world's second-largest continent express the energy of molten rock slowly roiling beneath the earth's crust. Visibly charged with the power of this subterranean energy, they awaken a profound sense of mystery and awe.

The highest peaks and greatest concentration of sacred mountains occur in East Africa. The forces that ripped open the Great Rift Valley uplifted the Ruwenzori, a mysterious range of alpine peaks covered with snow and hidden by mist. Elsewhere in the region of East Africa the fracturing of the earth's crust has produced clusters of isolated volcanoes, such as the Virungas and Ol Doinyo Lengai, the Mountain of

God. The highest of these volcanic peaks, Mount Kenya and Mount Kilimanjaro, have played particularly important roles as sacred mountains. The Kikuyu, the largest and most influential tribe in Kenya, have long revered the summit of Mount Kenya as the principal abode of their god Ngai. The mountain possesses the distinction of being the only peak in the world to have a country named after it. The Chagga who live on the slopes of Mount Kilimanjaro draw spiritual and material sustenance from the snows of its summit, the highest point on the continent. Just as the perfect cone of Fuji symbolizes Japan, so has the great dome of Kilimanjaro come to represent Africa in the eyes of the world.

Not far from Kilimanjaro, in northern Tanzania, lies Olduvai Gorge. There, in beds of volcanic ash and sedimentary clay, anthropologists have unearthed the primitive tools and skeletal remains of *homo habilis*, one of the earliest known precursors of the human species, nearly two million years old. Cave paintings scattered throughout the continent suggest that mountains may have served as shrines and objects of veneration for the prehistoric peoples who evolved out of these protohumans. In the middle of the Sahara, the mysterious peaks of the Ahaggar and Tibesti massifs rise up like the black bones of a fossilized monster, struggling to emerge from the desert sands. Figures inscribed on cliffs that lie between these mountains depict in graceful detail giraffes and other animals that inhabited the region when it was covered with jungle and grassland. Similar paintings appear in caves and under overhanging rocks in the Drakensberg Range of South Africa and the Tsodilo Hills of Botswana. Bushmen whose ancestors painted some of these

Point John, a peak of Mount Kenya, emerges from mist over the waters of Hut Tarn. (Clive Ward)

works of art venerate certain of the mountains in which they are found. The beliefs of these aboriginal people, who hunt and gather as the earliest humans did, suggest how prehistoric tribes may have regarded sacred peaks, both in Africa and other parts of the world where similar cave paintings exist, such as the Lascaux Cave in southern France. One of their legends tells us that the highest peak in the Tsodilo Hills, called Mount Male, was once a man with two wives. Because he showed more affection for his second spouse, his first wife bashed him with a stick and moved away with their children. When the great god found out what had happened, he turned the man into a mountain and his wives and children into the smaller peaks and hills of the region. A crack running up the face of Mount Male marks the scar inflicted by the jealous wife's blow to her husband's head. Like the Aborigines of Australia, the Bushmen of Botswana see mountains as the petrified forms of their primordial ancestors, a view that may have been widely shared by prehistoric peoples.[1]

The oldest clear-cut references to sacred mountains in Africa occur in European rather than African sources. The ancient Greeks looked to the mysterious continent for the far-off place where Zeus had sentenced Atlas to stand in endless anguish, supporting the enormous weight of the sky on his head and shoulders. In the fifth century B.C. the Greek historian Herodotus identified the Titan of Homeric mythology with the Atlas Mountains of North Africa—a range of rocky peaks that runs through the countries of Tunisia, Algeria, and Morocco, reaching an altitude of 13,665 feet. Rising above the straits of Gibralter, where the Mediterranean empties into the vast Atlantic, this range lay for the ancient Greeks at the extreme limits of the world they dared to navigate and explore. In the *Metamorphoses*, the classic poem describing the transformations of men and gods, the poet Ovid recounts the myth the Romans used to explain the creation of the Atlas Mountains. After slaying Medusa, Perseus showed the Titan the head of the Gorgon, which turned him to stone so that

Atlas was all at once a mountain, beard
And hair were forests and his arms and shoulders
Were mountain ridges; what had been his head
Was the peak of a mountain, and his bones were
   boulders.[2]

As the fossilized remains of a deity who held apart the earth and sky, the Atlas Mountains fulfilled one of the most important functions of a sacred peak: to stand as a pillar linking, yet separating, the various levels of the cosmos. Without such a pillar to give it form, the world would collapse on itself and return to the chaos from which it emerged.

Of the African mountains known to the ancients, the ones that most excited the imagination of the outside world were the elusive Mountains of the Moon—a range of fabled peaks that seemed to epitomize the primordial mystery of Africa itself. Around 500 B.C. the Greek dramatist Aeschylus made a brief reference to "Egypt nurtured by the snows." Fifty years later Herodotus stated his belief that the Nile issued from the waters of a bottomless lake set between two peaks. Aristotle wrote that the river had its source in a "Silver Mountain." In the second century A.D. the Greek geographer Ptolemy identified this mountain as a range and called it the "Mountains of the Moon," a name that resonated with the haunting connotations of another world—the divine realm of a lunar god. There among peaks glistening with silver snow lay the legendary source of the sacred river of the ancient world. During the latter part of the nineteenth century, the search for these mountains and the waters they concealed assumed the character of a religious quest.

A number of explorers risked their lives in arduous efforts to seek the solution to the outstanding problem of modern geography. In 1888, traveling over ground covered by his predecessors, Henry Stanley became the first European to discover, almost by chance, a range of snow peaks that fit Ptolemy's description of the Mountains of the Moon—the Ruwenzori, situated on the border of modern-day Uganda and Zaire. An almost perpetual cover of rain and mist had hidden them from the gaze of the earlier explorers. Looking toward them on a rare day when their snows emerged from the misted sky, Stanley nearly mistook the mountains for a range of clouds:

. . . while looking to the south-east and meditating upon the events of the last month, my eyes were directed by a boy to a mountain said to be covered with salt, and I saw a peculiar-shaped cloud of a most beautiful silver color, which assumed the proportions and appearance of a vast mountain covered with snow. Following its form downward, I became struck with the deep blue-black color of its base, and wondered if it portended another tornado; then as the sight descended to the gap between the eastern and western plateaus I became for the first time conscious that what I gazed upon was not the image or semblance of a vast mountain, but the solid substance of a real one, with its summit covered with snow.[3]

Even when discovered at last by Europeans, the Mountains of the Moon retained the aura of mystery that had enveloped them for thousands of years. Expeditions that set out to study and climb the peaks found in their rain forest and shrouded tundra an uncanny world of monstrous plants that seemed to belong to a prehistoric age of the primordial past. The

Perspiration gleams on the faces of Masai warriors who have danced themselves into a state of trance.
(Galen Rowell/Mountain Light)

chronicler of the Duke of Abruzzi's scientific expedition, which made the first ascent of the highest summit — 16,795-foot Margherita — in 1906, described the eerie, almost supernatural impression that the unearthly landscape made on him and his comrades:

The ground was carpeted with a deep layer of lycopodium and springy moss, and thickly dotted with big clumps of the papery flowers, pink, yellow and silver white, of the helichrysum or everlasting, above which rose the tall columnar stalks of the lobelia, like funeral torches, beside huge branching groups of the monster senecio. The impression produced was beyond words to describe; the spectacle was too weird, too improbable, too unlike all familiar images, and upon the whole brooded the same grave deathly silence.[4]

The quest for scientific knowledge had taken the Duke of Abruzzi and his companions into the mysterious realm of the sacred, hidden in the gray mist of the Ruwenzori.

Although the human race probably originated in East Africa, not far from the cloudy peaks of the Ruwenzori, the ancestors of most of the Africans now living in the region have come from other parts of the continent. A long, slow migration of Bantu-speaking tribes circled in from Central Africa, approaching the highlands of modern-day Kenya and Tanzania from the south and east. The Kikuyu, the largest of these tribes, reached the foot of Mount Kenya in the seventeenth century. About a hundred years later Nilo-Hamites moved into the area from the region of the Nile River valley to the north. The most warlike of these newcomers, the Masai, took control of the open grasslands lying beneath the forested mountains. Unlike the Kikuyu, who settled the land and cultivated crops, the Masai roamed the plains, grazing herds of cattle. Sometime before the nineteenth century, predominantly Bantu-speaking peoples with a sprinkling of Masai gathered on the green slopes of Kilimanjaro to form the farming and trading clans of the Chagga tribe. Other Bantu speakers and Nilo-Hamites displaced aboriginal inhabitants, primarily pygmies, from mountain ranges such as the Virunga Mountains and the Ruwenzori.

The Virungas, the last remaining habitat of the imperiled mountain gorilla, form a range of eight volcanic peaks that stretch along the border between Zaire, Rwanda, and Uganda. They include two dormant volcanoes whose divine power, still burning invisibly in the mist, attracts the spirits of the dead. According to Dian Fossey, the researcher who publicized the plight of the gorillas and was killed for her efforts to protect them from poachers, the Bahutu, the largest tribe in the region, believe that the souls of the good go after death to dwell forever on the summit of the highest peak, 14,782-foot Karisimbi. There, far above the somber jungle that shrouds the mountain's lower slopes, a soft ring of meadows and lakes, green and blue beneath a volcanic cone softened with a cover of white hail, provides a beautiful setting for a paradise of the dead. Rising over forests of thick bamboo, the nearby peak of 13,540-foot Muhavura serves as another heavenly resting place for the spirits of those whose lives have been pure. Its name means "He Who Shows the Right Way"—the way that leads to eternal happiness. Perhaps Fossey's spirit followed that way and now dwells on the heights of the mountain, happily watching over the gorillas for which she gave her life.[5]

To the south of the Virungas, in the dry savannah of northern Tanzania, rises a sacred mountain that plays an even more central role in the lives of the people who live within sight of its summit. The Masai who wander the open country around its base call it Ol Doinyo Lengai, the "Mountain of God." An active volcano nearly 10,000 feet high, it looms over the surrounding plains, thrusting up a gray cone wrinkled with gullies and covered with slippery ash. When it erupts, as it last did in 1967, it seems to draw its power, like thunder, from the sky itself. The Masai regard it as the special abode of Engai, the one and only God, who dwells in the blue heights of heaven. They believe that he withdrew to the mountain peak after a hunter shot an arrow at him, leaving their world a place of famine and death. Although he no longer lives among them, his universal presence remains: he still harkens to their prayers and sends blessings to relieve their sufferings. They offer sacrifices of lambs without spots to him upon his mountain in the sky, asking for cattle and children, the two things they treasure most. The men of the tribe still climb the volcano on mysterious missions to invoke his power.[6]

The Sonjo, a farming people who speak a Bantu language and share their territory with the pastoral

A gorilla peers out from his sanctuary in the mountain jungle of the Virunga Volcanoes. (Allen Bechky)

Masai, also revere Ol Doinyo Lengai. They call it Mogongo jo Mugwe, which means the "Mountain of God" in their language as well. The volcano occupies a prominent place in their lives: the silhouette of its symmetric cone can be seen from all their villages. According to their beliefs, in the distant past a supernatural being named Khambegeu came to the Sonjo and created for them a golden age of peace and prosperity. After he died, he rose from the grave and flew up to the sky to become one with God. A solar deity who lives in the sun, Khambegeu also dwells on the summit of Ol Doinyo Lengai. The Sonjo believe that he cares for their welfare and that when the world comes to an end and other peoples are destroyed, he will save those among them who have remained faithful to him.[7]

Ol Doinyo Lengai, the "Mountain of God," an active volcano in Tanzania sacred to the Masai and Sonjo. (Clive Ward)

Mawenzi veiled in shining mist, viewed from the snowfields of Kibo. A Chagga myth attributes Mawenzi's shattered appearance to a beating the mountain received from the larger Kibo. (William Frej)

Mount Kilimanjaro

Not far from the border with Kenya, the earth of northern Tanzania bulges up to form one of the most massive and isolated mountains in the world—Mount Kilimanjaro. Visible from more than a hundred miles in every direction, rising nearly 17,000 feet above the surrounding plains, the great snow-tipped dome of Kibo, the highest peak in Africa, spreads across the equatorial sky, white and cool above the dust and heat that envelop the lowlands. Its broad outline, polished to a smooth finish by the blurring effects of height and distance, soothes the eye and calms the mind, lifting the spirit with a reassuring glimpse of the sacred.

Kilimanjaro actually consists of three widely separated volcanic peaks linked together by an elevated plateau. The immense blue mass of 19,340-foot Kibo so dominates the view that most people identify it with the mountain itself. Fumaroles and a huge crater hidden in its summit attest to its relatively recent volcanic activity. The second-highest peak, 16,896-foot Mawenzi, set to the side, has a very different look from

that of its serene companion: black pinnacles of crumbling rock give it a dark and angry appearance. They form the sharp and brittle remains of a huge explosion that shattered its crater thousands of years ago. The oldest and lowest peak, 13,140-foot Shira, scarcely protrudes from the plateau. With its three peaks and their diverse climatic zones, ranging from dry scrubland through damp jungle to alpine meadows and arctic glaciers, Kilimanjaro forms a world complete in itself.

The people who inhabit this world, the Chagga, have a story that explains the disfigured appearance of Mawenzi. Long ago, before the coming of Christianity, when the peaks were still active volcanoes, Kibo and Mawenzi were wives of the great god Ruwa. One day Mawenzi went to the house of Kibo to get embers to light her fire. She arrived just before mealtime, which meant that according to Chagga custom she had to stay and eat. Mawenzi was lazy and did not want to cook. So after leaving, she came back, saying that she had tripped crossing a stream and had dropped the embers in the water. Since she had timed her return to arrive just in time for dinner, Kibo had to feed her again. Realizing that she was on to a good thing, Mawenzi came back for a third meal. This time Kibo had cooked just enough food to prepare a special dinner for her husband. She flew into a rage and beat Mawenzi with a wooden paddle she was using to pound bananas, giving her the battered look she has today. Some people say that Mawenzi is so ashamed of her scarred and ugly face that she usually hides herself in clouds so as not to be seen in public. According to others, Kibo faces away from Mawenzi out of disgust, so that all one can see is the back of her head—a smooth mountainside without any prominent features.[8]

The Chagga clans came to Kilimanjaro over the last few centuries, seeking streams to nourish their crops and forests to protect them from attacks by the warlike Masai. There on the lush slopes of the mountain, in an idyllic setting, perched above the heat and dust of the plains, they felt they had found an earthly paradise. Even today they refer to their homeland as *kari ko ruwa*—"God's backyard." People who live downslope toward the grasslands with their harsh, dry climate, are regarded by the Chagga as deprived, even cursed. A Chagga told me, "You are not a full human being if you don't come from Kilimanjaro. In fact, the higher up the mountain you live, the more fully human and blessed you are." The region above the inhabited zone they considered especially sacred: only men who had received the proper initiations were allowed to go there.[9]

When the Chagga arrived, they displaced the aboriginal inhabitants of Kilimanjaro, who migrated over the mountain and vanished toward the west. Reflecting tribal memories of their predecessors, who were probably pygmies, Chagga legends tell of little people who live on the heights of Kibo, hidden from view. They possess great wealth, and if approached correctly, will bestow it on others. In one of many stories about them, two brothers suffering from poverty decided to set out in search of a generous king said to dwell on top of the mountain. The elder brother, who went first, met an old woman with suppurating eyes. She offered to show him the path if he would lick the pus off her eyelids, but he refused in disgust and went on. Higher up the mountain, he came across some little people, whom he mistook for children. When he asked to speak with their fathers, they told him to wait and went away. When they failed to return, he gave up and came down the mountain. The younger brother went up to try his luck. When he met the old woman, he licked her eyes clean, and she told him to address the little people not as children, but as members of the royal council. He did so and they led him to the king, who gave him cattle and wealth so that he and his relatives prospered.[10]

Another Chagga myth, unique in its conception of mountain building, describes the creation of Kibo, Mawenzi, and other peaks and hills of the region. Long ago, before any of these mountains had come into existence, a skeptic named Tone challenged Ruwa, the supreme god of the Chagga, to demonstrate his power by inflicting a famine on the world, which he did. In revenge for the suffering he had caused them, the people tried to kill Tone. He sought refuge with a kindly old man, who put him in charge of two cows named Tenu and Meru—the latter being the name of a prominent volcano not far from Kilimanjaro. He warned Tone that if he ever opened the door to their stall he would die. One day Tone left it open, and the cattle escaped. He chased after them, crying out, "Tenu, wait for me!" The cow called back, "Come here, I wait for you." When Tone approached her, she threw up a hill and kept going. Tone had to run over it. In this way, she continued to lure him on, throwing up ever-higher and higher hills, culminating in Mawenzi. Nearly dead from exhaustion, Tone staggered to the top and croaked out, "Tenu, Tenu, my friend, have pity, for I die." Tenu replied one last time, "Come quickly, I am merciful and wait for you." Then she tossed up Kibo, the highest peak, and galloped up its slopes with Tone following her. They disappeared into the crater, never to be seen again.[11]

Before European missionaries converted them to Christianity, the Chagga regarded Kibo as the divine embodiment of all that was exalted, eternal, and nourishing. They honored their chiefs by calling them "Kibo." They also used the name of the mountain in expressions wishing each other the blessing of long life:

"May you endure like Kibo."
"As Kibo moves not, so may life not be removed from you."
"As Kibo ages not, so may you never be old."

Many of their customs reflected the high esteem in which they held the sacred mountain. On rising in the morning, the older men would stand outside their houses facing the volcano and spit toward the sky as a form of offering to Ruwa, their god. The Chagga regarded the side of the village oriented toward Kibo as the one of honor. There the men would assemble for feasts and councils. Whenever a person met a superior on the road, he would pay him respect by allowing him to pass on the side closest to Kibo. The Chagga buried their dead with their heads toward the mountain. After a year or two, they would remove the skeleton from its grave and place the skull in an ancestral grove, facing Kibo in silent communion with its forefathers. People today still dig up the bones of

The pink light of dawn glows softly on the summit glaciers of Kibo, highest peak of Kilimanjaro. (William Frej)

their dead and put them in ancestral groves. The power of the mountain has even influenced the practice of Christianity: most of the churches have their altars on the side closest to Kibo.[12]

Modern Europeans learned of the existence of Kilimanjaro at a fairly late date, in 1848 when a German missionary, Johannes Rebmann, sighted it during a journey to set up missions in the interior regions of East Africa. His report of snow on the top of a peak near the equator elicited mockery and disbelief among authorities back in Europe, who claimed the mountain would have to be huge, which indeed it was. Ten years later an expedition led by a flamboyant German baron confirmed Rebmann's observation — and aroused European interest in climbing the enormous volcano. After a number of attempts by succeeding expeditions, Hans Meyer finally made the first recorded ascent of Kibo in 1889. The Chagga themselves had been climbing high on the mountain to retrieve a white medicinal substance called *ikata*. The ascent of Kibo, a long slog up ash and snow slopes, required no technical skill, only endurance and the will to reach the top. Mawenzi, a much more difficult peak to climb, yielded its summit much later, in 1912.

Dr. Richard Reusch, a German missionary who climbed Kibo in 1926, reported a legend, current among Ethiopians of the region, that linked the peak to the seal of Solomon. According to a story told to Reusch, the King of Israel and the Queen of Sheba had a son named Menelik, who succeeded his mother to become the ruler of Ethiopia. After conquering a number of other countries and establishing a great empire in East Africa, he started home and found himself crossing the high plateau between Kibo and Mawenzi. A great fatigue came over him, and filled with a yearning to die, he said farewell to his generals and climbed up to the summit of Kibo to vanish into its crater, accompanied by slaves bearing his treasure. There, swathed in clouds and snow, he sleeps the eternal sleep of the hallowed dead. Someday, according to prophecy, a descendant of the ancient king will restore the Ethiopian empire to its former greatness. When the time comes for him to fulfill his destiny, he will climb to the summit of Kibo and find the seal of Solomon on the hand of Menelik. He will put the ring on his finger, and the wisdom of Solomon and the spirit of Menelik will fall upon him, and he will go out to conquer the nations of the world.[13]

Although Ethiopia lies to the north of Kilimanjaro, the legend has some basis in the national epic of that country. According to this ancient work of Coptic poetry, the *Kebra Nagast,* Solomon and Sheba had a son named Menelik who was supposed to have conquered a vast territory to the south of Ethiopia. He also obtained the seal from Solomon, although he later returned it to his father in Jerusalem. The rulers of Ethiopia believed themselves descended from the kings of Israel, and the grandfather of Haile Selassie, the last emperor, deposed in 1974, took the name of Menelik II when he reunited the country in 1889. The *Kebra Nagast* does not mention Kilimanjaro, but an Arab writing in the sixteenth century apparently referred to it when he spoke of the "Olympus of Ethiopia." Although the legend linking the mountain to Menelik was probably a later addition, it seems to have had a number of adherents by the twentieth century. When Reusch climbed Kilimanjaro in 1926, the Ethiopian Christians living in the region refused to believe he had actually reached the summit after they asked if he had seen the ancient king and his treasure there and he replied that he had not.

On that ascent Reusch discovered, near the highest point of the mountain, the frozen body of a leopard that inspired Ernest Hemingway to write a short story that made Kilimanjaro one of the most famous mountains in modern literature — "The Snows of Kilimanjaro." The story begins with a quote describing the bizarre remains on top of the mountain and the enigmatic question they pose:

Close to the western summit there is the dried and frozen carcass of a leopard. No one has explained what the leopard was seeking at that altitude.[14]

The end of the story gives an indication of the ultimate answer to the question of what the leopard, and Hemingway, were seeking. As the hero, a writer named Harry who has wasted his life, lies in his safari tent, dying of gangrene, he falls into a delirium and has a vision of a bush pilot coming to save him. They take off in the airplane and fly into a storm. When they come out, heading toward the mountain, Harry suddenly realizes that this is no ordinary flight:

Compie [Compton, the pilot] turned his head and grinned and pointed and there, ahead, all he could see, as wide as all the world, great, high, and unbelievably white in the sun, was the square top of Kilimanjaro. And then he knew that there was where he was going.[15]

For Hemingway, as for the Africans themselves, Kilimanjaro was an awe-inspiring place of the dead and a symbol of the mystery into which they vanish.

Converted to Christianity and involved in the pursuits of the modern world, the Chaggas today have abandoned most of the beliefs and practices directed toward Kilimanjaro, and few of the younger generation are aware of the significance that the mountain once had in the lives of their people. I asked Willy Makundi, a Chagga finishing a Ph.D. in forestry and natural resources at the University of California at Berkeley, what he knew of the sacred aspects of Kibo. "I really don't know much about that," he replied. "There are stories that people tell, but it's all part of the past that hasn't been handed on to my generation. Since we went to Christian schools, we never had the opportunity to learn that kind of thing from the old people. In a way, we're a lost generation."

However, as we talked, he began to realize that many seemingly unimportant things in contemporary Chagga life that he had not thought much about came from old beliefs and practices having to do with the sacred mountain. When they are about to leave on a journey, even young people with a modern upbringing spit in the direction of Kibo for good luck — a relic of the offering of saliva that their elders used to make every morning at dawn. Everyone in the Chagga country sleeps with his or her head toward the mountain. "I thought it was because of the slope of the land, but I guess there's more to it than that," Makundi observed. Thinking about that point, he noted that the Chagga always build their homes and shops on the side of their farm or plot of land closest to Kibo. His father wanted to set up a brewing club for the men in his village — a place where they could pay to distill their own liquor. Makundi's grandmother told him that he would have to build it on the upper end of the farm facing the mountain.

"My father thought that was the women's way of discouraging him from something they disapproved of because he would have to build the club right next to their house, where no one would want to have a brewery. So he built it on the lower part of the farm, away from Kibo. My grandmother said bad luck would come of it, and sure enough, in a few years people stopped coming and he went out of business.

"You know," he added, "Kibo is not a mountain."

Makundi explained that the Chagga language puts special words before place names, indicating their relation to the rest of the sentence and what kinds of places they are — flat areas, spaces, or points of land. With mountains they use a word that we would translate as "mount." So a Chagga will say, "I go to Mount Meru." But in the case of Kibo, they use nothing. They simply say, "I go Kibo." As far as their language is concerned, Kibo is not a place, much less a mountain. When I asked Makundi if any other places were treated in the same way, the only ones he could think of were home and heaven — two places of particular sanctity in cultures throughout the world. Although no longer conscious of the fact, through their language, the Chagga still set their sacred mountain apart, in a special realm outside of space and time.

Mount Kenya

Mount Kenya, the second-highest peak in Africa, has a very different appearance and feel from Kilimanjaro. Whereas the sight of the smooth outline of Kilimanjaro fills the mind with a quiet feeling of serenity, the first glimpse of the jagged silhouette of Kenya tends to startle the viewer with a sudden burst of excitement, piqued with fear. Left as the remnants of a volcanic core, its peaks shoot up like daggers of rock, stabbing the underbelly of the sky. The highest summit, Batian, reaches an altitude of 17,058 feet. The two other principal peaks, Nelion and Point Lenana, are 17,022 and 16,355 feet high. Clinging to ridge crests and plunging down precipices, glaciers of wind-polished snow hover in the mist or glisten in the sun, depending on the clouds that swirl about the mountain's heights. Situated some seventy miles northeast of Nairobi, the capital of the country, Mount Kenya rises in magnificent isolation, a dazzling pinnacle of rock and ice that dominates the landscape of central Kenya.

The Kikuyu, the largest tribe in Kenya, with more than two million members, call the mountain Kere-Nyaga, the "Mountain of Brightness," from which has come the modern name of Mount Kenya. They believe that their god, Ngai, the creator who dwells in the sky, chose this mountain as his principal resting place on earth. In their prayers and sacrifices those who still hold to the old ways address him as Mwene-Nyaga, the "Possessor of Brightness." In addition to Mount Kenya in the place of honor to the north, he has abodes of lesser importance on smaller mountains located in positions of symbolic significance to the east, south, and west. Traditional Kikuyu worship him beneath sacred trees that stand for these mountains and act as temples of their god, the supreme deity who gave them life as a people and sustains them to this day.

According to the story the Kikuyu tell of their origins, in the beginning, when Ngai created the earth and divided it among the peoples of the world, he took the first Kikuyu to the summit of Mount Kenya. From the heights of his holy resting place, he showed him the beauties of the land he was about to give him. Then he pointed to a grove of fig trees in the middle of the country below them and commanded the man, whose name was Kikuyu, to make his home and family there. When Kikuyu went down to the place Ngai had indicated, he found a wife waiting for him, and together they had nine daughters. Their descendants became the nine clans of the Kikuyu tribe, named for the primordial ancestor who had stood on the divine summit of Mount Kenya at the beginning of the world.

Before sending him down to the land he had given him, Ngai told the first Kikuyu that whenever he had need of assistance, he should make a sacrifice and raise his hands toward the sacred mountain. This act of ceremonially invoking the god of Mount Kenya became one of the most basic rituals of the Kikuyu religion. It occupied such a central place in their religious and secular life that when Jomo Kenyatta, the father of the modern nation of Kenya, wrote a book about his people, published in 1938, he titled it *Facing Mount Kenya*. Even today, this book remains the best-written source on traditional Kikuyu religion and society. Every event of any importance—consecrating a house, performing a marriage, holding a tribal council, settling a dispute—required the participants first to face Mount Kenya and invoke the blessings of Ngai.[16]

The mountain played an especially important role in initiation ceremonies. In rites of circumcision for children becoming adults, elders drew sacred symbols on the initiates' faces, throats, and stomachs with pieces of white chalk called "snow," brought down from the heights of Mount Kenya. The most important part of the ceremony announcing the engagement of a boy and girl involved the sacrifice of a fat sheep, whose blood would be sprinkled toward the sacred mountain. After his marriage a young man entered the third stage of manhood and underwent an initiation into eldership. During the ritual an elder would rise and face Mount Kenya; holding up a drinking horn filled with beer, he would beseech Ngai to grant the tribe peace, wisdom, and prosperity and to bless the new initiate and his family. Then he would give the young husband a staff of office and a bundle of sacred leaves, indicating that he had passed beyond the stage of war and had become a man of peace, no longer a bearer of spear and shield.[17]

The most elaborate rituals directed toward Mount Kenya occurred in ceremonies asking for rain, the source of life. The Kikuyu would choose a lamb of a single color, usually red or black, and have two children lead it in a procession to a sacred tree symbolizing the mountain. Taking a calabash of beer in one hand, milk in the other, an elder would lift up his arms toward Mount Kenya and pray in a loud, clear voice, "Reverend Elder who lives on Kere-Nyaga. You who make mountains tremble and rivers flood; we offer to you this sacrifice that you may bring us rain." Then the procession would circle the tree seven times, sprinkling the milk and beer on the trunk and branches. On the eighth circuit they would sit down in a ring, place the lamb on its back with its head facing the mountain, and strangle it. After roasting the meat and feasting on it, they would make burnt offerings toward Mount Kenya. The Kikuyu performed this ceremony at the beginning of the growing season and whenever the rains were late in coming.[18]

Like the Chagga on Kilimanjaro, the Kikuyu have discarded much of their traditional culture. Many of them have converted to Christianity and live in the urban environment of Nairobi, one of the most modern cities in Africa. Today only a few of the older Kikuyu continue to practice the old ceremonies directed toward the sacred mountain. During a severe drought in 1984, some local farmers brought red and black lambs to sacrifice at the foot of Mount Kenya. The god of the mountain took a long time to respond to their entreaties, and one of the men lamented, "Years ago rain would come only ten minutes after my people sacrificed to Ngai on the mountain. Now all our requests are bounced." He blamed the situation on the young who no longer followed the ways of their ancestors. With a worried look toward Mount Kenya, he added, "The mountain is still like a magnet to our eyes, but it is not helpful to us anymore. We are scared."[19]

During their struggles for independence from British rule in the 1950s, the Kikuyu warriors of the dreaded Mau Mau movement took refuge in the forests of Mount Kenya, where they felt protected by the god of their sacred mountain. One of their leaders, Kiboi Muriithi, described how he and his companions would face the peak and pray to Ngai to watch over them. On the eve of dangerous missions, they would offer sacrifices to the deity on the mountain, asking him to keep them safe and grant them success. Thirty years later, in the 1980s, Moses Muchugo, a veteran of the Mau Mau movement, still rose every morning before dawn to face Mount Kenya and thank Ngai for the protection granted him and his companions years before: "No warplanes' bombs could harm us, because we stayed near our god. Our god and our mountain were our only friends then."[20]

Followers of hybrid religious movements that blend Christianity with the old Kikuyu religion have adopted the practice of facing Mount Kenya. Lifting their hands toward the sacred mountain, members of one of these movements, a sect called the People of God, fall into states of possession, shaking with the spirit of the Holy Ghost. Simulating the cries of lions and leopards, they become more than human and believe themselves able to communicate directly with Ngai, whom they view as one with the Christian deity. On eight nights between 1953 and 1984, a farmer deeply influenced by the new religious movements heard a voice telling him to arise and ascend Mount Kenya.

Each time he obeyed and went up to the mountain wearing a white gown and carrying a Bible. There, scantily clad in the icy air of the heights, he prayed to his Kikuyu and Christian god. He described what he felt in the following words:

I move up the mountain as if I am being carried, like lightning. I cannot feel the altitude. Everything Ngai tells me up on the mountain happens below. I take the messages to the government of Kenya, but they ignore me. Something very big will happen soon and everyone will understand what has been revealed to me.[21]

In 1978 two British climbers encountered another holy man inspired by his god, this one on top of the mountain. They had just completed the ascent of an extremely difficult route on Nelion, the second-highest peak of Mount Kenya. There on the summit, standing in the snow, they found an African clad in a blanket, wearing plastic sandals. Assuming he was the caretaker of a nearby hut, they asked him to fetch them some water. To their surprise he informed them that he had just climbed their route to pray for the welfare of mankind. They offered to help him down an easier way, but he declined their invitation, saying he had a few more days to spend on the summit. After descending the mountain, they told the Mount Kenya park superintendent about the holy man trapped on top of Nelion. He mounted a rescue effort, but none of his rangers could climb up the imposing face. They were standing around on the glacier below it when the holy man came nimbly scrambling down the cliff in his plastic sandals.

One of the rangers called out to him, "Can I help you, my brother?"

"I'm not your brother, and I don't need your help. I do this three times a year," the man shouted back and ran off down the mountain, vanishing into the mist. The park warden, a climber himself, later called the holy man's exploit one of the most amazing ascents in the history of mountaineering: mountaineers regard the route he had climbed without any equipment as one of the most difficult in the world.[22]

Even more than Kilimanjaro, Mount Kenya has cast a spell over European climbers, luring them to attempt its much more demanding and exhilarating ascent. Ludwig Krapf—a companion of Johannes Rebmann, the missionary who "discovered" Kilimanjaro—first sighted Kenya in 1849. His report of seeing snow on its summit, which he described as "two horns . . . covered with a white substance," was also mocked in Europe. An Englishman by the name of Sir Halford Mackinder made the first ascent of Batian, the highest peak, in 1899. On the way to the mountain, he had to fight off hostile natives, losing two of his men in the process. He named the three major peaks of Mount

A Chagga porter carries a load up Kilimanjaro, heading toward the snow-capped dome of Kibo. (William Frej)

Kenya after M'batian, a prominent chief and medicine man of the Masai, and his brother and son, Nelion and Lenana. Thirty years later in 1929, Eric Shipton climbed Batian and made the first ascent of the much more difficult 17,022-foot Nelion. Overtaken by nightfall on his descent from the summit of Batian, Shipton had a strange experience:

I felt very tired and the phantom moonlight, the shadowy forms of ridge and pinnacle, the wisps of silvered mist, the radiant expanse of the Lewis glacier plunging into soundless depths below induced a sense of exquisite fantasy. I experienced that curious feeling, not uncommon in such circumstances, that there was an additional member of the party—three of us instead of two.[23]

A traditional Kikuyu would probably have said that for a brief time Shipton experienced the presence of Ngai.

One of the most interesting ascents—or attempted ascents—of Mount Kenya took place during World War II. Three Italian prisoners of war in a nearby British camp escaped on a lark to climb the peak as a break from the boredom of prison life. They fashioned crampons and other climbing equipment from

odd scraps of material and left a note saying they would be back. To the astonishment of the camp commandant, they returned as promised, having failed in their attempt to reach the summit. However, they had plenty of adventures, avoiding wild animals in the jungle and struggling up difficult faces in bad weather with primitive equipment. One of the Italians, Felice Benuzzi, wrote a delightful book about their escapade, titled *No Picnic on Mount Kenya*.[24]

The writings of the Scottish explorer Joseph Thomson, who saw the mountain in 1873, reflect the kind of spiritual feelings that the dramatic sight of Mount Kenya continues to inspire among people today. Having heard about the wonders of the mysterious peak, he developed such an obsession to see it that his journey took on the character of a religious pilgrimage. When he finally saw the object of his quest, he was, indeed, transported:

The sun set in the western heavens, and sorrowfully we were about to turn away, when suddenly there was a break in the clouds far up in the sky, and the next moment a dazzling white pinnacle caught the last rays of the sun, and shone with a beauty, marvelous, spirit-like and divine,

apparent in an observation made by an anthropologist who spent a year with the Koyukon in the mid 1970s:

In the past (if not still today), it was considered disrespectful to talk about the size or majesty of a mountain while looking at it. A child doing this would be told, "Don't talk; your mouth is small." In other words, this was "talking big," disregarding the need to be humble before something so large as a mountain.[2]

Hunters of the northern forest, the Koyukon depend for their survival on an intimate relationship with features of the natural landscape. Among these features, the most important are rivers, along which the Koyukon live and from which they derive their tribal identity. The importance of water in Koyukon life is reflected in an unusual flood myth that makes the highest peak in North America a wave of stone. Long ago in the distant time of the beginnings, the trickster deity Raven assumed the form of a young man. Having heard of a beautiful maiden living in a village across the water, he went in a canoe to ask her to marry him, but she refused. Another woman offered him her baby daughter. He took the child into his boat, and as he began to paddle away, the haughty girl came down to the water. To punish her for having rejected him, he made her sink into the mud and vanish. In revenge the girl's mother set two bears to beating up waves to drown him. As a result of their efforts, the waters rose and inundated the world so that everyone in the village perished.

Raven used his magic powers to smooth a path through the waves heaving up before him. Gradually he tired, and overcome at last with fatigue, he cast his harpoon at a wave and fainted. When he awoke, he found that the breaker he had hit had turned to stone, forming a small mountain. Glancing off its crest, the harpoon had struck an even larger wave, which had become Denali, the "High One." The missile had ricocheted up from the summit of that peak, the highest of all, to stick in the sky, where it remains to this day, visible only to shamans with supernatural sight. Raven looked around, and there to his delight stood the baby girl, now grown to womanhood. He took her in his arms, and they repopulated the world.[3]

White people have treated the mountain with much less reverence. Tired of companions who were boring him with arguments for silver, a prospector named the mountain after William McKinley, the American presidential candidate who in the election of 1896 favored retaining gold as the monetary standard. McKinley won and the name stuck. Fourteen years later, in response to a wager made in a saloon in Fairbanks, a group of prospectors nicknamed the Sourdoughs climbed the mountain in 1910, reaching its slightly lower northern summit, which they mistook for the highest point. Following the official first ascent in 1913, innumerable climbers have "conquered" the peak, most of them in the last thirty years. In 1917 the mountain became the nucleus of Mount McKinley National Park. In 1980, in a belated show of respect for Native American views of the peak, the United States Congress renamed the preserve Denali National Park. The official name of the mountain itself, however, remains Mount McKinley.

Mount Saint Elias

Mount Saint Elias, the second-highest mountain in the United States and Canada, rises to an altitude of 18,008 feet on the border between the two countries, right where the Alaska Panhandle begins. In contrast with McKinley, the modern name of Saint Elias betrays a sensibility for the extraordinary, even sacred, nature of its object. Vitus Bering, the Danish navigator who discovered the straits separating the North American and Asian continents, named the mountain for

Climbers at a camp beneath the north face of Mount Saint Elias on the border between Alaska and the Yukon Territory of Canada. (Edwin Bernbaum)

the saint on whose holy day he first spied its white peak rising over the gray waters of the Gulf of Alaska. It seems quite probable that he also had in mind Saint Elias's close association with mountains in the Eastern Orthodox faith of his Russian employers. Because of Saint Elias's role in the transfiguration of Jesus on top of a mountain, shrines dedicated to him appear on numerous hills and peaks in Greece and eastern Europe.

The great pyramid of Mount Saint Elias, soaring up from the ocean in one smooth sweep of snow and rock, can have a powerful effect on the observer—on the rare occasions when the mountain breaks free from the mist and cloud that normally envelop its flanks. In 1965, long before I knew anything about its significance as a sacred peak, I came to climb Saint Elias on a mountaineering expedition. Early one morning in the darkness before dawn, after a night hauling up loads through the Arctic twilight, I walked off alone to the edge of an enormous drop. A blue glow, smooth and clean, lit the sky to the north. Just in front of my boots, the ground fell away to a glacier six thousand feet beneath my toes. Flanked by a row of little peaks, their summits far below me, it stretched off straight and wide, a gray highway a hundred miles long, to blur and vanish in the azure shadows of the distance. For a moment, alone on the mountain, I felt as though I were standing on the blue edge of infinity.

Surrounded on all sides by a sea of glaciers, Saint Elias looks as though it were emerging from the waters of a frozen flood. The Yakutat Tlingit, a tribe living in the vicinity of Yakutat Bay on the southeastern coast of Alaska, believe, in fact, that a deluge did cover the world and that only the summits of Saint Elias and two other mountains remained above it. Some of them point to tales of skin robes found among rocks up on the mountain as evidence that their ancestors survived the flood by finding refuge high on the peaks. Others say that the great white peak acted as a signpost guiding their forebears on their historical migrations down the Alaska coast from the mouth of the Copper River to their present homeland in Yakutat Bay. As one woman so beautifully put it, "When the migrators were coming to Yakutat across the ocean, offshore, they saw Mount Saint Elias ahead, looking like a seagull on the water."[4]

The mountain played such an important role in stories of their origins that one clan of Yakutat Tlingit even made it their crest, or totem. According to one account, those of their ancestors who traveled along the coast by land were so happy at finding their new home that they danced down the slopes of Mount Saint Elias. In gratitude they built a special house, or lodge, named for the sacred mountain. On festive occasions, some of the Yakutat Tlingit perform ceremonial dances in which they wear shirts stitched with designs representing the peak as a pyramid surrounded by clouds depicted as circles. A song composed by a man in 1931 tells how Mount Saint Elias "opened the world" with sunshine, making the people happy.[5]

The mountain itself did not always share in this happiness. According to Yakutat Tlingit belief, Saint Elias was once the husband of Mount Fairweather, a prominent, but lesser, peak a hundred miles away. The two had a rocky marriage, eventually shattered by bickering. After a particularly violent quarrel, Saint Elias moved away to the west, taking many retainers and slaves with him. Stretching out after him in a long line, they became the range of smaller mountains that links the two great peaks and marks the border between Alaska and Canada's Yukon. One peak, a slave belonging to both, acted as a go-between, continuing to carry messages between the estranged spouses. The couple's children remained with their mother, forming the mountains to the east of Fairweather.[6]

The Yakutat and other Tlingit tribes to the south believe that spirits inhabit the mountains and glaciers that frame the world in which they live. Their shamans go up into the forests, toward the high peaks, to obtain from them their supernatural powers. Ordinary people take special care to avoid offending the powerful spirits who dwell inside glaciers and occasionally come out, looking like frost upon the ground. When the Tlingit have to travel near them, they put on their best clothes and address them with respect. They also smear black pitch over their faces to conceal their eyes: mountains and glaciers find the gaze of humans offensive. If a stranger on a mountaineering expedition has the bad manners to look at them directly without dark goggles, they hide themselves in clouds and cause bad weather.[7]

Glaciers come in sexes, like people. Males tend to be more aggressive and bothersome than females. If someone tries to cook near a male glacier, the spirit will come out and cause trouble. It will not go away until its harried victim burns everything in the fire. Female glaciers, on the other hand, allow people to cook and eat their food in peace. They are usually retreating and have large medial moraines that look like the parted braids of a woman's hair. If provoked to anger, however, they can join with males to advance and destroy everything in their paths. In 1986 the Hubbard Glacier surged across a fjord, blocking a major river and threatening an ecological catastrophe. One of its tributaries, the Valerie Glacier, apparently triggered the event by starting to move at the amazing speed of 112 feet a day. The local Tlingit surmised that someone may have angered the glaciers. One person identified the Hubbard as a male glacier and the Valerie as a female. "And," he added, "females are known for their unpredictable behavior."[8]

Cloud caps over the summit of Mount Rainier create an unearthly effect. (William B. Bryan)

Mount Rainier

Gracefully draped with the largest glaciers in the United States outside Alaska, Mount Rainier, the highest peak in the Cascades, soars up in a majestic dome of blue-white snow, its broad summit gleaming in the sky, 14,408 feet above the waters of Puget Sound. An enormous dormant volcano, unrivaled in height by any of its neighbors, the mountain dominates the landscape of Washington State south of Seattle. The sight of its massive, yet beautiful, form so moved one writer, John H. Williams, that in 1910 he published a book about the peak titled *The Mountain That Was "God."* The title caught on as an epithet for Rainier itself, leading some to mistake it for a translation of one of the native names of the volcano, Tahoma or Tacoma, which probably means the "Great White Mountain." Williams captured in one short phrase the powerful religious sentiments that the peak has awakened in numerous other writers, such as Theodore Winthrop, who wrote of his impressions on seeing Rainier in 1853:

Only the thought of eternal peace arose from this heaven-upbearing monument like incense, and, overflowing, filled the world with deep and holy calm. . . . And, studying the *light and majesty of Tacoma, there passed from it and entered into my being, to dwell there evermore by the side of many such, a thought and an image of solemn beauty, which I could thenceforth evoke whenever in the world I must have peace or die. For such emotion years of pilgrimage were worthily spent.*[9]

Even today the local residents of Washington State refer to Mount Rainier as "*the* mountain," putting it in a special category set apart from all other mountains, no matter how high or impressive they may be. In 1963 some of the most proficient mountaineers in America were returning from an expedition in which they had made the first ascent of one of the highest and most difficult faces of Mount McKinley. When they stopped off to climb Mount Rainier, a much lower mountain, the park ranger told them they would have to hire a guide. "But we've just come back from climbing the Wickersham Wall on McKinley," they protested. "Doesn't matter—it's not *the* mountain," the ranger replied and refused to give them permission to go up it on their own.

The Native American tribes who shared the region before white settlers took over—Puyallup, Yakima, Nisqually, and others—also regarded the mountain with awe, believing that no one could climb it and survive the power of the spirits inhabiting its summit. On the occasion of its first ascent in 1870 by General Hazard Stevens and P. B. Van Trump, their Yakima guide, Sluiskin, pleaded with them not to go, saying:

Your plan to climb Tacoma is all foolishness. No one can do it and live. A mighty chief dwells upon the summit in a lake of fire. He brooks no intruders. . . . You make my heart sick when you talk of climbing Tacoma. You will perish if you try to climb Tacoma. You will perish and your people will blame me. Don't go! Don't go![10]

The great size and solitary splendor of Mount Rainier made it a focal point for Native American myths embodying the two themes of flood and family strife that we saw attached to prominent peaks in Alaska. According to one set of these myths, the mountain used to stand on the other side of Puget Sound, where she was the wife of a peak in the Olympic Mountains. One of a number of wives, she quarreled with the others and moved away in disgust, taking up residence on the far side of the water. Filled with anger, she swelled up to her present, enormous size. Some accounts say that the Great Changer punished her and her husband by changing them into the snow-capped peaks of Mount Rainier and Mount Olympus. Others assert that the mountains on the Olympic Peninsula became too crowded, forcing Rainier to move across the sound, where she grew so big that she turned into a monster who had to be subdued by Changer in the form of a fox. When he killed her, the blood that gushed from her sides turned into the rivers of water that flow from the mountain today.[11]

According to a myth based on the flood theme, long ago the Great Spirit dwelled on the summit of Tacoma. Angered by the beings he had created, he decided to get rid of them all, except for one good man and his family, along with the animals who had not become evil. He told the man to shoot a series of arrows into a cloud floating over the mountain. The first one stuck into the cloud and the others into each other, forming a rope that hung down from the sky. Followed by his family and all the good animals left in the world, the man climbed up this rope. When they reached the top, they looked down to see the others scrambling up after him. The man shook the rope of arrows, and it broke, so that the bad animals and people fell back to earth. Then the Great Spirit made the rain to fall, causing a flood to cover the world and drown them all. When the waters reached the snow line on Tacoma, he commanded them to recede. The man and his family, accompanied by the good animals, lowered the rest of the rope and climbed down to the summit of Tacoma. As dry land emerged from the water, they walked down the mountain to repopulate the world. The myth adds that since that time Tacoma has been free of bad animals and snakes.[12]

Mount Shasta

The mountain in North America that arouses the greatest excitement among contemporary spiritual and mystical groups is Mount Shasta, the second-highest peak in the Cascades. An enormous snow-capped volcano 14,160 feet high, it looms like a Himalayan giant over the hills and ranges of northern California, just south of the Oregon border. Unlike many other mountains, Shasta stands by itself, awesome in its isolation. Because of its location near the Pacific Ocean, frequent storms and strange clouds create eerie effects around its summit, making it seem charged with supernatural power. Somewhere far beneath its sealed crater, the clash of tectonic plates melts rock and makes it roil in preparation for some future eruption. Awed by its size and power, many people regard the mountain as a fountainhead of cosmic energy.

Mount Shasta has such an imposing appearance, rising more than 10,000 feet above its base, that it makes a profound impression on anyone who sees it, whether mystically inclined or not. Once, on a family trip driving back to California from Oregon, I pointed out the great white pyramid of shimmering snow to my son, David, not yet two years old. The sight so impressed him that a month later he produced as his first work of representational art a drawing of the peak. For the next two years every mountain he saw, no matter what size or shape, whether a photograph or an actual view, elicited the same excited cry, "Look, Daddy, Mount Shasta!" Such responses help to give life to a mountain and make it a sacred place, a focal point of myth and legend.

Mount Shasta stands out as a sacred peak of particular power and significance for a number of Native American tribes in northern California and southern Oregon, including the Wintu, the Shasta, the Pit River or Ajumawi, the Modoc, the Karok, and the Hupa. Over the centuries they have traditionally looked to it for healing, strength, wisdom, and guidance. The Wintu call the mountain Boyem Puyuik and consider it the most sacred of their sacred places, the one that "has it all." They believe that prayers made on its slopes have great power and efficacy. The Pit River, who know Shasta as Ako-Yet, revere the mountain as the abode of a powerful spirit who keeps the universe in balance. Whichever view they hold, Native Americans object to development on the sacred mountain—in particular, the expansion of a ski area that threatens to desecrate a pristine meadow on its southern flank.

The last pink light of sunset touches the summit of Mount Shasta, imbuing it with an aura of mystery, enhanced by the presence of the moon. (Robert McKenzie)

Joaquin Miller, a nineteenth-century author and poet, lived among the Modoc of northern California and recorded a Modoc myth about the creation of Mount Shasta and the origin of the human race. According to the story as he understood—and probably embellished—it, the Great Spirit formed the mountain from snow and ice that he pushed down through a hole he made in the sky. Stepping down from the clouds onto its summit, he created plants and trees, birds and animals, and appointed the grizzly bears masters over all. Then, in order to remain on earth and complete his work of creation, fashioning the land and sea, he hollowed out Mount Shasta in the form of a lodge and took up residence inside it. The fire and smoke that used to issue from the volcano came from the cozy fireplace he built in its center to warm himself and his family.

Once a great storm blew in from the ocean, threatening to topple Mount Shasta. The Great Spirit asked his youngest daughter to stick her arm out of the top of the mountain and signal the tempest to lessen its force. When out of curiosity she stuck out her head to see the ocean, the wind grabbed her by the hair and hurled her down the mountain, where a grizzly found her shivering in the snow. He took her back to his wife, who decided to rear the child along with her own brood of cubs. When she grew up, the mother bear had her marry her oldest son. The children of this couple, who resembled neither of their parents but shared some of their characteristics, were the first human beings. Up to this time, the bears walked on two legs, like people. However, when the Great Spirit discovered that a new race had been created without his authorization, he became angry with the grizzlies and ordered them down on their hands and knees. Ever since that time, bears have gone about on four legs, only getting up on two when they have to fight. The unauthorized children of the couple, the ancestors of the human race, the Great Spirit scattered throughout the world.[13]

A number of contemporary cults and mystical groups find in another creation myth material to bolster one of their favorite beliefs about Mount Shasta. According to this myth, which falls into a pattern of flood stories reaching from Alaska to California, Coyote, the wisest and wiliest of all the animals, angered an evil water spirit by shooting it with a bow and arrow. In retaliation the deity flooded the world so that only the summit of Mount Shasta protruded above the waves. Coyote escaped the rising water by running up to the top of the mountain, where he lit a fire to warm himself. Seeing him there, various animals swam over to seek refuge with him. After the flood subsided, they all came down from the summit to renew life on the earth below.[14]

In 1931, the Rosicrucian Order, a world-wide mystical society with roots in Europe and headquarters in San Jose, California, published a book that popularized the belief that Mount Shasta used to stand on the eastern edge of Lemuria, a vast continent that once spread over much of the area now covered by the Pacific Ocean.[15] The people who lived there, a race of advanced beings, belonged to an extraordinary civilization related to that of Atlantis. When the continent sank in a great cataclysm of volcanic activity, a number of survivors escaped the ensuing flood by finding refuge on Mount Shasta, as Coyote did in the Native American myth. The Rosicrucians — and many other groups influenced by them — claim that the Lemurians have remained there ever since, living in cities within the mountain, where they preserve a knowledge far exceeding that known to modern science. Continuing reports of odd-looking people, some of them dressed in white and wearing sandals, others appearing out of nowhere to pay for everything with gold, have convinced many residents of the area that a colony of advanced Lemurians does indeed exist, even today, hidden inside the mysterious recesses of Mount Shasta.[16]

Stories of strange phenomena — wavering lights, balls of fire, eerie sounds, flying saucers — have led many to view the mountain as a magnet of cosmic forces, a sacred center of psychic and spiritual power. In the summer of 1987, thousands converged on Mount Shasta as part of the Harmonic Convergence, a worldwide effort to draw on such energies to initiate a new age of universal peace and harmony. Over the years, the mystical aura surrounding the peak has attracted numerous religious groups and spiritual seekers. The mountain has assumed importance for as many as a hundred different sects and centers, each with its own view and interpretation of Shasta's significance. These include, to list only a few, the Brotherhood of the White Temple, the Radiant School of Seekers and Servers, Gathering of the Ways, the Association Sananda and Sanat Kumara, the Zen monastery of Shasta Abbey, and the I AM Foundation.

The I AM Foundation, one of the oldest and most influential sects associated with Shasta, owes its existence to a mysterious encounter on the slopes of the sacred mountain. In 1930 its founder, G. W. Ballard, wandered up the peak to commune with nature and enjoy the views. According to his account, as he bent over a stream to dip his cup in the water, he felt an electric current pass through his body. He looked around to see a young man standing behind him. The stranger said, "My Brother, if you will hand me your cup, I will give you a much more refreshing drink than spring water." When Ballard drank the liquid offered him, he gasped with surprise at its marvelous taste and effect. "That which you drank," the young man explained, "comes directly from the Universal Supply, pure and vivifying as Life Itself. In fact it is Life — Omnipresent Life — for it exists everywhere about us." Then he revealed himself to Ballard as the master Saint Germain: "He stood there before me — A Magnificent God-like figure — in a white jeweled robe, a Light and Love sparkling in his eyes. . . ."

In the course of a number of subsequent encounters described in his book, *Unveiled Mysteries*, Ballard received the teachings that became the basic doctrines of the new religion he founded. The I AM, or Saint Germain, Foundation, as it is also known, spread around the world, with centers in numerous countries. Some years after Ballard's experience, his followers purchased land and built a center at the foot of the mountain. Each summer members of the sect perform a religious pageant in an amphitheater beneath the awesome backdrop of Mount Shasta.[17]

Despite a history of efforts to destroy and ignore their ways of life, Native Americans who live in the area have managed to maintain traditional beliefs and practices centered on Mount Shasta. Each summer in August, members of various tribes led by the Wintu spiritual doctor Florence Jones gather in a meadow southeast of the peak to perform ceremonial dances and invoke the spirit of the mountain. Then they climb the lower slopes of the peak to pray at a sacred spring from which Jones draws her power and knowledge as a healer and teacher. On the full moon of that month, a Karok medicine man named Charles Thom conducts a sweat ceremony of ritual purification at 7,000 feet on the slopes of the mountain. After following the white man's ways for many years, Thom felt himself called to Mount Shasta. Now he lives within sight of the peak and looks to it for inspiration and meaning in his life and practice.

Sierra Nevada

From the coastal mountains of California south of Shasta, on a clear day in spring one can look a hundred miles across the Sacramento Valley to the luminous wall of the Sierra Nevada, a long white line gleaming above bands of purple forest and golden grass. In 1868 such a view from a pass not far from San Francisco moved the well-known naturalist John Muir to write his often-quoted words about this mountain range, the one he loved above all others:

Then it seemed to me the Sierra should be called not the Nevada, or Snowy Range, but the Range of Light. And after ten years of wandering in the heart of it, rejoicing and wondering, bathing in its glorious floods of light, the white beams of the morning streaming through the passes, the noonday radiance on the crystal rocks, the flush of the alpenglow and the irised spray of countless waterfalls, it still seems above all others the Range of Light, the most divinely beautiful of all the mountain-chains I have ever seen.[18]

His writings make it clear that Muir was not speaking about an ordinary light, but a divine radiance such as mystics see in visions of glory infusing the world around them and transforming it into a reflection of a higher, more perfect reality. In the same passage he says of the Sierra, "It seems to be not clothed with light, but wholly composed of it, like the wall of some celestial city." And elsewhere he writes, making his meaning perfectly clear, "Never had I beheld a nobler atlas of mountains. A thousand pictures composed that one mountain countenance, glowing with the Holy Spirit of Light!"[19]

For Muir and many others who followed his lead, the Sierra Nevada was a range of sacred mountains. Its peaks, valleys, forests, meadows, and glaciers were his church and his paradise, the shrines where he communed with nature and spoke with God—as spirit, truth, and beauty. He went so far as to say that the Sierra Nevada was to him as holy as Mount Sinai.[20] The religious connotations of the form and the name of Cathedral Peak, a magnificent spire of granite towering over Tuolomne Meadows, fascinated him: before making its first ascent in 1869, he wrote that he hoped to climb the mountain not to conquer it but "to say my prayers and hear the stone sermons."[21] Yosemite Valley with its walls of glacier-sculpted rock he regarded as a temple more magnificent than any made by human hands. And whereas others might use such metaphors for literary effect, he really meant it. He even found in his sacred mountains the mystical experience of self-transcendence sought as the goal of many religious traditions:

Brooding over some vast mountain landscape, or among the spiritual countenances of mountain flowers, our bodies disappear, our mortal coils come off without any shuffling, and we blend into the rest of Nature, utterly blind to the boundaries that measure human quantities into separate individuals.[22]

Indeed, in accounts like this one, Muir gave expression to the tenets of a philosophical tradition to which he felt he did belong—the transcendentalism of Ralph Waldo Emerson and Henry David Thoreau, who believed in the existence of a spiritual reality underlying and unifying the world revealed by the physical senses. When Emerson visited him in Yosemite in 1871, Muir compared the old philosopher to a giant sequoia and later named a mountain after him.

A miraculous experience of deliverance from death on the heights of Mount Ritter in the Sierra Nevada gave Muir a special feeling for the peak as a divine embodiment of all that he revered in the range as a whole. He was creeping up a cliff close to the top when he ran out of holds and found himself clinging to the smooth rock, unable to move in any direction. He suddenly realized there was nothing he could do to keep from falling. He was going to die. At that same moment something seemed to take over—"the other self, bygone experiences, Instinct, or Guardian angel,—call it what you will"—and with incredible power and precision his limbs carried him up the rest of the cliff. "Had I been borne aloft upon wings," he wrote, "my deliverance could not have been more complete." When he reached the top, emerging from the shadow of fear that had darkened his mind, Muir found Mount Ritter and the world around it glowing in a special light:

How truly glorious the landscape circled around this noble summit!—giant mountains, valleys innumerable, glaciers and meadows, rivers and lakes, with the wide blue sky bent tenderly over them all. But in my first hour of freedom from that terrible shadow, the sunlight in which I was laving [bathing] seemed all in all.[23]

A Paiute Indian who lived on the eastern slope of the range not far from Mount Ritter also had a miraculous experience of power and deliverance that gave him a feeling of intimate relationship with a particular mountain in the Sierra Nevada. One night in the mid-nineteenth century, Birch Mountain, a 13,000-foot peak that rises in a dark triangle above the Owens Valley, appeared to Hoavadunaki in a dream and said to him, "You will always be well and strong. Nothing can hurt you and you will live to an old age." A short time later, walking alone across the desert, he fell ill, poisoned, he believed, by a witch doctor. With no one to help him, he crawled under a bush and collapsed on the verge of death. But then a thought came to him: *Since my mountain has spoken and told me that I shall not die, why should I die here?* Revived by this thought, he rose and went on to a village, where a medicine man completed the restoration of his health. Hoavadunaki attributed his deliverance to the power of Birch Mountain. For the rest of his life, he felt that it remained with him, helping him in times of need. Often when he had trouble hunting, he would ask

the peak for aid, and without fail he would get a deer. As he put it, "My mountain is always good to me."[24]

Although they both felt a sense of intimate relationship with their mountains, Hoavadunaki and Muir related to them in very different ways. Hoavadunaki experienced Birch Mountain as a supernatural person who appeared to him in a dream and assumed the role of his guardian spirit, telling him that it would protect and aid him for the rest of his life. Muir had a much more abstract relationship with Mount Ritter. He viewed the mountain not as a divine being capable of personally acting on his behalf, but rather as a temple filled with the Holy Spirit of light. The power that carried him up the cliff and saved his life he attributed not to the peak itself, but to some other, more mysterious agency — "the other self, bygone experiences, Instinct, or Guardian angel." Living in a culture that considered spirits a part of the natural order of things, Hoavadunaki had no difficulty accepting Birch Mountain as a sacred person who could reach out to help him. Given his religious and scientific background, which emphasized monotheism and discouraged anthropomorphic views of physical objects, Muir saw Mount Ritter as the divine expression of something more abstract — Nature, Beauty, or God.

ROCKY MOUNTAINS AND BLACK HILLS

East of the Sierra Nevada, across the deserts of the Great Basin, rise the rugged peaks and ridges of the Rocky Mountains, the longest and widest range in North America, stretching from Canada in the north to Mexico in the south. Seen from the Great Plains on their eastern side, they appear as an enormous wall blocking the western horizon. Here and there a peak rises above this wall, like a turret on a fort or castle: the granite mass of Pike's Peak in Colorado, the alpine spires of the Tetons and Wind Rivers in Wyoming, the layered faces of the Rockies above Banff in Alberta. Just to the east of the Rockies, emerging as an upwelling of rock, like an island in a sea of grass, stand the granite peaks and volcanic buttes of the Black Hills of South Dakota.

Most of the Native Americans who roamed the eastern slope of the Rocky Mountains until the end of the nineteenth century migrated there from the woodlands of Minnesota and the Ohio Valley to the east. With the help of horses, introduced by the Spanish in the sixteenth century, various tribes of the

Sioux, such as the Oglala and the Hunkpapa, moved onto the Great Plains around 1700. As nomadic hunters and warriors, they cultivated visionary experiences to help them in finding and killing their enemies as well as their prey — vast herds of buffalo that once speckled the land. Like other tribes of the plains, such as the Cheyenne and the Crow, they climbed sacred hills and peaks on vision quests, seeking the kind of power and protection that Hoavadunaki obtained from his dream of Birch Mountain. Most of these tribes made such a quest an obligatory rite of passage leading to adulthood and the attainment of a full and responsible life.

After receiving religious instruction from a medicine man and ritually purifying himself in the sacred steam of a sweat lodge, a youth would be taken up to a high place and left there to pray and fast alone, exposed to wind, rain, sun, and stars. Awed by the wild grandeur of his surroundings, his senses quivering with anticipation and fear, he would gradually open himself to the mysterious forces of the world around him. A rustling in the grass, the sudden flight of a bird, the sparkle of a star, any of these could trigger a vision — or induce a dream. He might find himself flying up through the clouds to a rendezvous in the sky, or a sacred being might appear directly before him, sometimes in the shape of an animal, sometimes in the form of a man.

What he saw and heard in the magic realm of vision gave him the power and guidance he needed to live in the world of everyday life. Often, but not always, he would get a guardian spirit — the sacred being or animal that had appeared to him — to instruct and protect him after his quest. In some cases the power and guidance might come directly from the visionary experience itself, or the memory of it. Crazy Horse, a famous Sioux war chief whose profile has been sculpted into a cliff in the Black Hills near Mount Rushmore, got his name from a vision in which he saw himself riding a horse that was dancing like a shadow in the shimmering world of the spirits. Whenever he rode into battle, he had only to remember this vision, and no bullet or arrow could touch him — so he and his people believed.

After a youth came down from his quest, a medicine man would help him to interpret and understand what he had experienced. To bring the power of his vision into the world, he might have to sing the songs he had heard and act out the things he had seen in a ritual play performed before the members of his tribe. He would have the rest of his life to comprehend

A vision of Half Dome emerging from clouds reveals the kind of view that inspired John Muir to extol Yosemite Valley as a temple fashioned by God. (Galen Rowell/Mountain Light)

and assimilate the full significance of his most important youthful experience.[25]

Although vision quests might take place in forests or other places, depending on the nature of the local terrain, the tops of hills and mountains were the sites that most Plains Indians preferred. They provided an ideal environment for engaging in such quests. The awesome view of a vast horizon circling around him dwarfed a person to insignificance, putting him into a suitably humble and receptive frame of mind. Standing on a high and lonely summit, exposed to the power of the elements, he could expose himself to the mercy of the spirits, calling on them to take pity upon him and grant him a vision. In a typical prayer a Sioux would cry out, "Wakantanka, pity me! I want to live; that's why I am doing this." The Sioux word for a vision quest means, in fact, "crying for a vision." The harshness of the environment on the summit of a mountain also contributed to the ordeal of self-inflicted suffering that some tribes made an essential part of the undertaking. To demonstrate their commitment and willingness to sacrifice everything for a vision, the Crow would even cut out strips of their flesh and chop off joints of their fingers.

The Crow, who put a particular emphasis on the positive role of suffering, ranked sites for vision quests in order of increasing altitude. The higher the site, the greater the severity and solitude of the quest and the more powerful the vision it could produce. The easiest and least significant places were down in sheltered valleys close to human habitation; the most difficult and sacred lay high on the windswept peaks of the Bighorn Mountains of Wyoming and Montana. Two Leggings, a warrior chief obsessed with the quest for personal power and status, deliberately chose the top of an imposing mountain that most people feared as the home of the terrifying Thunderbird. "We decided to fast there because we wanted a stronger medicine," he told his biographer. On the fourth day of his vigil, just before dawn, as he slipped into a dream, a huge man appeared on the horizon with a hawk perched on his head. He told him that the name of the bird was "The Bird Above All the Mountains" and that in the future he, Two Leggings, would be known all over the earth. The man also showed him a vision of four small sweat lodges and instructed him to build copies of them whenever he wished to go on the warpath.[26]

The Stoney, or Assiniboine, a Siouan people living in Canada near Calgary, still go on vision quests on

Moraine Lake and the Valley of the Ten Peaks in the Canadian Rockies. The Stoney Indians go to these mountains on vision quests. (Galen Rowell/Mountain Light)

peaks in the Rocky Mountains above Banff and Lake Louise. Like the Sioux tribes to the south, who also maintain the practice, they purify themselves in sweat lodges and smoke peace pipes, offering prayers to the four directions. Then they go off alone into some of the highest and most beautiful mountains in North America to seek power and guidance in a vision, dream, or sign of nature, such as the sight of an animal acting in an unusual manner. The importance of the Canadian Rockies and the feelings of reverence they inspire are revealed in the words of a Stoney leader, Chief John Snow, who wrote a book in 1977 about his people titled *These Mountains Are Our Sacred Places:*

These mountains are our temples, our sanctuaries, and our resting places. They are a place of hope, a place of vision, a very special and holy place where the Great Spirit speaks with us. Therefore, these mountains are our sacred places.[27]

Pike's Peak and Harney Peak

In addition to being sites for vision quests, mountains might appear in visions themselves. Black Elk, an Oglala Sioux medicine man who lived near the Black Hills, fell suddenly ill at the age of nine and had a spontaneous vision in which two spirit warriors in the form of geese took him up into the sky to meet the six Grandfathers in a tipi made of rainbows roofed with clouds. These powers of the six directions, who together represented Wakantanka, the Great Spirit of the Oglala religion, showed him the future of his people and how he would heal and lead them when he grew up. At the end of the vision, as he prepared to return to his body, the home of the Grandfathers underwent a startling transformation:

I heard a voice saying: "Look back and behold it." I looked back and the cloud house was gone. There was nothing there but a big mountain with a gap in it.[28]

That mountain, he told John Neihardt, the poet who recorded his life story in a beautiful book called *Black Elk Speaks,* was Pike's Peak — one of the highest and most prominent mountains in the Colorado Rockies. Like the heavenly house of the Grandfathers, it lay off to the west of his world, looming over the edge of the Great Plains like a billowing cloud of rock and snow.

Earlier in the vision, spirit guides took Black Elk to a mountain that had even more significance for him and his people. He had gone with them to the four directions, where he had received the powers to heal and lead his nation in the times of trouble that lay ahead. Now, as the culmination of all that they had revealed, the spirits said, "Behold the center of the earth for we are taking you there." And this is what Black Elk saw at the high point of his vision:

As I looked I could see great mountains with rocks and forests on them. I could see all colors of light flashing out of the mountains toward the four quarters. Then they took me on top of a high mountain where I could see all over the earth.[29]

Two men came with the daybreak star and gave him an herb that bestowed on him the power to accomplish all the things he had been given to do. From his vantage point he looked down and saw people suffering and rejoicing in the world below. The spirits told him that all the sick that he could see he would restore to health and happiness:

They had taken me all over the world and showed me all the powers. They took me to the center of the earth and to the top of the peak they took me to review it all. . . . I was to see the bad and the good. I was to see what is good for humans and what is not good for humans.[30]

The mountain at the center of the universe represented the place from which Black Elk could behold the totality of all things, both good and evil. As he remarked to Neihardt, this center, this point of total awareness, can be found anywhere — if a person can recognize it in the spot where he finds himself.

The peak to which Black Elk went in his vision he later identified as Harney Peak, the highest summit of the Black Hills, an isolated range of mountains sacred to the Sioux. The center of an island of forest in the middle of the Great Plains, it rises to 7242 feet in crags of gleaming granite to culminate in a pinnacle of rock suspended above a landscape of dark pines and bright meadows. In his old age, having witnessed the death of many of his people and the destruction of most of his culture, Black Elk returned to the mountain of his vision. Standing on the summit of Harney Peak, convinced that he had failed in the sacred task entrusted to him, he extended his arms and cried out in despair:

In sorrow I am sending a voice, O six powers of the earth, hear me in sorrow. With tears I am sending a voice. May you behold me and hear me that my people may live again.[31]

On the way up the peak, Black Elk had told his son that if he had any power left, when he offered his prayer on the summit, a little rain and thunder would come. Although the day had been sunny, just before he prayed to the Grandfathers, wisps of cloud materialized, and as he made his plea, a fine drizzle fell, accompanied by a faint rumble from the sky. For the Sioux this was no accident: Black Elk had chosen to offer his prayer on top of the sacred peak where the Thunderbird makes its nest. According to their beliefs, when this huge, eaglelike bird blinks its eyes and flaps its wings, lightning flashes and thunder roars.[32]

Bear Butte

Just to the east of the Black Hills rises another prominent mountain, Bear Butte, sacred to both the Sioux and the Cheyenne. A great monolith of volcanic rock streaked with gullies and speckled with pines, it stands in impressive isolation, surrounded by rolling prairie, visible from miles away. The Sioux named it Mato Paha, "Bear Mountain." They would climb it on vision quests, leaving stones in the branches of trees to mark their efforts. Crazy Horse was said to have obtained remarkable powers from one such quest that he undertook on Bear Butte. A Sioux legend holds that Custer died at the hands of Crazy Horse and his warriors because he had offended the spirit of the mountain by climbing it just before the battle of Little Big Horn.[33]

Bear Butte has had a particular importance for the Cheyenne. They regard it as the most sacred place on earth, the holy source of the spiritual power that has nourished and sustained them as a people. They call it Noahavose, the "Good Mountain." According to their accounts of their history, many generations ago, Sweet Medicine, their greatest leader and prophet, went up to Bear Butte with his wife and found a cave inside the mountain. There they met Maheo, the Great Father and Creator of All. He gave them four commandments and four sacred arrows to take back to the Cheyenne. After four years of spiritual instruction inside the cave, they emerged from the mountain and returned to their people. Through the sacred arrows Sweet Medicine and his wife brought down from Bear Butte, the blessings of Maheo flowed into the Cheyenne. Two of these arrows gave them power over buffalo, the other two over men, supplying them with nourishment and protecting them from their enemies. Their greatest disaster as a people occurred around 1830 when Pawnee warriors stole the sacred objects in a raid. Eventually the Cheyenne recovered the four arrows, but only after they had suffered defeat and incredible hardship at the hands of other Native Americans and the white men who followed them.

Before passing on, Sweet Medicine appointed a successor to keep the sacred arrows safe in his possession. In each generation the person entrusted with them has been regarded as the greatest holy man in the tribe. In 1945 the keeper of the arrows took them back to Bear Butte to recharge them with the power of the sacred mountain. Since that time groups of Cheyenne have gone to the peak on vision quests to fast and pray for peace whenever the need for it has arisen—at the end of World War II, in the Korean War, and during the fighting in Vietnam. Like their ancestors in the past, they have offered themselves up in suffering for the benefit of their people. After days and nights of torment, frozen by the wind and roasted in the sun, tortured by insects and distracted with doubts, they have seen something in the dim light of dawn that gives them the hope and assurance that a power will emerge from the mountain to renew the world and flood it with life.[34]

SOUTHWEST

Today Native American beliefs and practices devoted to sacred mountains retain their greatest vitality in the desert regions of the southwestern United States—Arizona, New Mexico, and southern Utah and Colorado. The spacious views and incredible clarity of the air cause distant peaks to stand out with a sharpness of outline that gives them a supernatural appearance. The tribes living within sight of them, such as the Navajo, Hopi, and Rio Grande Pueblos, have had the most success in preserving traditional ways of life that encourage the veneration of mountains. Unlike their neighbors in the plains to the north, however, they

The Oglala Sioux medicine man Black Elk in his old age. (Joseph Epes Brown, courtesy of National Anthropological Archives, Smithsonian Institution, Washington, D.C.)

do not climb their peaks to cry for visions; they go up them, instead, to perform rituals and gather substances for ceremonial and medicinal purposes. As one Hopi put it to me, "We go to the mountains to pray quietly." Sometimes in the course of a pilgrimage to the San Francisco Peaks, a ring of summits left as the shattered remnants of an ancient volcano in northern Arizona, the Hopi will see the *katsina* spirits who dwell on the sacred massif, but rather than prize such a vision, they regard it as a great misfortune. *Katsinas* only reveal themselves to those who doubt their existence: if a person sees one, it indicates that something bad will happen to him as punishment for his lack of faith.

The oldest inhabitants of the Southwest, the Pueblo tribes, are descended from the Anasazi, an ancient culture that left impressive ruins throughout the region, some of them dating back to as early as A.D. 900. Present-day Pueblos such as the Hopi of Arizona and the Tewa of New Mexico have inherited highly developed ceremonial religions based on the practice of agriculture. They live concentrated together in compact villages centered around ritual structures called *kivas*. The open spaces of desert and mountains that surround them are inhabited by the Navajo and Apache, nomadic tribes that wandered into the Southwest from Canada sometime after A.D. 1300. Speaking Athapaskan languages related to those of the Koyukon and Tlingit in Alaska, they follow a different way of life from the Pueblos, herding sheep and cattle and living in isolated communities.

For most of these Native Americans, mountains serve as cosmic pillars defining the limits of the world in which they live. Gazing out from the centers of their villages, the Tewa Pueblos of the Rio Grande Valley near Albuquerque have singled out four peaks spaced around the horizon to mark the boundaries of their territory: Conjilon to the north, Tsikomo to the west, Sandia Crest to the south, and Truchas to the east. In their system of beliefs, these sacred mountains support the sky and divide the earth into quarters, conferring order and stability on the land that lies between them. On their summits dwell the deities who guided the ancestors of the Tewa people up from the underworld into the light of the sun. Altars of stone laid out in the pattern of keyholes mark the sites of earth navels that lead down into the supernatural realm of the gods hidden beneath the ground. When a person dies, he or she goes to the top of the mountains to return to the underworld from which his or her ancestors came. Each peak is associated with a color corresponding to the direction in which it lies: blue or green for north, yellow for west, red for south, and white for east.[35]

Other tribes in the Southwest have similar sets of mountains surrounding their territories, some of them overlapping with each other. Because it rises to the north of one tribe and to the west of the other, Mount Taylor, for example, is the northern mountain for the Acoma Pueblo and the western one for the Laguna Pueblo. The Apache and Navajo adopted this scheme of four sacred peaks from the Pueblos. The four *ga'an*, or mountain spirits, play a particularly important role in the ceremonial life of the Apache. Male dancers wearing masks and waving sticks emblazoned with lightning bolts impersonate these deities in the coming-of-age ceremony of an Apache girl. They protect her during the ritual and assure her of well-being in her future life as a woman. According to Apache belief, when the world was created, the *ga'an* stabilized it at the points of the four directions. After living for a while among humans, the *ga'an* retired to four sacred mountains, where they dwell as divine protectors of the Apache people, giving them power and wisdom and sustaining them in times of need. Associated with the colors of the four directions, the *ga'an* represent the principles of order and balance on which the Apache world depends.

Mountains of the Navajo

The Navajo, the largest tribe in the United States, have expanded the scheme of four mountains to cover an enormous area, extending from the San Francisco Peaks of Arizona in the west to the Sangre de Cristo Range of Colorado and New Mexico to the east. According to various versions of their creation myth, when First Man and First Woman emerged from a hole in the earth, they brought with them the soil of the sacred peaks that had given light to the worlds lying beneath ours. With this soil they fashioned the sacred mountains of the four directions, along with two or three others, depending on the version of the myth. One account describes the creation of the western peak, Doko'o'slid, or the San Francisco Peaks, in the following way:

The mountain of the west, they fastened to the earth with a sunbeam. They adorned it with abalone shell, with black clouds, he-rain, yellow corn, and all sorts of wild animals. They placed a dish of abalone shell on the top, and laid in this two eggs of the Yellow Warbler, covering them with sacred buckskins. Over all they spread a blanket of yellow evening light, and they sent White Corn Boy and Yellow Corn Girl to dwell there.[36]

First Man and First Woman created the mountains of the other three directions in a similar manner, fastening them to the earth with lightning, a flint knife, and a rainbow and spreading over them coverings of white dawn, blue sky, and darkness. Sis Najini, which some identify with Blanca Peak in the Sangre de Cristo Mountains, they placed in the east; Tso Dzil, or Mount Taylor, in the south; and Dibe Ntsa, which seems to be Mount Hesperus in the La Plata Range of southern Colorado, they put in the north. Each one they adorned with its own distinctive jewel, rain, plants, animals, and birds. Most importantly First Man and First Woman dispatched to the mountains the holy persons that give them life and make them sacred.

Each mountain has its inner form, a deity or spiritual essence in human shape, who acts as its soul or spirit, imbuing it with a power and intelligence that makes the peak itself a supernatural being. One Navajo with whom I spoke at the college where he teaches used the analogy of an instrument and its music to describe the relationship between the mountains and their indwelling deities: "If you ask a person where music is in a violin and he takes it apart, he will find nothing. In the same way, if we excavate and take apart a sacred mountain, we will also find nothing. But with belief we can find the holy person and his power in the mountain."[37]

The inner forms of the sacred mountains even partake of the nature of music itself: they come to life in songs sung by medicine men as they invoke their powers in rituals of healing and blessing. In Blessingway, one of the most important of these ceremonies, the inner form of the San Francisco Peaks sings this song of exultation as he climbs the mountain and takes his place on its summit:

To the summit of the San Francisco Peaks I have ascended, ascended
Now to the summit of chief mountain I have ascended, ascended
To an abalone footprint figure I have ascended.
. . . .
Now I am long life, now happiness as I have ascended, ascended.
Before me it is blessed as I have ascended, behind me it is blessed as I ascended, I have ascended, o ye.[38]

The song expresses not only the power of the deity, but also the feelings of the Navajo people for the sacred mountain and what it means to them.

The mountains arranged and inhabited by the gods enclose and protect the sacred homeland of the Diné, the People, as the Navajo call themselves. "In the midst of these four Sacred Mountains that were placed, there we live. With that, we who are the People are the heart of the world."[39] Within the boundaries established by their sacred peaks, the Navajo feel at home, safe and secure. They view the mountains of the four directions as the four crossed beams of a fork-sticked hogan, their oldest kind of traditional dwelling—a tipi-shaped structure covered with mud. The remaining two, or three, peaks form the chimney and entrance of the house, whose opening faces east, toward the dawn. In this way the vast expanse of their land, with its stark buttes and deserts exposed to the icy winds of the open sky, takes on the more inviting and intimate quality of a home made cozy by a cheerful fire.

Since the hogan defines the sacred space par excellence in Navajo culture, the correspondence between mountains and beams is more than an intellectual construct. It has a visceral effect. Like a city dweller leaving the safety of his apartment for the dangerous realm of streets at night, traditional Navajo feel anxious when they go outside the peaks that enclose the sanctity of their home. There lies a chaotic world of evil forces that may spring forth to attack them at any moment. On their return they have to go through rites of purification to drive out evil influences and restore them to a sacred condition. Even a modern Navajo woman like Theresa Boone, who works as a nurse in a Flagstaff hospital, feels a quiver in her stomach when she goes on trips to cities outside the boundaries of the sacred mountains. "When I come back and see the San Francisco Peaks, I feel at home, safe and protected," she says.

The Navajo also view the mountains as people and relatives for whom they have a deep personal affection. According to one man, "These mountains are our father and mother. We came from them; we depend on them. Between the large mountains are small ones which we made ourselves. Each mountain is a person. The water courses are their veins and arteries. The water in them is their life as our blood is to our bodies."[40] The trees and grass growing on the slopes of a mountain are the clothing covering its body. The Navajo even use their mountains as models for the way they see and dress themselves: "We adorn ourselves as they do, with bracelets of turquoise and precious jewels about our necks. From them, and because of them, we prosper."[41] Another person told me, "We don't pray to the sacred mountains: we talk to them. We address them as relatives, just as we say to the earth, Mother Earth, give us your blessings."[42] Some Navajo regard the San Francisco Peaks as a woman seated with her drawn-up knees represented by two ridges enclosing a basin that opens out to the east. From that basin, as from the womb of a mother, issue her children—a line of small volcanic hills and cones running out toward the Hopi mesas.[43]

The four mountains play a major role in the chantways—intricate ceremonies of healing and blessing that form the soul of the Navajo religion. As markers defining the boundaries of the world, they establish the setting for the sacred events of the mythic past narrated and sung in the course of the rituals. The

The four sacred mountains of the Navajo represented as different-colored hogans with doors framed by rainbows.
A holy person stands on top of each mountain. Ceremonial painting by Jeff King.
(Maud Oakes, reprinted by permission of Princeton University Press)

medicine men, called singers, who perform these ceremonies, usually for the benefit of a sick person, draw paintings of sand—or pollen and other substances—on the ground that represent the mountains with abstract symbols such as circles and squares of different colors. Occasionally they depict the peaks in the form of little hogans, arranged at the four points of the compass. The mountains provide the sense of order needed to restore the spiritual harmony and balance that a person requires for physical health and well-being. They appear with the greatest frequency in the songs of Blessingway, the ceremony that confers the fundamental blessings and happiness from which all the other chantways derive their power.

An object essential to most of these ceremonies is a medicine bundle with pinches of soil from each of the sacred mountains. With appropriate song and ritual, the singers travel to the peaks to gather the earth and place it in leather pouches, which they wrap, together with jewels and other objects, in a buckskin hide. During the ceremonies they hold the bundles in their hands or place them beside the paintings drawn on the ground. According to the mythology of

Blessingway, these bundles symbolize the primordial bundle that First Man used to create the inner forms of natural phenomena, culminating in those of the earth itself—hence their importance in the ceremony today. Every twelve years the singers are supposed to return to the peaks from which they collected the soil and gaze at the summits above them to renew the power of their songs. Individuals keep mountain earth bundles in their homes for protection, prosperity, happiness, and long life.

Medicine men also go to the sacred mountains to gather medicinal herbs. Before traveling to a mountain like the San Francisco Peaks, they purify themselves with ritual sweatbaths, making their journey a pilgrimage. On reaching their destination, they stand before a plant and make offerings to its inner form, reciting the myth that tells of its origin. Then, to avoid injuring the particular one they have just addressed, they pick another herb. The inner form, the life and soul of the plant, gives it its power to heal, making the medicine that comes from it more than a concoction of dead leaves and twigs.

In addition to the four that enclose their land, the

Navajo venerate a number of other mountains. The two that First Man and Woman created along with the peaks of the four directions are neither as high, nor, some say, as important. Gobernador Knob, the mountain of the center, located in New Mexico, was the birthplace of Changing Woman, a major deity who embodies the principles of renewal and restoration on which the world depends for its continued existence. She grew up on the nearby summit of Huerfano Mesa, where she had her sacred hogan. Unlike the peaks of the four directions, which reach up to 14,000 feet in altitude, neither mountain rises more than 8000 feet above sea level. However, both have an important place in ceremonies as sites of divine origin and markers of the sacred center.

North of the San Francisco Peaks, not far from the Grand Canyon, rises the long, dome-shaped ridge of 10,388-foot Navajo Mountain, which the Navajo compare in appearance to a loaf of blue cornbread. They did not recognize it as a sacred peak until 1863 when a group of them found refuge there from U.S. cavalry troops who were pursuing them. Then it revealed itself to their singers as the Head of Earth Woman, the place where Monster Slayer, the holy warrior of heroic deeds, was miraculously born and raised in a single day. Although only recently recognized, Navajo Mountain is today a major place of pilgrimage. After growing up, Monster Slayer slew the dreaded Cliff Monster, a winged demon whose corpse turned into Ship Rock, one of the highest and most spectacular buttes in the Southwest. The Navajo regard it as a sacred peak and objected when rock climbers made the first ascent of its difficult cliffs in 1939. Today they refuse to give anyone permission to climb it.[44]

San Francisco Peaks

Unlike their neighbors, who venerate a number of different mountains, the Hopi focus most of their reverence on the San Francisco Peaks, a cluster of discrete summits regarded as a single mountain. Spaced around the rim of an exploded crater, the peaks of this extinct volcano culminate in 12,633-foot Mount Humphreys, the highest point in Arizona. Viewed from the Hopi mesas, eighty miles away across the open expanse of a vast plateau, the mountain appears to float in the sky, a cloud of blue crystal frosted with snow. Like Mount Kailas in Tibet, the San Francisco Peaks stand alone on the horizon, opening the mind to a mysterious sense of fathomless depth and limitless space. The Hopi call the massif Nuvatukya'ovi, the "Snow Mountain Higher Than Everything Else." They revere it as the source of the clouds that bring them rain, enabling them to survive in the awesome world in which they live.

Although they share a veneration of the San Francisco Peaks with the Navajo, the Hopi relate to them in a very different way, reflecting profound differences in their societies and religious traditions. The Navajo lead nomadic lives, living in isolated hogans and herding sheep. They perform rituals invoking the power of sacred mountains mostly for the benefit of individuals whenever they get sick or need help. The Hopi, on the other hand, live together in houses perched on mesas, farming the land beneath their villages. The ceremonies that involve the San Francisco Peaks follow a rigid schedule set by the agricultural calendar. Priests perform them for the well-being of the entire community rather than for the welfare of individual members. Since knowledge belongs to the priesthood and comes from initiations, the Hopi tend to be much more secretive than the Navajo about religious practices having to do with the San Francisco Peaks. In addition, since each village and clan has its own beliefs and traditions, they disagree among themselves as to what many of those practices are, making it difficult to say anything definitive about them.

The Hopi view the San Francisco Peaks as one of the principal dwellings of the *katsinas*, supernatural beings on whose goodwill they depend for their livelihood and survival. Sometimes regarded as messengers of the ancestors, sometimes as ancestral spirits themselves, *katsinas* take the shapes of clouds, swelling up from the mountain to soar over the desert with rain. They also appear in semihuman forms, impersonated by masked dancers in elaborate ceremonies performed on the Hopi mesas. Through these dances the Hopi invoke the aid of the *katsinas* and call on them to take messages to their ancestors and other deities who dwell within the sacred slopes of the San Francisco Peaks.

Although they also live in other places associated with moisture, such as Kiisiwu spring to the east and Betatakin ruins to the north, legend says that the *katsinas* first appeared to the Hopi at the foot of the San Francisco Peaks. Anxious to find out the identity of these mysterious beings, seen prowling near their villages, the priests sent a young warrior to climb the mountain with an offering of feathered prayer sticks called *paahos*. Near the summit the youth came across a kiva, an underground chamber used in Hopi ceremonies. A voice invited him to enter, and he descended the ladder protruding from the hatchway that served as its entrance. Inside he found a friendly man who identified himself as an immortal spirit living in the underworld beneath the mountain. A frightening creature with a black face and a long snout came out, its white teeth gleaming in the firelight, and the man said that it was a *katsina*, one of the mysterious beings the Hopi had seen. When the youth gave him the *paahos* he had brought, the creature was very pleased with the gift and told him the *katsinas* would form rain-bearing clouds over the mountain whenever his people prayed and offered them prayer feathers.[45]

The Hopi see the San Francisco Peaks as an enormous kiva, similar to the ones they use in their villages. Like the Navajo, who view the peaks as a hogan, or the beam of one, the Hopi have projected the image of their ceremonial center onto the mountain, making it their equivalent of a church or temple. In 1978, during legal hearings to block the expansion of a ski resort, Hopi elders described the San Francisco Peaks as a kiva and compared the construction of ski runs on its slopes to the desecration of a Christian place of worship. Like the kivas found in pueblos, the mountain has a *sipapuni*, or hole that leads to the heavenly underworld where the ancestors dwell. There everything exists as an inverse image of the world above: when it is night here, it is day there; when it is summer here, it is winter there; while people live, they stay here; when they die, they go there. On their return from missions to the Hopi mesas, the *katsina* spirits use the *sipapuni* in the kiva of the San Francisco Peaks as a doorway to enter the underworld and convey their messages to the powers below.

A Hopi doll representing a nuva or snow katsina, a rain deity closely associated with the San Francisco Peaks. (Edwin Bernbaum, courtesy of the Heard Museum)

According to the story of the young warrior, the *katsinas* came down from the mountain and dwelled among people. But after a number of years, when men and women ceased to show them the proper respect, they withdrew and no longer appeared in bodily form. At that time the Hopi began the practice of making masks and impersonating the *katsinas* as a means of embodying their spirits and invoking their aid. A Hopi told me how it feels to become a *katsina*: "When I put on the mask and start to dance, I feel something enter me, and I am no longer myself: I become someone else. A power comes, but often I don't see its effects until much later."[46] Although nearly everyone receives an initiation into the *katsina* cult, the priests who actually perform the ceremonies come from only a few clans, such as the Badger, who have their ancestral shrines on the San Francisco Peaks and other places where the *katsinas* dwell.

Each year, at the time of the winter solstice when the sun sets behind a depression beside the San Francisco Peaks, the *katsinas* leave their spirit homes and move to the Hopi mesas to enter the bodies of dancers and initiate the ceremonial season of masked dances. There they remain until July, bringing the Hopi the rain and blessings needed to prepare the ground and start their crops. Then, when the corn is high and their work is done, they return to the underworld from which they came. Shortly after the summer solstice the Hopi give them messages to take back to the ancestors and send them off toward the San Francisco Peaks with the final ceremony of the *katsina* season, the Niman or Going Home dance. Some say that the *katsinas* go first to the sacred mountain and then disperse from there to their various homes in springs and other sites located at the points of the four directions.

A number of *katsinas* have close associations with the San Francisco Peaks. In the story of the youth who found a kiva on the mountain, the fearsome creature who appeared to him was a Cheveyo Katsina. This figure came down to teach the Hopi how to perform the proper ceremonies. The Nuva, or Snow, Katsinas come from the snowfields of the San Francisco Peaks, where they have their home. Despite their icy demeanor, they have a positive, nourishing role: they bear the winter snows that bring the water that young shoots need in spring. The Hemis Katsinas, whom dancers impersonate in the Niman ceremony, wear collars of sacred spruce brought from the San Francisco Peaks. The moist green needles of their costumes help to call forth the cool mist of dark clouds loaded with rain.

The Hopi go on numerous pilgrimages to the sacred mountain to leave prayer sticks and collect spruce and fir for ceremonies. Sometimes an individual will run from the mesas to the San Francisco Peaks and back

The San Francisco Peaks loom over the Colorado Plateau behind ancient Native American ruins at Wupatki National Monument. (Edwin Bernbaum)

to encourage the *katsinas* to come quickly with life-giving rain, but most people travel there in groups, going slowly and taking the time to perform rituals along the way. In the old days, when the Hopi walked to the mountain, they would go by way of Wupatki ruins and the ice caves on Sunset Crater, where they would stop to place prayer sticks. Today most groups drive directly to the Pavasiuki ruins at the foot of the San Francisco Peaks. There, according to Hopi tradition, all the Indian nations of North America gathered for a last meeting before dispersing across the continent to the places where they now live. After performing rituals at this ancestral shrine, where undergrowth covers the barely discernible traces of an ancient village, the pilgrims climb up to the evergreen forests on the slopes of the mountain. Most groups go no farther: stopping here, they make offerings at shrines marked with weathered prayer sticks and cut boughs of fir and spruce to take back with them. Only a few elders ascend all the way to the summit of Mount Humphreys to commune with their ancestors and pray to the *katsinas* who represent them.

Visible from every Hopi village, suspended in the sky, floating on the horizon, the San Francisco Peaks make a deep impression on the Hopi, giving them a strong sense of place and eliciting powerful emotions. Every morning on rising they look toward the mountain and throw cornmeal in its direction. The sight of its luminous peaks fills them with energy and gives them a feeling of well-being, similar to what they experience in watching the *katsina* dances. One Hopi summed up his people's feelings about the San Francisco Peaks by saying, "They house the divine source of our sustenance in all respects—physical as well as spiritual." Before telling me anything about the mountain, another person took me to the edge of the mesa and said, "Let's just look at it first." He wanted me to see that the sheer, physical presence of the San Francisco Peaks has a powerful spiritual effect—an effect that I felt that morning when I rose to look at them at dawn and a coyote glided by and stopped to look at me.

As a place of powerful deities, the mountain also inspires fear. A Hopi woman told me, "You must be careful to look at the San Francisco Peaks with good thoughts. If you look at them with anger, thinking something bad, the spirits who live there, not just the *katsinas*, may strike you." People who go on pilgrimages to the San Francisco Peaks have to maintain the proper attitude. If they quarrel among themselves or question what they are doing, the spirits of the mountain will punish them. Some pilgrims go singing songs to remove all hurtful thoughts from their minds, replacing them with kindness to keep from desecrating the sacred place with anger. Members of certain clans whose ancestors offended the *katsinas* of the mountain must avoid going near the San Francisco Peaks.

A striking story told me by the mother of a girl from one of these clans illustrates the consequences of treating the mountain lightly. On the understanding that the group was not going to the San Francisco Peaks, she gave permission for her daughter to go on a high school camping trip. Despite having made assurances to the contrary, the counselor, a white man ignorant of Hopi traditions, took the students to the sacred mountain. It was bad enough that the girl had gone to the peak forbidden to women from her clan, but some Hopi men who came along made matters even worse by taking the group up to a secret shrine and forcing her to make offerings of cornmeal. Trouble started right away. A couple of white friends of the counselor who joined the party brought along marijuana and give it to the Hopi boys and girls to smoke that night. The police heard about it and came to question the students. As a result, the girl lost her job at the tribal council. Then she began to have terrible nightmares. Unable to sleep, she lost weight and grew weaker and weaker. Finally her mother took her to a Hopi priest, who told her that she had come just in time: the girl had offended the *katsinas* and they were about to take her life. After he performed a ceremony to propitiate the wrathful spirits of the mountain, she ceased to have nightmares and completely recovered her health and peace of mind.

The Hopi feel that it is dangerous to tamper with the San Francisco Peaks. A transgression against the powerful spirits who inhabit the mountain can have consequences far worse than the punishment of one individual. It may effect the welfare of the community and even the future of the world itself. Offending the *katsinas* by thinking bad thoughts on a pilgrimage to the San Francisco Peaks may, for example, cause them to withhold the rain, endangering the crops of everyone in the tribe. The Hopi view themselves as custodians of the land, entrusted with ritual practices required for its care and protection. They believe that if they do not perform their ceremonies correctly, maintaining the sanctity of natural shrines like the San Francisco Peaks, drought and famine will result. They fear that they may even lose their right to the land they were given to protect; legal deeds of ownership according to the laws of secular society mean little or nothing in comparison to the entitlements and obligations of sacred ceremony. Their prophecies add that if the Hopi allow their ritual practices to deteriorate sufficiently, evil will increase and earthquakes, floods, and chaos will destroy the world.

Concern over the need to maintain the efficacy of their ceremonies drove the Hopi to join with the Navajo in an attempt to block the expansion of a ski resort desecrating the slopes of the San Francisco Peaks. In 1972, without consulting any of the Native Americans involved, the United States Forest Service formulated a master plan for the development of the sacred mountain. Up to this time the Hopi and Navajo had had no recourse to legal representation and had been unable to prevent the original construction of the Arizona Ski Bowl on the San Francisco Peaks in the 1950s. When he heard about the plan, Ben Hufford, a lawyer with the recently formed Legal Services Corporation, offered to represent them if they wanted to contest it. They did. Surprised by their unexpected intrusion into the process, the Forest Service agreed to ban development on the higher parts of the San Francisco Peaks. At that point the company that owned the ski resort applied for permission to build condominiums on the lower slopes of the mountain. When the zoning commission approved the construction, Hufford went to court, where he argued that the development violated the Native Americans' constitutional right to practice religion without obstruction and that the application for rezoning had been made improperly. He won the case on the superficial technicalities of the second point rather than on the deeper issues of religious freedom raised in the first.

Two years later the company returned with a properly completed application. This time, when the zoning commission held its hearing, two thousand people attended: the Native Americans had gained the support of many local residents who opposed the development of an unspoiled area. Navajo medicine men and Hopi priests and elders who had been reluctant to speak publicly about sacred matters stood up to voice their peoples' feelings about the San Francisco Peaks and why they must not be desecrated, arguing that the well-being of their tribes depended on their ability to maintain a harmonious relationship with the sacred mountain. Impressed by the sincerity of the speeches and the support of the audience, the commission turned down the application for rezoning.

In 1978 a new company, which had bought out the old one, decided to expand the Arizona Ski Bowl within the boundaries of the original permit. Once again the Hopi and Navajo tried to prevent any further desecration of the sacred mountain. But this time the Forest Service ruled against them, pointing out that the San Francisco Peaks lay on public lands off Indian reservations. The Federal Court of Appeals upheld the ruling on the grounds that while the ski area did interfere with ceremonial practices on the mountain, those practices, in its opinion, were not central to the Hopi and Navajo religions and did not, therefore, violate their constitutional rights. When the elders and medicine men appealed the decision to the Supreme Court, the justices refused to hear the case, and the company went ahead with construction of the ski runs, leaving white scars that can be seen from far away.

A Hopi spokesman had foretold his people's reactions to this act of desecration by noting that the Hopi derived much of their sense of spiritual well-being from prayers and songs in which they saw "a perfect mountain with perfect beings in perfect balance with each other." If they knew the San Francisco Peaks were scarred by ski runs, they would no longer be able to visualize them properly in their religious practices. "And, if I am not able to achieve this kind of spiritual satisfaction because of that," he had concluded, with an eloquence that speaks for many other Native Americans today, "I have been hurt, I have been damaged."[47]

EAST COAST

The mountains of eastern North America lack the impressive height and spectacular appearance of their western counterparts. Much older ranges than the Rockies and the Sierra Nevada, rounded down by millions of years of erosion, the granite knobs of the Appalachian and Adirondack mountains no longer leap up with blades of rock to slash the sky. With a few exceptions a heavy cover of forest softens their summits and obscures the spacious views that open the mind to ecstatic experiences. Blending with the terrain around them, they encourage a quiet kind of peaceful contemplation. Rather than seek out the exposed heights of stormy peaks, Native Americans of the Eastern woodlands tended to pursue their vision quests in the secluded depths of forests where spirits dwelled among the shadows of silent trees.

Although they did not make it a common practice to go to summits in search of visions, eastern tribes such as the Cherokee in the south and the Iroquois in the north regarded certain mountains as places of special power and significance. According to a myth held so sacred by the Cherokee that one could only hear it after purifying oneself in a ritual bath, Mount Mitchell—the highest peak east of the Mississippi, rising to 6684 feet in the Appalachians of North Carolina—had a sacred cave from which game animals magically issued like food from the cauldron of plenty in the fairy hills of Celtic mythology.[48] The Seneca, one of the tribes making up the Iroquois Confederacy, maintained a special relationship with a mountain in upper New York state. According to one of their traditions, their ancestors emerged from the earth of a bare-topped hill at the head of Canandaigua Lake near Rochester. As a consequence of this view of their tribal origins, the Seneca called themselves the "People of the Mountain." Because of the sanctity of the hill, they would go there to hold important council meetings on its summit. The tradition linking the Seneca to their mountain reflects the intimate connection they felt with the earth, the mother who had given birth to their people.[49]

Mount Katahdin

Near the northern end of the Appalachians stands the most imposing mountain in the eastern part of the continent. A lonely massif of frost-shattered granite, 5267-foot Mount Katahdin rises steeply over the wilderness of northern Maine. The Penobscot and Passamaquoddy tribes of the region regarded the mountain as the awesome abode of various deities. There, close to the spirit of the night wind, itself a manifestation of Katahdin, perched a storm bird of terrible wrath. Within the mountain itself, as in a lodge, lived a benevolent giant with eyebrows of stone. According to legend, he sometimes took native women for wives and fathered sons with supernatural powers. For fear of the storm bird and the reach of his claws, the Penobscot and Passamaquoddy stayed away from the summit of Katahdin.[50]

The mountain also made a profound impression on white people. In 1846, while living at Walden Pond, Thoreau took a trip to Mount Katahdin. Climbing the mountain alone, he emerged from the forest into a harsh and eerie realm of misted boulders that stunned him with an overwhelming experience of otherness:

It was vast, Titanic, and such as man never inhabits. Some part of the beholder, even some vital part, seems to escape through the loose gratings of his ribs as he ascends. He is more alone than you can imagine. . . . Vast, Titanic, inhuman Nature has got him at disadvantage, caught him alone, and pilfers him of some of his divine faculty. She does not smile on him as in the plains. She seems to say sternly, why came ye here before your time? This ground is not prepared for you. Is it not enough that I smile in the valleys? I have never made this soil for thy feet, this air for thy breathing, these rocks for they neighbors. . . . Shouldst thou freeze or starve, or shudder thy life away, here is no shrine, nor altar, nor any access to my ear.[51]

So completely alien, so utterly sacred, had Nature become on Mount Katahdin that it lay beyond all reach of any prayer or appeal that any mortal could ever hope to make. There, on the heights of the mountain, Thoreau confronted an aspect of wilderness far more awesome than anything he had ever experienced in the friendly woods of Walden Pond.

Mount Marcy

In 1898 the well-known philosopher and psychologist William James went camping on Mount Marcy, the highest peak in the wild Adirondack Mountains of New York State. Lying awake one night in a condition of "spiritual alertness," he had a visionary experience of something incredibly meaningful and perplexing, which he attempted to describe in a letter to a friend:

I spent a good deal of [the night] in the woods, where the streaming moonlight lit up things in a magical checkered play, and it seemed as if the Gods of all the nature-mythologies were holding an indescribable meeting in my breast with the moral Gods of the inner life. . . . The intense significance of some sort, of the whole scene, if one could only tell *the significance; the intense inhuman remoteness of its inner life, and yet the intense* appeal *of it; its everlasting freshness and its immemorial antiquity and decay. . . . In point of fact, I can't find a single word for all that significance, and don't know what it was significant of, so there it remains, a mere boulder of* impression.[52]

James had this experience only a short time before delivering a series of lectures that became one of his most famous and influential works—*The Varieties of Religious Experience.* It was the closest that he himself ever came to having the kind of mystical experiences that he explored and analyzed in others.

An Iroquois youth, for whom Mount Marcy was sacred, would not have come away from such an experience with "a mere boulder of impression." His tradition would have put it in a context, making it understandable as an encounter with the kind of spiritual powers sought in a typical vision quest. An older, wiser man would have explained to him the meaning of what he had experienced and what it portended for his life. James lacked a cultural background that understood and encouraged such experiences, particularly in the physical realm of "nature-mythologies." As a consequence, the events of his night on Mount Marcy remained for him something of inexplicable import charged with intense emotional significance.

It is vital to understand, or at least appreciate, the significance of the feelings that William James experienced on Mount Marcy. On such an understanding our lives and the future of our world may well depend. Native American views of sacred mountains are not just matters of limited cultural interest. They raise issues of environmental and spiritual importance that concern us all.

While doing research on the San Francisco Peaks in Arizona, I decided to drive to New Mexico and see Mount Taylor. As I pulled off the highway to go around a building for a better view of the peak, I was stopped by the sight of a weathered sign standing beside a wooden cross. I got out of the car to look at it. In bleak letters printed in black, it read:

> *Caution Radiation*
> *Air, Water, and Land Contaminated*
> *by Homestake Milling Co.*

I looked down at my feet and shook the dust off my shoes, regarding the earth around me with a sudden twinge of fear and suspicion.

Mount Taylor, the sacred mountain of the south for the Navajo, has suffered one of the worst forms of desecration imaginable. As I learned the next day in the town of Grants, from the 1950s through the 1970s the peak was the center of the most concentrated uranium mining in the United States. Waste left from this mining—and the milling that went with it—has contaminated the mountain with radioactive soil and dust. On the east side of Mount Taylor, the largest open-pit uranium mine in North America has polluted the drinking water of the Laguna Pueblo, another tribe for whom the peak is sacred, with chemical and nuclear poisons. In a grim twist of irony, the mountain to which Navajo singers have traditionally gone for medicinal herbs and the power to heal has become a source of sickness and death.

Problems arising from the mining have compounded the desecration of the sacred peak. When demand for uranium dropped off in the late 1970s, the mines around Mount Taylor closed down or curtailed their activity. Severe unemployment struck Grants, a town at the foot of the mountain that had advertised itself as the uranium capital of the world. In a desperate quest for jobs, the people of Grants agreed to accept what no other community in the state was willing to take—a female penitentiary. In 1984 the New Mexico Western Correctional Facility for Women was built on the lower slopes of Mount Taylor. A mountain that served as a symbol of happiness and well-being had become a place of punishment and bondage. Efforts by the Grants Chamber of Commerce to build a ski resort near the summit as an additional source of employment would exacerbate the desecration of Mount Taylor—as such a resort has already done on the San Francisco Peaks. If such a resort should be constructed, skiers who come to play on the mountain could unwittingly expose themselves to the invisible menace of radiation.

The problems posed by the desecration of the sacred peak extend far beyond the immediate vicinity of Mount Taylor. The mountain rises next to a major interstate highway that handles much of the traffic crossing the southern part of the United States. A number of important tourist sites, including Acoma Pueblo, one of the oldest continually inhabited towns in the country, lie within sight of the peak—and reach of the radioactive dust that blows from its slopes. Albuquerque, the capital of New Mexico and a major population center, is only sixty miles from Mount Taylor. Wastes from uranium mining throughout the Southwest constitute a much greater threat to public health than waste from nuclear reactors. Ignorant of the invisible hazards of their work, many miners—mostly Native American and Mexican American—have contracted lung cancer and died from breathing radon gas and radioactive dust. Children play on deadly tailing piles strewn like sandboxes around the countryside. In places like Durango, Colorado, people have discovered to their horror that they live in radioactive houses built with cement made from these tailings.

Despite all these problems, Chevron continues to operate its Mount Taylor Mine high on the western slope of Mount Taylor. With interest in nuclear energy returning as a result of worries over the green-house effect, pressures are mounting to revive uranium mining throughout the Southwest, including a proposal to exploit Red Butte, a sacred site on the southern rim of the Grand Canyon just outside the boundaries of the national park. When Anaconda Minerals leased land from the Laguna Pueblo to excavate its open-pit mine on the eastern side of Mount Taylor, the company promised to restore the land to a safe and usable condition. In 1989, after years of delay in negotiations over costs and plans, partial restoration work began, aimed at filling in and covering up the worst of the damage. Nobody knows how to dispose completely of the radioactive wastes that have accumulated at Laguna Pueblo and elsewhere in the Southwest. Any solution to the problem will require massive efforts by business and government in response to public pressure. Such efforts would have the side benefit of providing jobs for a long time to come.[53]

To generate the interest and energy required to reclaim and preserve the environment, we need to feel the kind of relationship to the land expressed in Native American views of sacred mountains. When I asked a woman from Acoma Pueblo how she felt about what had happened on Mount Taylor, she said, very simply, "It hurts." Perhaps, when it hurts us enough, we will finally do something about it—and make sure it never happens again.

LATIN AMERICA
MOUNTAINS OF VANISHED EMPIRES

From Mexico in the north to Argentina and Chile in the south, Latin America includes within its far-flung boundaries some of the highest and most mysterious mountains in the world.[1] Like fumes slowly dissipating from past eruptions, memories of little-known civilizations linger around the cones of Mexican volcanoes such as Popocatepetl and Iztaccihuatl. The surrounding highlands of Mexico and Central America hold ruins with ancient pyramids whose forms reflect the shapes of mountains still revered by native peoples. From the warm Caribbean shores of Colombia and Venezuela to the icy fjords of Tierra del Fuego, the Andes stretch 4500 miles down the west coast of South America to form the longest mountain range in the world. Soaring over the driest desert and the largest jungle on earth—the Atacama and the Amazon—these mountains reach their greatest height in the rugged peak of Argentina's 22,834-foot Aconcagua, the highest point outside of Asia. Like the ranges of Mexico and Guatemala to the north, the Andes conceal among their heights the ruins of ancient civilizations of which we know only shards of information. Elsewhere, to the east of the snow-capped Andean wall, mysterious mesas of primeval rock rise out of the jungles of Venezuela and Brazil to preserve on their isolated summits remnants of prehistoric plants and animals. Some of the tribes who live near them regard these plateaus as the stumps of primordial trees cut down by the gods.[2]

The earliest inhabitants of Latin America followed mountain ranges down from North America around twenty thousand years ago. As they progressed through Mexico to Central and South America, they developed complex societies that between 1000 B.C. and A.D. 1500 created some of the most extraordinary civilizations known to history—the Olmec, the Maya, the Aztec, the Chavín, the Tiahuanaco, and the Inca, among others. As these civilizations rose and fell, their memories obscured or obliterated by their successors, they left behind the ruins of magnificent cities, temples, and works of art, many of them located near or among the highest peaks in Latin America. The placement of these ruins, some of them as high as 22,000 feet above sea level, reveals that the worship of mountains played an important role in the lives of the peoples who built them. When the Spanish conquistadors arrived in the sixteenth century, they found the last of the pre-Columbian empires, the Aztec and the Inca, centered around the lofty cities of Tenochtitlán at 7000 feet in the Valley of Mexico and Cuzco at 11,000 feet in the Andes of Peru.

Because they were the last to fall, the pre-Columbian civilizations of which we know the most are the Aztec and Inca. The Spanish who conquered their empires left eyewitness accounts of their religious beliefs and practices, albeit biased. The Dominican friar Diego Durán, one of our principal sources, begins his description of the gods and rites of the Aztec religion with the following passage:

I am moved, O Christian reader, to begin the task [of writing this work] with the realization that we who have been chosen to instruct the Indians will never reveal the True God to them until the heathen ceremonies and false cults of their counterfeit deities are extinguished, erased. Here I shall set down a written account of the ancient idolatries and false religion with which the devil was worshiped until the Holy Gospel was brought to this land.[3]

The Towers of Paine in Patagonia. According to native beliefs, they are the fossilized remains of ancestral warriors killed by an evil spirit when he flooded the earth with water. (Gil Roberts)

Despite the efforts of the Catholic church to eradicate them, remnants of pre-Columbian beliefs and practices, especially those concerning mountains, have survived among native peoples descended from the Aztecs and the Incas. These contemporary beliefs and practices afford additional windows opening onto views of sacred summits that once held sway over cultures throughout much of Latin America. This chapter will accordingly focus on the highest and most important mountains revered by the Aztecs and Incas and their descendants—the volcanoes of Mexico and the peaks of the Andes.

MEXICO

A series of high plateaus and valleys bordered by mountain ranges occupies most of the land surface of Mexico. The slow convergence of tectonic plates has thrown up the peaks and ridges of the Sierra Madre Occidental and the Sierra Madre Oriental, two long cordilleras that run down either side of the country from its border with the United States in the north. At their southern ends they intersect with the Sierra Volcánica Transversal, a range of volcanic peaks that extends from the Gulf of Mexico in the east to the Pacific Coast in the west. Here, where the grinding together of tectonic plates has generated immense amounts of subterranean heat, the mountains of Mexico shoot up to their greatest altitudes in the volcanoes of Orizaba, Iztaccihuatl, and Popocatepetl. In a broad flat valley beneath the snows of the latter two peaks lies Mexico City, the site of the Aztec capital of Tenochtitlán.

The earliest pre-Columbian civilization to create a highly developed society, the Olmec, rose around 1000 B.C. in the lowlands bordering the Gulf of Mexico. Its sophisticated culture, expressed most powerfully in enormous heads of stone, influenced later civilizations that developed in the highlands of Mexico and Central America. During the first half of the first millennium A.D., the mysterious city of Teotihuacán with its great pyramids of the sun and moon took form in a valley northeast of Mexico City. With a population of more than 100,000 people, this city became the center of what may have been the largest empire of pre-Columbian Middle America. During the same period, another great civilization, that of the Maya, rose in the lowlands of the Yucatán Peninsula and Guatemala. After the collapse of Teotihuacán around A.D. 500, the Maya, who constructed cities of pyramids in the jungle, and the Zapotec, who founded a great pyramid complex on top of Monte Alban, became the predominant powers of the region. When their rule came to an end around A.D. 900, the fierce Toltec gained ascendancy in the mountains of Mexico with their capital at Tula, not far from Mexico City.

In the middle of the fourteenth century, the Aztecs, a warlike people who spoke a language related to that of the Hopi in the United States, wandered in from the north to establish their capital of Tenochtitlán on an island in a lake where Mexico City stands today. There they built a huge metropolis with more than 150,000 inhabitants, far more populous than any European city of the period. With great brutality the Aztecs defeated their enemies and created in the fifteenth century an empire whose existence depended on human sacrifice. From the civilizations that preceded them—such as the Toltec, the Maya, and that of Teotihuacán—they adopted various deities, in particular, Tlaloc, the god of rain and storm associated with mountains.

Perhaps no civilization other than the Chinese or Japanese elevated mountain worship to such heights of ritual importance as did the Aztec—or Mexica, as they called themselves. Like the emperors of China, Aztec rulers would ascend sacred hills and peaks to make offerings and perform sacrifices to the gods. They made the veneration of mountains an integral part of state ceremony and religion. The Aztecs regarded prominent peaks, especially those around whose summits rain clouds tended to gather, as gods whom they worshipped as sources of water and fertility. In their ceremonies they fashioned images of mountains in human forms, which they treated as they did statues of other deities. They also viewed high peaks as powerful originators of diseases that came down with the wind and cold that swirled off their icy summits. Believing the mountains capable of curing the very ailments they inflicted, the sick would commission priests to make elaborate offerings to the peak that stood closest to the village where they lived—images of deities made of dough, embellished with eyes of black beans and teeth of pumpkin seeds, adorned with streamers of paper, and surrounded by lavish gifts of food and drink.[4]

But the most distinctive and horrifying feature of Aztec worship of sacred mountains was the practice of human sacrifice. On many of the feast days that filled the ceremonial calendar, priests would offer up men, women, and children to mountain deities, high on the peaks and hills themselves or down in temples dedicated to them in villages and cities throughout the land. In the beginning of the first month of the year, the Aztecs celebrated one of their most important ceremonies—a great festival devoted to the gods of rain and water. In preparation for the day of this feast, they would seek out infants with double cowlicks in their hair and buy them from their mothers. Dressing them in fine clothes and placing them in litters adorned with plumes and flowers, they would bear the doomed children, amid singing and dancing, up to the tops of hills and mountains. There the priests would tear out the hearts of the infants and distribute pieces of their bodies to be cooked and eaten by the people present.[5] Bernardino de Sahagún, a sixteenth-century Franciscan friar who wrote a detailed account of the Aztec religion, described one of the most pitiable aspects of these sacrifices in the following words:

And if the children went crying, if their tears kept flowing, it was said, it was stated: "It will surely rain." Their tears signified rain. Therefore there was contentment; therefore one's heart was at rest.[6]

Human sacrifices to mountains were only one form of Aztec sacrifice. The greatest sacrifices of all took place after victories over rival kingdoms: the blood of hundreds, even thousands, of slaves and prisoners captured in battle would flow in streams down the steps of the main temple in the capital city of Tenochtitlán. Although this temple stood on a pyramid shaped like a mountain, the rituals themselves focused on a powerful god of war who embodied the might of the Aztec rulers. As in this sacrifice, where victims would receive the name of the deity to whom they were offered, people assumed the names of the mountains on which they were slain. In each case the person sacrificed took on the nature of the god to whose realm he or she was dispatched. The blood of the victim was offered in exchange for water, the blood of the earth. Without such offerings the Aztecs feared that streams would cease to flow from the mountains and the people would die of famine.

Popocatepetl and Iztaccihuatl

On rare days when murky smog thins to a translucent haze, two snow-capped peaks materialize like clouds, swelling up over the mountains southeast of Mexico City. Rising 10,000 feet above the valley floor to altitudes of 17,887 feet and 17,343 feet respectively, the steep cone of Popocatepetl and the white mass of Iztaccihuatl appear to belong to a higher and purer world, not yet polluted by the industrial wastes of man. Viewed from Tenochtitlán in the time of the Aztecs, the two peaks must have sparkled in a clean blue sky—when they were not gathering black clouds about their summits in preparation for releasing violent storms. Because of their location, close to the center of the empire, Popocatepetl and Iztaccihuatl played a more important role in Aztec religion than did 18,700-foot Orizaba, the highest mountain in Mexico and more than a hundred miles away.

The name Popocatepetl means "Smoking Mountain." When Hernando Cortés, the Spanish conqueror of the Aztecs, crossed the pass between Popocatepetl and Iztaccihuatl and first saw Tenochtitlán, the volcano was much more active than it is today. Durán tells us that it shot up smoke, and sometimes flames, two or three times a day. Slippery ash, steep slopes, and poisonous fumes made its ascent extremely hazardous, seemingly impossible. Before the arrival of the Spanish conquistadors, the Aztec emperor Moctezuma had sent ten of his strongest warriors to discover the mysterious source of the smoke issuing from the summit of the mountain. Two of them died on the ascent. The others managed to reach a place from which they could see smoke roaring out from a latticework of cracks in the rock. Only two, however, returned to report this sight to the emperor: the others perished on the descent. And those two, according to the Aztecs, never recovered the health they lost on the perilous ascent of the Smoking Mountain.[7]

After conquering the Aztecs in 1521, Cortés found himself in need of sulfur to replenish his supply of gunpowder. In 1522 he dispatched a party of soldiers led by Francisco Montaño to climb Popocatepetl and fetch the substance from its crater. Partway up the mountain they came to a halt on an icy slope, nearly paralyzed with cold. A red-hot boulder fortuitously ejected by the volcano rolled to a stop next to them, and they were able to warm themselves with its heat and proceed to the top. To reach the sulfur, Montaño had to be lowered on a rope down the cliffs ringing the crater, convinced he was going to fall into the fires seething below him and go straight to hell. Despite his fears, he and his men managed to obtain a good supply and return alive, laden with bags of the

A human sacrifice performed on the steps of an Aztec temple. A priest is tearing out the victim's heart. Illumination from the Codice Magliabechiano, Biblioteca Nazionale, Florence. (Scala/Art Resource, N.Y.)

Popocatapetl, revered as a source of water and fertility by the Aztecs, rises behind a Christian graveyard in Mexico. (Robert Wenkam)

yellow powder.[8] Their ascent of Popocatepetl marks the highest summit climbed by Europeans up to that time. In keeping with the bloodthirsty nature of mountain worship among the Aztecs he conquered, Cortés used a material embodiment of the power of the sacred volcano to make gunpowder for killing people—human sacrifices for the greater glory of the Spanish empire.

The Aztecs revered Popocatepetl as one of the most important of their many sacred mountains. Its rich volcanic soil combined with a congenial climate and abundant water to make its lower slopes one of the most fertile and heavily populated areas of pre-Columbian Mexico. The closer maize and wheat grew to the peak itself, the earlier they ripened and the better they tasted. Despite the wrathful nature of its eruptions, the mountain represented to many Aztecs a divine embodiment of all that was good and desirable. Pilgrims would come from distant regions to fulfill vows and throw offerings of gold and precious stones into the mountain's springs, streams, and deep-cut ravines. As a major source of water, the volcano had a particularly close association with the goddess of springs and rivers.

Tepeilhuitl, the Feast of Mountains, celebrated in the thirteenth month of the ceremonial calendar, centered on Popocatepetl and the peaks around it. People would fashion images of the volcano out of dough made from amaranth and maize flour. These they would decorate with eyes and mouths and place in their homes, surrounded by figures representing the other mountains of the region. Dressing the volcano images with paper caps and tunics, the Aztecs would perform ceremonies and make offerings to them. After two days they would decapitate and eat them, as they did victims of human sacrifice. This ceremonial meal they called Nicteocua, which means "I Eat God." During the Feast of Mountains, people also climbed to the summits of nearby hills and peaks to light fires and burn incense.[9]

Right beside the image of Popocatepetl in the center of each household shrine prepared for the festival, worshippers would place a dough image of Iztacci-huatl, the "White Woman." A goddess of high repute, she resided in permanent images kept in various city temples and inside a cave high on the mountain itself. A wooden statue showing her robed in blue with a crown of white paper stood in a shrine in Tenochtitlán.

The image had the face of a young woman with hair clipped short in front and hanging long to her shoulders on the sides. On the day of Iztaccihuatl's special feast, a female slave would be arrayed in green with a white crown set on her head, representing the snows above the forests of the peak. Having given her the name of the goddess, the priests would sacrifice her in front of the image in the capital. Accompanied by nobles, two little boys and two little girls would be borne in litters up to the cave on the sacred mountain, there to be sacrificed before another statue, amid lavish offerings of jewels, feathered headdresses, clothing, and food. The retinue of aristocrats would remain for two days on the heights of Iztaccihuatl, fasting and performing ceremonies in honor of the White Woman, who embodied the divine power of the peak itself.[10]

Both Iztaccihuatl and Popocatepetl are covered with the ruins of shrines dating back to the Aztec period—and possibly even earlier. The Mexican archaeologist José Luis Lorenzo has identified at least ten sites at about 12,000 feet, not far from the snowline of the two peaks. Religious activities continue at some of these shrines, even today. Every year on May 3, people come from distant villages to climb Iztaccihuatl to attend a ceremony performed at the shrine of Alcalican. There, at night, high on the slopes of the sacred mountain, they seek to make contact with their *nahuals*, guardian spirits who take the form of animals to protect their human charges with supernatural power and insight. Anthropologist Carlos Castaneda popularized contemporary Mexican conceptions of the *nahual* in books he wrote allegedly describing his experiences with a Yaqui sorcerer named Don Juan. The idea itself and the practices associated with it go back to the time of the Aztecs—and the civilizations that preceded them.[11]

The physical proximity of Iztaccihuatl to Popocatepetl has brought the two peaks together in a popular legend that originated after the Spanish conquest of Mexico. According to this legend, Popocatepetl was a warrior of an Aztec tribe who fell in love with Iztaccihuatl, the daughter of the tribal chief. When the lovers went to her father, he told them that he would agree to their marriage only if Popocatepetl would first conquer an enemy tribe and bring him the head of its leader. The young warrior succeeded in his mission, but he took so long that Iztaccihuatl, thinking he had perished, succumbed to sorrow. When Popocatepetl returned in triumph to find his lover dead, he picked her up and carried her to the top of a nearby mountain. Overcome with grief, he laid her body to rest on the long summit ridge, which assumed the form of a sleeping woman—the shape that many people see in the outline of Iztaccihuatl today. There

Popocatepetl has remained, standing with a torch lit to watch over his lover, which accounts for the smoke that issues from the volcano that bears his name.[12]

Tlaloc

Just north of Popocatepetl and Iztaccihuatl rises another mountain, lower and less impressive than the two snow-capped peaks, but closer to Tenochtitlán and more prominent in the state ritual of the Aztec emperors. The Aztecs regarded 13,615-foot Tlaloc as a great storehouse of clouds, mist, rain, and snow. There, on or within its rocky summit, dwelled one of the greatest deities of the Aztec pantheon. God of rain and lightning, thunder and storm, Tlaloc shared the main temple of Tenochtitlán with Huitzilopochtli, the mighty lord of war. This temple stood on top of a great pyramid located at the center of the city. Its placement there, characteristic of similar temples throughout Mexico and Central America, suggests that the pyramids of earlier civilizations, such as the Maya, functioned, in part, as symbolic mountains whose summits represented the dwelling places of the gods.[13]

Diego Durán gives us a vivid description of the image of Tlaloc that once stood carved in stone on top of the pyramid that enshrined the sacred power and secular might of the Aztec empire:

Its horrendous face was like that of a serpent with huge fangs; it was bright and red like a flaming fire. This was a symbol of the brilliance of the lightning and rays cast from the heavens when he sent tempests and thunderbolts; to express the same thing, he was clad totally in red. His head was crowned with a great panache of green feathers, shining, beautiful, rich. From his neck hung a string of green beads of jade in the form of a necklace and hanging from it a round emerald set in gold. . . . In his right hand [Tlaloc] carried a purple wooden thunderbolt, curved like the lightning which falls from the clouds, wriggling like a snake toward the earth.[14]

Another image of Tlaloc, similar to the one in Tenochtitlán, stood in a temple built on the summit of the mountain that bears his name. A wall eight feet high, plastered with stucco and crowned with spikes, enclosed a square courtyard with a wooden chamber that contained the statue of the god. Like a crown on the head of a king, its outline stood out sharp against the sky, visible from miles away. No other peak in Mexico had a higher or more impressive shrine situated on its very summit. The ruins of this temple still stand on top of Mount Tlaloc, weathered walls of stone now inhabited by the whispering spirits of mist and wind. An archaeological expedition that climbed the mountain in 1953 found the site remarkably untouched compared to other sites near Mexico City: the height and difficulty of the ascent had inhibited the destructive activities of the Spanish conquistadors and the treasure seekers who followed them.[15]

Each year in spring, accompanied by a vast retinue of nobles and princes, the Aztec emperor would ascend to the summit of Tlaloc to beseech the god of the mountain to grant his people a good harvest. A great procession would bear a closed litter containing a six-year-old child into the courtyard of the temple. There, in the red light of dawn, concealed behind a curtain of cloth, the throat of the young victim would be slit to the wild music of trumpets, conches, and flutes. Following the sacrifice the emperor would advance to the great statue of Tlaloc and place a feathered crest on its head and fine garments on its body. After he and the other rulers had spread a banquet of food before the god, priests would sprinkle a portion of the child's blood on the offerings and pour the remainder over the idol itself. Then the emperor and his retinue would return to their quarters to feast and dine before descending to the lowlands. Guards would remain behind to watch over the offerings as they slowly rotted away in the thin cold air of the mountain heights.[16]

Although the Spanish put an end to the Aztec practice of human sacrifice in the sixteenth century, child sacrifices on Mount Tlaloc continued to take place until the end of the nineteenth century. Recounting stories heard in their youth, old people in the vicinity of Tetzcoco near the foot of the peak tell of villagers taking babies up to the summit to offer them as sacrifices to obtain good crops. According to their accounts, only unbaptized infants would do as sacrificial offerings — which must have encouraged parents to baptize their children as early as possible. The last child sacrifice reported on Tlaloc occurred in 1887. A woman told a researcher that her mother remembered people whispering and making secret preparations in Purificación, a village near Tetzcoco, before departing for the summit with an unbaptized baby, who never returned. A Mexican anthropologist reported that in 1950 people were still being initiated into the rites and chants of ancient deities in ceremonies taking place in a cave high on the slopes of Malinche, a sacred peak that the Aztecs worshipped as the husband of Mount Tlaloc. Although himself an Indian, the anthropologist was not invited to attend the secret rituals and felt it would be too dangerous to risk coming to observe them without permission.[17]

THE ANDES

Like a raised binding stitched along the edge of a blanket, the Andes form a sinuous line of folded rock that traces the western coast of South America. The immense pressure of the continent riding up over the Nazca Plate of the Pacific Ocean has buckled the earth's crust, producing a narrow range of mountains 22,000 feet high, overlooking one of the deepest submarine trenches in the world, the Peru-Chile Trench, 25,000 feet deep. The geological forces responsible for this enormous disparity of elevation — 47,000 feet within a horizontal distance of only one hundred miles — have riddled the Andes with the highest volcanoes in the world. Because of the equatorial bulge, one extinct volcano, Ecuador's 20,577-foot Chimborazo, is even higher than Mount Everest — if one measures its altitude by the distance of its summit from the center of the earth. Interspersed among groups of massive volcanoes, the Andes include alpine ranges that contain some of the sharpest and most beautiful mountains in the world: sheer pinnacles of glistening granite, like Fitzroy and the Towers of Paine in Patagonia, and delicate wisps of ice and snow, such as the peaks of the Cordillera Blanca in Peru. It is not surprising that the Incas and their predecessors regarded the Andes as the home of powerful deities.

The first major civilization known in South America appeared at Chavín de Huántar in the Cordillera Blanca of Peru around the end of the second millennium B.C. The orientation of the ceremonial center of the site suggests that the people who built it venerated high snow peaks as sacred sources of water. Distinguished by the representation of pumas, animals associated with mountain deities, the culture of Chavín influenced a number of other civilizations, such as the Moche and Nazca, that developed along the Peruvian coast between 400 B.C. and A.D. 600. From A.D. 600 to 1000, the Tiahuanaco civilization in Bolivia spread its influence north to Peru through the Huari empire. Like Chavín, the site of Tiahuanaco, set in a valley near Lake Titicaca, is oriented with respect to mountains revered by native Andeans today, most notably the impressive snow peak of 21,185-foot Illimani.[18]

Until the fifteenth century a multiplicity of cultures, each in its own valley or local area, characterized the history of South American civilizations. In 1438 the Incas, a tribe that had established itself at Cuzco in the thirteenth century, marched out from their mountain stronghold to conquer and unify the Andean re-

Inca ruins on the summit of 22,057-foot Llullaillaco mark the highest known archaeological and religious site in the world. (Johan Reinhard)

gion in an elongated empire that eventually stretched from southern Colombia in the north to central Chile in the south. Over the more than 2500-mile length of this rugged empire, the Incas established a remarkable system of roads and administrative centers. As a means of reinforcing their rule, they also carried out a campaign to eradicate all memory of previous civilizations. This action, combined with a lack of a written language, makes it very difficult to know much about the beliefs and practices of the people who preceded them. The cultural destruction wrought by the Spanish, who in 1532 conquered the Inca empire in search of gold, compounded the problem, making our knowledge of pre-Columbian views of sacred mountains in the Andes sketchy at best.

Whereas the practice of human sacrifice characterizes the worship of Mexican volcanos by the Aztecs, the construction of religious structures at incredibly high altitudes distinguishes the veneration of Andean peaks by the Incas. In 1952, believing themselves to be among the first people ever to climb the mountain, two Chilean mountaineers, Bíon Gonzalez and Juan Harseim, reached the barren snow-streaked summit of Llullaillaco, the eighth-highest peak in the Andes. There, to their amazement, at an altitude of 22,057 feet, they came upon a low rock wall and the remains of a leather bag half-buried in the frozen ground. The Incas had beaten them to the top—by more than four hundred years. Although Gonzalez and Harseim had failed to make a pioneering ascent of the mountain, they had accomplished something of far greater significance: they had discovered the highest archaeological site in the world.[19]

Subsequent expeditions uncovered an altar on the very summit and close to it the foundations of a structure with two rooms. Only two hundred feet below the top, ruins were found of a much more extensive complex with living quarters for people and an enclosure for llamas, along with seeds, corn, bits of charred wood, and pieces of pottery and cloth. The remnants of a path with rock supports led up to the site, situated at an altitude of 21,800 feet, overlooking the brown desert mountains of northern Chile and Argentina. In 1971, while exploring the north side of Llullaillaco, Argentine climbers stumbled on six graves with skulls and skeletons at 18,000 feet.

In the fifteenth and sixteenth centuries, hundreds of years before European mountaineers managed to reach comparable altitudes in the Himalayas in the nineteenth century, South Americans had not only climbed up to 22,000 feet in the Andes, but had stayed

to build altars and living quarters on the summits they "conquered." Climbers and archaeologists have discovered about one hundred sites with Inca ruins on mountains higher than 17,000 feet, opening up the fascinating new field of high-altitude archaeology. All of these sites lie well to the south of Cuzco, the capital of the Inca empire, in the drier regions of southern Peru, Bolivia, Chile, and Argentina where the existence of elevated snowlines made the ascent of rounded volcanoes easy and the construction of permanent structures on their heights feasible. Many of them, including the very highest, rest on great volcanic peaks that rise like enormous piles of sand heaped up beside the arid wastes of the Atacama Desert.

Because they have remained relatively untouched, high-altitude ruins constitute some of the most important archaeological sites in the Andes. Unlike most of their counterparts at lower elevations, whose contents have been thoroughly ransacked, high-altitude sites continue to provide finds of rare objects dating back to the time of the Inca empire. These objects include figurines of gold and silver, images fashioned from sea shells, pieces of clothing, remnants of ancient food, and offerings of coca leaves. Researchers have speculated that the statues, depicting humans and animals, may represent deities, things that people wanted, or substitutes for sacrifices of living beings. Those who offered them up have left no written records to describe what they were doing in the mysterious ceremonies they performed at the uppermost limits of physical endurance.

One of the largest groups of high-altitude ruins lies perched on the circular rim of Licancabur, a 19,421-foot volcano in northern Chile with a lake set in its extinguished crater. Since Incas are known to have tossed offerings into sacred bodies of water at lower elevations, several expeditions climbed up to this strange lake to do the highest scuba diving in the world—in search of Inca artifacts. Underwater research at over 19,000 feet, where the air pressure is less than half that at sea level, presented some unusual problems. Coming up from a depth of 15 feet below the surface was equivalent to flying up from sea level to 20,000 feet: if the divers emerged from the water out of breath, they ran the risk of blacking out from oxygen deprivation. To add to the difficulties, they had to carry extra loads of lead weights up the mountain to compensate for the increased flotation of their wet suits, which ballooned up at high altitude. Unfortunately, a layer of sediments prevented them from finding any artifacts on the lake bottom.[20]

An Inca burial image discovered on the summit of 17,815-foot Cerro El Plomo in the Chilean Andes. (Ianiszewski/Art Resource, N.Y.)

Andeans on a ridge with Salcantay, a sacred mountain near Machu Picchu, seen in the background. (Sara Steck)

The archaeological discoveries that have attracted the most attention have been a few rare finds of frozen bodies preserved on the heights of Andean peaks—victims of human sacrifice. In 1954, a climber encountered two locals hurrying down from near the summit of El Plomo—a 17,815-foot dome of ice visible from Santiago, the capital of Chile—lugging a heavy bag with something big in it. They had just dug up the perfectly preserved corpse of an eight-year-old boy sacrificed on the mountain during the Inca period. He was dressed in fine clothes and wore a headdress of condor feathers. They had found him buried beneath a stone enclosure, surrounded by offerings of sea shells and statues. A few years later a couple of climbers came across a human head sticking out of the summit of Cerro El Toro, at 20,952 feet on the border between Chile and Argentina. They disinterred the rest of the body and took it down to a museum in Mendoza, where analysis revealed it to be the remains of a twenty-year-old youth who had been intoxicated with corn liquor before being sacrificed on the desolate heights of the peak. Another climber, Antonio Beorchia, a pioneer in the field of high-altitude archaeology, found a child encased in ice on the summit of Quehar, a 20,106-foot mountain in northern Argentina, but when he returned to the site to dig it out, he found that treasure seekers had blasted the body to bits with dynamite, looking for things they considered more valuable.[21]

What drove the Incas to climb to such extreme heights to build religious structures and offer human sacrifices? Whereas Durán and Sahagún have left us detailed accounts of Aztec rituals performed on mountains such as Tlaloc and Iztaccihuatl in Mexico, we have no comparable written records describing Inca practices on high peaks of the Andes. Most observers have assumed that these practices were devoted to sun worship, which we know to have played a major role in Inca religion, but the archaeological sites themselves have yielded no convincing evidence of this. Nor have they given much support to other conjectures that the Incas and their predecessors climbed high peaks to construct signaling stations or to pray to deities guarding gold and other sacred metals hidden inside the mountains.

Taking a completely different approach drawing on observations of contemporary practices among native Andeans, Johan Reinhard, an American anthropologist who has done much of the recent work on high-altitude archaeology, has proposed a hypothesis that seems to account for the incredible efforts devoted by the Incas to climbing and constructing religious sites on the summits of some of the highest peaks in the Andes. Each year the people of Socaire, a village in the dry Atacama region of northern Chile, perform a ceremony in which they make offerings to more than twenty mountains to invoke their aid in bringing rain. Researchers have climbed fifteen of these mountains, and on thirteen of them, including Licancabur and Llullaillaco, they have discovered Inca ruins. Reinhard also observed that the relative importance of the peaks in the ceremony correlates with the size and significance of the archaeological sites found on their summits. Archaic features of the ritual and other deities invoked, such as Pachamama, Inca goddess of the earth, indicate a tradition reaching back to the time when the Incas constructed the now abandoned ruins. All this suggests that they built and used these ruins for the same reason that their descendants in Socaire now make offerings to the peaks on which the sites are found—to ask the mountain deities who control the weather to send them water to irrigate their fields.

Elsewhere in the Andes, people still climb up to summits as high as 18,000 feet to perform ceremonies asking mountains for rain, and on the Island of the Sun in Lake Titicaca, Aymara Indians regularly sacrifice llamas to Illimani and Illampu, two of the highest and most sacred peaks in Bolivia, in return for water to irrigate their crops. Such practices would also appear to explain the presence of the mummified bodies found on the lofty summits of El Plomo and Cerro El Toro. In 1942 and 1945, some villagers in Peru sacrificed children, with the approval of their parents, to prevent a drought, and in 1958, another human sacrifice for rain reportedly occurred on top of a mountain near Lake Titicaca. Offerings of sea shells—symbols of the ocean, mother of waters—found with the ancient mummies, and in numerous other high-altitude sites, provide archaeological evidence that the Incas regarded high peaks as sacred sources of water to be obtained by building religious structures and performing ceremonies on their summits.[22]

The survival of ancient beliefs and practices lends added support to Reinhard's hypothesis that the pre-Columbian cultures of the Andes worshipped mountains as weather gods responsible for dispensing water and ensuring fertility. Along with the sun god, the Incas focused much of their religious attention on a deity of thunder and lightning known as Illapa, the "Flashing One." Lord of storms, he controlled the natural forces of wind, rain, hail, and snow. In times of drought, people would climb up to high places to make offerings to him. The Aymara of the Bolivian Altiplano had a similar deity, called Tunupa, whom they associated with Illimani, the highest and most important mountain in the vicinity of Lake Titicaca. As lord of storms, Illapa was also a powerful god of war: the great Inca emperor Pachacuti carried his image into battle when he went forth to found the Inca empire in 1438.

Many of the mountain deities or spirits worshipped by Andeans today share the attributes of this ancient god of war and weather. Called *apus*, *wamanis*, *aukis*, *achachilas*, and *mallkus*, they inhabit the high peaks of the Andes, where wrapped in clouds they control the impressive meteorological phenomena of rain, hail, snow, thunder, and lightning. Often regarded as the most powerful deities in the region, these spirits are looked upon as warlike protectors of fields and livestock, as well as sources of life-giving rain and water. Dependent upon their goodwill for survival, villagers climb up to high places to placate them with offerings of coca leaves and sacrifices of animals such as llamas.

The term *apu*, the title of the highest mountain deities in much of Peru, comes from the ancient Inca word for "chief or lord." Like rulers of men, *apus* have hierarchies of subordinates arranged on human models of political organization complete with governors, judges, and other functionaries. They also possess palaces concealed within mountains and lands filled with livestock—the wild animals who frequent the heights. Foxes are an *apu*'s dogs, pumas his cats, and vicuñas and guanacos his domesticated llamas. People regard condors as the special messengers of mountain deities, the intermediaries through whom they communicate with shamans and other religious experts, such as healers and diviners. During one festival they feed the great birds food and drink and then release them with messages to take back to their masters in the clouds.

If angered, the mountain deities and their retinues of spirits can wreak havoc on the people who live in the valleys below them, destroying their crops with barrages of hail. Only shamans who have amassed supernatural powers dare to venture up to the rarefied realm of the gods, sometimes leaving their bodies behind to journey there in trances of spiritual exaltation to heal the sick and rescue the dead. Others stay away from the heights, regarding the summits of sacred mountains as dangerous places impossible to reach—if one should be crazy enough to want to try. When Reinhard told a villager that he had just climbed a nearby mountain, the man thanked him for the fanciful tale and introduced himself as Jesus Christ.[23]

Myths and legends from various parts of the Andes reveal sacred mountains acting in a number of different roles—not just as weather deities. The theme of a great flood found as far north as Alaska extends down South America to Patagonia. According to a tradition recorded in Peru at the beginning of the seventeenth century, a llama learned that in five days the ocean would cover the earth. When his herder became angry at him for not grazing, the animal reproached him, saying, "How can you worry about such things when the world is about to end?"

Alarmed by the news, the man asked what he could do to save himself. The llama suggested that he and his wife take five days' food and climb a mountain called Huillcacoto. Clustered together on top, they found a number of different animals, including a fox. As the llama had predicted, the ocean rose and covered all the other mountains, except for the summit of Huillcacoto. Because the fox let his tail dip into the water, foxes have black-tipped tails. After five days the flood receded, and the herder descended with his wife and the animals to repopulate the world. According to the Catholic priest who recorded this myth, the non-Christian natives of the region revered the mountain as the savior of the human race. In other traditions from the Inca period, the peak chosen as a place of refuge keeps growing in height in order to stay above the rising waters.[24]

In Patagonian versions of the flood myth, people become mountains. According to the Araucanians of southern Chile, an evil serpent named Cai Cai created a flood to exterminate the human race. Despite the efforts of a friendly snake, who saved many people by helping them reach the tops of mountains, a number of warriors perished. When the waters receded, Cai Cai turned their bodies to stone, and they became the spectacular spires of Fitzroy and the Towers of Paine. The Araucanians regard themselves as the descendants of the lucky people helped by the friendly serpent. Few of the modern climbers who come to make some of the most difficult rock climbs in the world realize that they are jabbing their pitons and ice axes into the backs and bellies of mythic warriors.

Elsewhere in the Andes, people turn into mountains in myths and legends that have nothing to do with floods. A story about the Cordillera Blanca tells of an Inca warrior named Huáscar who came to the range and fell in love with Huandi, the beautiful daughter of the local chieftain. Her father, an avowed enemy of the Incas, disapproved of the alliance and forbade her ever to marry the handsome young man. When the two lovers attempted to run away, he set out in pursuit with his warriors and captured them. Enraged at what they had done, he had them taken up to the heights and bound to rocks, where they froze to death in sight of each other. Huáscar became Huascarán, at 22,205 feet the highest mountain in Peru,

and Huandi was transformed into the 20,981-foot neighboring peak of Huandoy. The tears the lovers shed for each other turned into the glaciers and streams that flow down from the heights where they died. And, indeed, viewed in the right way, the great white bulk of Huascarán has the broad and muscular appearance of a man, whereas the graceful pyramid of Huandoy possesses the more slender and feminine qualities of a woman.[25]

If people can turn into mountains, then mountains can have human descendants. A number of myths make prominent peaks the ancestors of the tribes or groups that live near them. Chimborazo, the highest mountain in Ecuador, married Tungurahua, a shapely female volcano. Jealous of his bride's affections and suspecting her of infidelity, he attacked Carihauirazo and Altar, two lesser peaks who had also courted her. Their shattered ridges and truncated appearance bear witness to the blows that the jealous husband inflicted upon them. The Puruhá who inhabited the region between Chimborazo and Tungurahua at the time of the Spanish conquest believed themselves descended from the union of the two volcanoes. The belief in ancestral links with sacred peaks has given many Andeans a particularly intimate sense of relationship with the mountains around them.[26]

Mount Ausangate

A large number of religious beliefs and practices, both pre-Columbian and contemporary, have collected around the snow-fluted precipices of Ausangate, an impressive 20,945-foot massif that dominates the Vilcanota Range southeast of Cuzco. One of the most important sacred mountains in the Andes today, Ausangate has a recorded history that goes back to the pre-Columbian period. Because of its height and location near their capital, the Incas regarded it as a major shrine, and the emperors would make offerings of gold and silver to it. When Pachacuti, the founder of the Inca empire, was campaigning to the north, two comets were said to have shot forth from Ausangate to announce the impending death of his father.[27]

The people of the Cuzco region still regard Ausangate as the seat of the supreme *apu*, greatest of all the mountain gods. In their rituals they always invoke it before the other mountains of the area. Some believe that the influence of the peak extends to the distant capital of Lima, where the *apu* caused legislators to pass a national law prohibiting the killing of vicuñas, the llamalike animals that graze wild on the slopes of high mountains under his protection. The deity himself keeps to the heights of Ausangate, aloof from the petitions of ordinary mortals. He interacts with the world below through his principal servant,

Mount Ausangate, the seat of the chief apu or mountain deity of the Cuzco region of Peru. (Sigmund M. Csicsery)

a mythical cat known as the *ccoa*, who flies through the sky, wrapped in clouds, spitting hail, urinating rain, and flashing lightning from its gleaming eyes. People describe it as having the appearance of a black puma with gold or gray stripes running down its back and a whiplike tail of brilliant colors.

The *ccoa* of Ausangate has his lair in a magical palace hidden inside the mountain. There he stores crops that he has taken by force from villagers' fields. The way to this magnificent palace leads through the doorway of four lakes found at the foot of the peak. Sometimes the *ccoa* leaves food floating on their waters—colored red, green, milky white, and clear. Anyone who tries to retrieve this food is sucked beneath the surface, never to be seen again. Only great shamans who have the patronage of the *ccoa* may enter the palace and return alive.

According to a story that the people of Kauri, a local village, hold firmly as fact, a shaman came across a woman weeping by the body of her husband, who had just been struck by lightning, the *ccoa*'s favorite weapon. He told her that he could bring the man back to life if she would bring him six of the most proficient thieves in the area and clothes to cover the nakedness

of the dead person's spirit. She complied with his instructions, and they all set out for Ausangate and entered the supernatural palace within the mountain. While the shaman kept the *ccoa* occupied with offerings, the widow and the thieves found the husband's spirit, dressed the captive in the garments they had brought, and rescued him unnoticed. On their return to the world of humans, the man's body came back to life, after being dead for two days. The story presents us with a typical example of a shamanic cure accomplished through a magic journey to retrieve a sick or dead person's spirit from the deities who have carried it away.[28]

Ausangate plays an important role in a contemporary religious festival that blends Inca practices with Catholic beliefs—the mountain pilgrimage of Qoyllur Rit'i, the "Star of Snow." Around the beginning of June, thousands of people gather in a high valley beneath glacier-laden peaks in the Ausangate range fifty miles east of Cuzco. Many of them come as dancers playing flutes and dressed in feather headdresses to represent lowland Indians from the jungle. Others have white knit masks embroidered with the faces of llamas; called *qollas*, they play the llama

*A procession of ukukus, or bear people, places a cross on an Andean glacier during the pilgrimage of Qoyllur Rit'i,
the "Star of Snow." Combining Christianity with indigenous religious traditions, the ukukus worship Jesus and mountain deities
of pre-Columbian origin. (Johan Reinhard)*

herders of the highlands. The smallest, but most important, group wears sack cloths hung with shaggy
black wool; they are the *ukukus*, the bear people,
descended from the mythical marriage of a bear and
a woman.

Shortly after midnight the *ukukus* rise to the shrill
sound of whistles and start off in a long procession,
climbing up through a silver landscape of moonlit
mountains. In the sharp cold of the icy air, their
breathing forms little gray clouds that waver like
ghosts before their faces. Only they, the *ukukus*, have
the power to overcome the spirits of the dead who
haunt the heights, condemned for their sins to carry
burdens of ice. Climbing in sinuous lines, they enter
the realm of the eternal snows, reaching the glaciers
with the first glimmer of dawn. There, at over 16,000
feet, they place a wooden cross in the ice and light
candles on the snow, and as the run rises toward the
dark horizon, they pray to Jesus and the mountain
gods.

Their ritual done, the *ukukus* cut chunks from the
glacier and carry them down, back to the villages from

which they have come. The ice, the people believe,
has medicinal properties that derive from the mountain itself—the mountain that draws moisture up from
the jungles below and sends it down in rivers to water
the land and keep humans and animals nourished and
healthy. The pilgrims regard the snow peak at the
head of the glaciers as an integral part of Ausangate,
the most important mountain deity east of Cuzco.
There, according to local belief, dwell the spirits of
the dead, condemned to remain within Ausangate's
eternal snows, awaiting salvation. As the pilgrims
climb up to the glacier, they worship the god of the
sacred mountain for the fertility and nourishment he
bestows on the living.

The Catholic church attributes the origin of the
festival to a miracle that took place in 1780. A lonely
Indian boy herding llamas found himself befriended
by a beautiful youth of mixed blood, whose presence
caused the flock to flourish and multiply in a miraculous fashion. When church officials came to investi-
gate the mysterious stranger, they were blinded by
the light radiating from his face. One man reached

out to touch him and found a wooden crucifix in his hand. The youth had vanished. Believing that he had lost his friend, the Indian boy died of grief and was buried beneath a nearby rock. The festival involves a pilgrimage to worship this rock, now painted with an image of Jesus known as the Christ of Qoyllur Rit'i. A temple built over it marks the focal point of the festivities preceding the ascent to the glaciers at the head of the valley.

Considerable evidence points to the persistence of an ancient rite overlain with a recent veneer of Christian belief. The name Qoyllur Rit'i, the "Star of Snow," suggests an astronomical connection, and the festival does indeed coincide with the emergence of the Pleiades from beneath the horizon after an absence of several months. Robert Randall, a researcher who has studied the pilgrimage in detail, notes that the Incas regarded the reappearance of this cluster of stars in the night sky as an important event marking the return of fertility to fields that had lain fallow since the last harvest. They called the period when the Pleiades were hidden below the horizon the time of "sickness." This would accord with the other part of the festival's name, Rit'i or "snow," referring to the medicinal ice brought down from the glaciers—ice whose healing water brings fertility and nourishment to the land and its people. As a supernatural mixture of human and animal, the *ukukus* who gather this ice have the ability to enter the dangerous realm of the spirits and bring back the blessings of the mountain gods, led by the supreme *apu* of Ausangate.[29]

Mount Kaata

Living in a remote region relatively uninfluenced by Christian missionaries, the Qollahuaya people of northeastern Bolivia have developed an especially intimate relationship with the mountains on which they live, viewing them as living beings with human bodies like their own. The Qollahuaya divide themselves into groups called *ayllus*, each one of which identifies itself with the particular mountain on which it lies. In 1972, nearly three months after he had come to do research on the isolated *ayllu* of Mount Kaata, a villager took an American anthropologist named Joseph Bastien aside and revealed the key to understanding his people's religious life and social organization. Standing high on the slopes of Kaata, he said:

The mountain is like us, and we're like it. The mountain has a head where alpaca hair and bunchgrass grow. The highland herders of Apacheta [the upper region] offer llama fetuses into the lakes, which are its eyes, and into a cave, which is its mouth, to feed the head. There you can see Tit Hill on the trunk of the body. Kaata [the main village,

in the middle] is the heart and guts, where potatoes and oca grow beneath the earth. The great ritualists live there. They offer blood and fat to this body. If we don't feed the mountain, it won't feed us. Corn grows on the lower slopes of Niñokorin [the lower settlement], the legs of Mount Kaata.[30]

Rather than pray to their mountain, making offerings to its spirit, the people of Kaata feed its body to keep it vigorous and healthy. The strength and well-being of the *ayllu* and its individual members depends, they feel, on the figure and health of Mount Kaata. The sacrificial blood that the ritualists pour on earth shrines embodies the life principle that animates the mountain so that it can produce food for the people who farm its slopes. The llama fat that they stuff into holes and caves gives Mount Kaata the energy it needs to remain strong and productive, ensuring the health of humans and animals. The diviners of Kaata, renowned for their healing powers, treat sickness as the consequence of mudslides, erratic water flow, and other problems afflicting the body of the mountain itself.

The view of Mount Kaata as a human being underlies and unifies almost all aspects of Kaatan religious and social life. During the year that he lived in the *ayllu*, Bastien took part in twelve rituals having to do with birth, marriage, sickness, death, and farming. Every one of them was based, in one way or another, on the underlying concept of the mountain as a human person intimately involved with the individual and collective life of its people. Diviners cured sick patients, for example, by performing rituals that symbolically healed Mount Kaata by putting its bodily parts back together in a healthy whole. People would choose husbands and wives from communities at different levels of the mountain to preserve and embody its integrity in their marriages. Agricultural rituals fed the body of Mount Kaata so that it would, in turn, nourish the potatoes and corn planted in its soil and feed the herds of llamas grazing on its grass.

The people of Kaata view the cycle of life and death as a circular journey through the mountain. The souls of the newly born emerge from the high lakes that form the eyes of Mount Kaata. Reflections shimmering on the surface of their water are the spirits of the dead coming back to life. As a person grows up and becomes old, he or she travels down the body of the mountain, to die and be buried beside the rivers at its feet. From there the dead climb up through underground channels of water to emerge once again from the eyes in the head of Mount Kaata.

The identification of the *ayllu* with the human body of Mount Kaata has bound the various communities of the mountain together into a single cohesive unit based on a deep sense of intimate relationship with the natural environment. Over the centuries the people of Kaata have drawn on the power of this conception of themselves and their place in the world to resist repeated efforts to break up the unity of their land and society. In 1592 a Spanish governor took over the lower level of Mount Kaata and made it his personal estate. Arguing that the various levels belonged together as parts of the mountain's body, the Kaatans finally convinced the authorities to return the expropriated land to them in 1799. Their struggle to maintain the integrity of Mount Kaata and its *ayllu* has persisted into the twentieth century. Even after modern political and economic developments have split up their communities, they continue to perform rituals that feed the sacred mountain and bring the people of Kaata together in a human body invisible to outsiders.[31]

Beliefs and practices directed toward sacred mountains provide important keys for understanding the past and the present. They help to explain the two most intriguing enigmas of South American archaeology—the Nazca Lines and Machu Picchu. Along the southern coast of Peru, spread over an area of eighty square miles, extends a mysterious network of animal figures, geometric shapes, and long straight lines, traced on the surface of a desert plain. Many of these figures are so large that they can only be seen in their entirety from high in the air, leading to fanciful theories that they functioned as signs for alien spacecraft. Based on his study of current Andean practices, Reinhard has proposed that the Nazca Lines, which date back to 200 B.C., were laid out to draw water down from mountain gods. Many of the straight lines point directly toward sacred peaks, and in parts of the Andes people follow similar pathways up to the tops of mountains to invoke deities for rain. The figures themselves appear to represent birds and animals that act as messengers of mountain gods, such as condors and the supernatural *ccoa* of Ausangate. According to Reinhard's hypothesis, the Nazca people would have created the representations to attract the attention of such messengers flying overhead on their way back to their masters on sacred mountains.[32]

Fifty miles northwest of Cuzco, on a narrow ridge suspended between snow-capped peaks and the Amazon jungle, lies the most spectacular archaeological site in the Andes: the abandoned city of Machu Picchu. Since its discovery in 1911, scholars have wondered

A double rainbow arches over the ruins of Machu Picchu, an Inca ceremonial site surrounded by four sacred mountains. (Mark Tuschman)

why the Incas chose to build an extensive complex of houses, temples, and terraces in such a remote and difficult location. What purpose could it have served? As Reinhard notes in a recent study, Machu Picchu sits on a ridge that descends from 20,574-foot Salcantay, one of the two most important sacred peaks in the Cuzco region—the other being Ausangate. Around the ruined city, situated at the four points of the compass, stand the rock peak of Huayna Picchu and three snow mountains revered by the local populace—Pumasillo, Veronica, and Salcantay. In the gorge below, the Urubamba, a sacred river carrying the waters of Salcantay, nearly encircles the site of Machu Picchu, completing the prerequisites needed to make it a ceremonial center of major importance. The main altar of Machu Picchu, a blade of rock called the Intihuatana, mimics the distinctive shape of Huayna Picchu, suggesting that it represents a mountain deity invoked in Inca rituals. The summit of the peak overlooking the city itself has artificial platforms similar to those found on higher mountains, such as Llullaillaco in Chile, where the Incas performed ceremonies to control the weather. All these factors provide powerful evidence that the worship of mountain gods played a major role in determining the location and purpose of Machu Picchu.[33]

In many parts of South America, pre-Columbian beliefs and practices, especially those concerning mountains, have survived by assuming guises acceptable to ecclesiastic authorities. Gods of sacred peaks have become Christian saints. The people of Peru and Bolivia who speak the Quechua language of the Incas worship Illapa, the Inca god of storm and war associated with mountain heights, as Santiago or Saint James. Practices devoted to Pachamama, goddess of earth and mother of mountains, continue in the name of the Virgin Mary. According to beliefs of Inca provenance, the souls of the dead go to dwell in idyllic villages blessed with pleasant weather inside 21,696-foot Mount Coropuna, one of the highest and most important sacred peaks in southern Peru. Today the Quechua people of the region believe that Saint Peter stands guard over the entrance to the spirit world concealed within the slopes of the pure white mountain.[34]

A similar blending of pre-Columbian and Christian beliefs underlies some of the most important shrines in Mexico and Central America. In 1531, ten years after the fall of the Aztec empire, a Catholic convert had a vision of the Virgin Mary on a hill sacred to Tonantzin, Our Lady Mother, goddess of earth and fertility. There, on the site of a temple dedicated to the Aztec deity, the church built a chapel that became the holiest shrine in Mexico—that of the Virgin of Guadalupe, just north of Mexico City. Like Our Lady of Czestochowa on the Shining Mountain of Jasna Góra

in Poland, the image of the Virgin of Guadalupe on the hill of Tepeyac functions as a national symbol, unifying the people of Mexico in a spirit of freedom and religious faith. Each year hundreds of thousands of pilgrims come to her shrine to invoke her love and help.[35] Elsewhere, in mountainous areas such as the Chiapas Highlands of southern Mexico, Christian crosses placed on the summits of sacred mountains mark the entrances to heavens ruled by Mayan gods. Beneath these crosses, the people believe, lie the subterranean realms of the earth lords and ancestral spirits responsible for the fertility of their crops and the fate of their souls.[36]

Even Catholic priests feel drawn to the heights of Latin American mountains, but for reasons that have little or nothing to do with pre-Columbian beliefs and practices. When I began climbing in Ecuador in the 1960s, one local cleric had developed a certain renown for ascending high peaks on Sunday and saying Mass on their summits. He would go up mountains like Cotopaxi, the highest active volcano in the world, with a portable altar strapped on his back. Standing in the snow, surrounded by clouds at 19,347 feet, he would put on his vestments, set out the wafers and wine, and conduct a service for anyone who happened to arrive on the summit.

Inspired in part by biblical images of Moses on Sinai and the transfiguration of Christ, a number of priests find spiritual uplift in Andean mountaineering. Once, at 19,000 feet in a windstorm on the upper slopes of Chimborazo, I encountered a group of Jesuit novitiates who needed help descending. We tied ourselves together on a rope, and I led the way down. Coming to an icy section, I yelled at them to stop and put on their crampons. I sat down to strap mine on, but they kept coming down. "There's no problem, it's easy," said one, just before he hit the ice and went shooting past me, to be caught by the rope. When I asked him why he had not stopped to put on his crampons, he replied, "We didn't bring any."

Assuming that God would take care of them, he and many of his colleagues went up into the mountains without the proper equipment or knowledge. Sometimes their quests to reach the heights of physical and spiritual experience ended in disaster. One group of priests and novitiates wandered up the glaciers of Antizana, a high peak east of Quito, without a rope. One of them stepped through the snow and fell into a hidden crevasse. He landed unhurt on a ledge, but his companions had no way of hauling him out, so they all left him there and went down to get help. When they returned with a party of rescuers, they could not find the crevasse: they had not thought to mark it, and a snowstorm had completely transformed the surface of the glacier. The priest died alone with his God, and, like the mummies left by the Incas, his body remains entombed in the mountain.

The mountains of Latin America continue to cast a mysterious spell. Popocatepetl, Iztaccihuatl, and Ausangate, these and other peaks draw tourists and climbers from all over the world. Few of these foreign visitors realize that many of the mountains they come to admire and climb are still worshipped as abodes of pre-Columbian gods. Whereas Buddhists and Hindus openly revere the snow peaks of the Himalayas, native Latin Americans keep their ancient beliefs and practices to themselves. The mysterious views of the Aztecs and the Incas remain alive, hidden on the heights of Mexico and the Andes. The persistence of pre-Columbian beliefs and practices, despite centuries of efforts to eradicate them, attests to the power of mountains to inspire a sense of the sacred under the most difficult and hostile circumstances.

OCEANIA
ISLANDS OF THE SKY

THE MOUNTAINS OF OCEANIA rise over flat expanses of sea, desert, and jungle. As the only points to break the monotony of a horizontal world, their vertical shapes, silhouetted against the sky, excite the imagination. Emerging from the deserts of central Australia, monoliths of ancient stone evoke visions of a primordial world that existed at the beginning of time. Elsewhere, in moister regions of Oceania, green crags and peaks of volcanic rock burst forth from the Pacific, streaming with vegetation, the symbol of life. Fuming in the sky, the cones and clouds of erupting volcanoes make visibly manifest the awesome power of the sacred. Far off over sea and jungle, at the limits of sight, alpine peaks with glaciers gleaming in the mist stir the spirit with glimpses of strange and unexpected beauty.

Oceania is made up of four major regions—the continent of Australia and the island groups of Melanesia, Micronesia, and Polynesia.[1] The sacred mountains of greatest importance and interest lie in Australia and Polynesia. Of the two remaining regions, only Melanesia contains land masses large enough to support peaks worthy of note. The largest island in Melanesia, New Guinea, hides deep within its jungles the highest mountain in Oceania—the mysterious 16,503-foot Mount Jaya, formerly known as the Carstenz Pyramid. This tropical snow peak lies too far from inhabited valleys to carry much sacred significance for the nearest people, the stone age Dani of Irian Jaya, who live at least five days' walk away. Other mountains of Melanesia, however, have assumed important roles in the most distinctive religious phenomenon of the region—the emergence of millenarian movements called cargo cults.

Beginning in the nineteenth century in response to the devastating impact of their power and wealth on indigenous cultures, Europeans began to notice the proliferation of native sects predicting the end of white rule and the advent of a golden age characterized by an abundance of material goods. When during World War II American forces arrived with planeloads of supplies to fight the Japanese, a widespread belief developed that airplanes would return with cargo for the people, announcing the beginning of the millennium. On the island of Tanna in Vanuatu, formerly the New Hebrides, followers of the John Frum church still climb the active volcano of Yasur to throw offerings into its crater as a means of speeding up the arrival of the blessed day. They believe that when their prophet, John Frum, comes from America with airplanes laden with goods, an army of fifty thousand soldiers will emerge from within the mountain to establish the golden age. The charismatic person who founded the sect in 1940, claiming to be John Frum himself, drew his spiritual authority from the deity of the highest peak on the island. In 1971, six thousand members of a cargo cult in Papua New Guinea filed up Mount Hurun to remove military markers from its summit, believing them to be demons placed there to keep airplanes from landing with cargo from heaven.[2]

Like a number of other contemporary practices in Melanesia, cargo cults mix modern ideas with traditional beliefs in ancestral spirits, many of them said to reside in sacred mountains. A friend of mine, an American lawyer named Jeffrey Falt, went to Papua New Guinea in a government program to train native people in the practice of modern law. During a vaca-

The Halemaumau firepit on Kilauea, abode and body of the goddess Pele. (Robert Wenkam)

tion one of Falt's students took him to visit his home in the jungle. Pointing to a peak above the village, the student said, "That is our sacred mountain."

"What makes it sacred?" Falt asked.

"There is a special cave in the mountain. When someone is accused of doing something wrong, we go to meditate inside it. There we enter a trance, and the spirits tell us whether the person is guilty or innocent."

"But what about due process and the right to a trial? What about giving the accused a chance to defend himself?"

"Oh, yes, that works too," the student replied.

AUSTRALIA

Of all the continents Australia is the lowest and flatest, as well as the driest. Its highest mountain, Kosciusko, located in the Snowy Mountains of the southeast coast, reaches an altitude of only 7310 feet. The low ranges that rim the continent and undulate through its flat interior reflect the extreme antiquity of Australia's rock, much of it laid down in the Precambrian

An Aborigine of the Pitjantjantjara contemplates the primordial landscape of central Australia. (H. R. Wenk)

period before life began. The long ages of erosion that have worn down these ranges into stubs of mountains have at the same time exposed some of the most spectacular geological formations in the world—great monoliths of ancient sandstone that loom out of the deserts of central Australia, most notably the Olgas and Ayers Rock.

The native people and culture of the continent mirror the primordial character of the land. The ancestors of the Aborigines migrated from Asia to Australia more than forty thousand years ago. Until British settlers, prisoners sent to a penal colony, arrived at the end of the eighteenth century and began to impinge on their culture, the Aborigines preserved in pure form traditions that went back to the hunting and gathering way of life followed for thousands of years by prehistoric peoples throughout the world. Wandering back and forth across the continent, following the tracks of their ancestors, they maintained networks of sacred sites that linked them physically and spiritually to the land. These places of ritual importance included, as one among many, the great dome of Ayers Rock.

Ayers Rock (Uluru)

Deep in the flat and desiccated interior of Australia stands the most impressive mountainous feature of the continent. An enormous monolith of reddish sandstone, Ayers Rock looks like a huge red pebble half-buried in the bed of a dried-up pond. Rising to 2820 feet above sea level, it has a length of $2\frac{1}{3}$ miles and a width of $1\frac{2}{3}$ miles. The sight of this extraordinary piece of stone, the largest isolated rock in the world, disorients the viewer: the eyes move back and forth, trying to ascertain the nature and proportions of what they see. Is it a boulder or is it a mountain? The extreme flatness of the desert, looking like a sheet of water shimmering in the noonday sun, accentuates the abruptness with which the walls of Ayers Rock burst out of the earth to rise in one clean motion more than 1000 feet above the surrounding plain.

Scoured clean of soil and vegetation, the impermeable surface of the monolith sheds nearly every drop of rain that falls on it, forming a ring of waterholes around its base. One of them, Mutijilda or Maggie Springs, is the most reliable waterhole for thousands of square miles. Nourished by the water that runs off the rock, a fertile fringe of trees, shrubs, and edible plants, such as fruits and tubers, supports a large and diverse population of wallabies, emus, kangaroos, bandicoots, dingos, and other kinds of wildlife. Numerous caves and overhangs pitting the foot of Ayers Rock provide shelter and protection, making the imposing monolith a gathering place for wandering

Uluru, or Ayers Rock, glows with a supernatural light, expressing the power and sanctity of the Aboriginal Dreamtime.
(Bernadette Joyner)

bands of Aborigines, who find food and water in plentiful enough supply to conduct elaborate rituals that require them to stay in one spot for an extended period of time.

The smooth curves and simple outline of the monolith evoke the timeless world in which the Aborigines used to live—a world of unchanging repetitive patterns, ceaselessly echoing events of the primordial past. According to Aboriginal traditions, in the mythic period of the beginning, known as the Dreamtime, ancestral beings with supernatural powers roamed the earth, creating, as a by-product of their lives, the features of the landscape that we see today, including Ayers Rock itself, which the Pitjantjatjara and Yankuntjatjara tribes of the region call Uluru. Living like the humans and animals who came after them, subject to the same emotions of love and hate, these ancestral beings banded together, gathered food, fought, loved, and died. In so doing, they established the sacred laws and ways of life that the Aborigines followed unchanged for thousands of years. At the end of the Dreamtime, the features that the ancestral beings gouged out of the earth in the course of their activities hardened to stone, and their bodies turned into dis-

tinctive boulders and piles of rocks that lie scattered like bones across the landscape they shaped—and continue to animate. The Aborigines who continue to maintain their traditions believe that the supernatural heroes of the Dreamtime live on in present-day people whose individual lives are the timeless dreams of their primordial ancestors.

The Pitjantjatjara and Yankuntjatjara read the features of Uluru as Jews and Christians do the pages of the Bible. They see written in its cliffs and gullies a record of Dreamtime stories that express the beliefs and practices of their ancient religion. In initiation ceremonies the men go to sacred caves and overhangs at the foot of the monolith to chant the primordial myths they find inscribed in natural shapes of living stone. Like the syllables of a Buddhist or Hindu mantra, the features of Uluru shimmer with supernatural power: by rubbing particular rocks, the Aborigines believe they can awaken the life force of the ancestral beings preserved within them and draw directly on the magical potencies of the Dreamtime. If, for example, they feel hungry, they can rub the stones that embody the primordial ancestors of certain snakes, and the serpents will come for them to catch and eat.

The original creation of Uluru itself the Aborigines attribute to the play of children. Two mythical boys shaped it out of soft earth left by a rainstorm. Then they went off to amuse themselves elsewhere, leaving behind a featureless mound of mud, which after further shaping hardened into the rock we see today. The significance of this seemingly trivial event lies not so much in what was done as in who did it and when: the fact that ancestral beings created it during the Dreamtime. By recounting the story, Aborigines evoke the reality of the sacred past and bring its power into the present, giving life and meaning to the world in which they live.[3]

Events related in other myths account for the particular features imprinted in the soft mound after the two boys fashioned its general form. A small red lizard named Kandju or Tjati was living in an open area west of Uluru. One day, as he was experimenting with a boomerang he had made, it twirled off course and imbedded itself deep in the mound. Quite upset over losing his new possession, Kandju scampered over to Uluru and began digging in its soft sand—it had not yet solidified into hardened rock. According to one version of the myth, he recovered his boomerang, but according to another, he could not find it and became so distressed that he died of worry. Long lines of scooped-out holes and deeply cut grooves on the western side of the monolith mark the place where the poor little lizard dug frantically in search of his precious stick.[4]

The two major myths accounting for most of the features of Ayers Rock both involve deadly struggles between antagonistic groups of ancestral beings, a feature characteristic of human society since prehistoric times. One tells the story of the demon dingo who attacked the hare wallaby people and drove them away from Uluru. This group of ancestral heroes had come to perform initiation ceremonies in sacred caves hollowed out of the northern side of the monolith. While the older men were engaged in circumcising the younger, a bird arrived with a message from the mulga seed people, inviting them to attend a ceremony they were holding off in the Petermann Ranges. The hare wallabies sent back a haughty reply, saying they were too busy to come. Offended by the scornful tone of this rebuff, the medicine men of the mulga seed tribe decided to create a demonic being to avenge the insult to their people. They arranged sticks on the ground in the skeletal form of a dingo with the teeth of a marsupial mole, the tail of a bandicoot, and the hair of their women to make the fur on its back. Then they chanted it to life, filling it with the spirit of hatred and evil.

When the dingo had filled out with flesh and grown to full size, they dispatched it to Uluru. The first people it encountered were the hare wallaby women and children who were staying apart from the men engaged in the initiation ceremonies. With savage fury, the dingo leapt into their midst and slaughtered them all. Piles of boulders near the north face of Uluru mark the bodies of the women and children trying to flee from the demonic being. Tufts of grass growing out of the rock are the pubic hairs of some of the older females. Alerted by the attack, many of the men managed to escape, climbing over the summit and running out into the desert. The younger initiates dragged a great pole used in the ceremonies with them, and it—along with their feet—left great furrows across the top of Uluru. The pole itself now stands as a great semidetached slab of rock propped up against the northern face of the monolith. As if photographed in a sequence of frames from a movie, the demon dingo moves through a series of angular boulders, each one showing him in a different position as he advances to leap on his victims.[5]

The other major myth deals with features on the south side of Uluru. The carpet snake people had been peacefully living there for some time when venomous snake warriors, who enjoyed making trouble, decided to attack them. They came slithering from the Olgas, a cluster of sacred domes to the west. The ensuing battle centered around Mutijilda. A carpet snake woman managed to hold off the advancing warriors—some of whom appear as desert oaks marching across the plain toward Uluru—with a digging stick and poison she spit at them until she was forced to retreat into Mutijilda gorge. The leader of the venomous snakes slashed her son with a stone knife, and he crawled up over the waterhole to bleed to death. The Aborigines regard the pool and the stream feeding it as his blood, clarified and fit to drink. In times of drought they will call to him to release some water from the source where he rests in death. Roused to fury, the carpet snake woman struck the leader of the venomous snakes such a great blow with her digging stick that she knocked off his nose, which fell to the ground and now stands on its own as a pointed boulder some seventy feet high. Her mouth, wailing in grief for her son, appears as a wide cave on the side of the gorge. The bodies of other carpet snake people slain by the poisonous serpents lie in the forms of cylindrical stones scattered over the top of Ayers Rock.[6]

Depressions gouged out of Uluru by an ancestral lizard in search of his boomerang. The curved hole near the base represents the boomerang itself. (H. R. Wenk)

The myths recorded in the natural features of the monolith continue to have significance for the Aborigines today, even though much of their culture has withered under the influence of modernization. The refusal of the hare wallaby men to attend the ceremony of the mulga seed tribe explains, for example, why the descendants of the two groups use different body decoration in their rituals. Individuals attribute their personal temperaments to events that happened in the Dreamtime, many of them on Ayers Rock. After noting that he was identified with a mythical ancestor who had speared another, one genial old man remarked to a researcher, "I'm a proper cranky bugger."[7] The cave where the hare wallaby men conducted their most important ceremonies still serves as a place of male initiation forbidden until recently to women under pain of death. Aborigines would go there to cut their arms and smear streaks of blood on its sacred walls. Paintings on the rock depict in highly symbolic form stories of things that happened during the Dreamtime, events that young men must learn and understand in order to become adults.

The Aborigines of the region regard themselves as custodians of the sacred sites on Uluru, entrusted with maintaining the ancient traditions connected with the monolith. The encroachment of the outside world has interfered with their ability to carry out these responsibilities, which give meaning and significance to their lives. After white Australians hunted them away from the vicinity of Ayers Rock in the first half of the twentieth century, the national government expropriated the monolith itself, removing it from the preserve it had established for the Aboriginal people. As one of the two best-known natural features in Australia—the other being the Great Barrier Reef—it had become a major tourist attraction, bringing the country international attention and foreign currency.

Concerned that growing numbers of tourists would desecrate sacred sites banned to all but initiates, the Aborigines tried to regain conrol of the monolith. In 1985, after years of frustration dealing with a legal system whose illogical complexities they had little interest in comprehending, they finally succeeded: the government returned ownership of Uluru to its traditional custodians, the Pitjantjatjara and Yankuntjatjara tribes of the area. They then leased it back to the state as a national park over which Aborigines maintained nominal control by having six of the ten positions on its board of management. Although tourists are coming in ever greater numbers, swarming over the rock with cameras clicking and babies crying, at least Uluru has a chance of remaining a sacred place, a natural monument expressing the primordial power and traditions of the Dreamtime.[8]

POLYNESIA

East of Australia and Melanesia, the islands of Polynesia lie scattered across the ocean, ranging from Hawai'i in the north to New Zealand in the south. Like breakers flecked with foam or purple masses of clouds banked along the horizon, the mountains that give these islands their form rise up over the blue swells of the dark Pacific. As they come into view, poised between sea and sky, they reveal the presence of bits of land that seem to drift on the waves like pieces of flotsam left by the catastrophic collision of unknown continents. The highest billow up into the enormous, isolated domes of great volcanoes, such as Mauna Loa and Mauna Kea in Hawai'i. Qthers in temperate regions to the south merge into each other to form the icy ranges of the New Zealand Alps.

The Polynesian people who migrated across vast expanses of the South Pacific used these mountains to guide them to their final destinations. Having come from Samoa, they reached the Society Islands of Polynesia around the end of the first millennium B.C. There, in central Polynesia, they split up into various branches. One sailed north to discover Hawai'i about A.D. 100, while another journeyed southwest around seven hundred years later to settle New Zealand. In myths about their legendary place of origin, the Polynesians appear to have taken with them the memory of a sacred mountain, an extinct volcano, located on the island of Raiatea, just west of Tahiti.

In ancient times Polynesians throughout the Society Islands believed that the souls of the dead journeyed to Raiatea, the religious and political center of the region, to climb the crater peak of Mount Temahani. There, on a ridge running along the rim of its summit, they came to a junction guarded by a god, who silently indicated the path destined for each to follow. Those of evil fate veered left and plunged from a heap of slag into the extinct crater of the volcano to enter the darkness of the underworld and suffer the torments of hell until purged and fit to become gods of the earth. The souls of the blessed, however, turned to the right and took a path that rose to a promontory on the heights of the mountain, from which they ascended to a fragrant paradise hidden in the sky.[9]

The ancient Tahitians called Raiatea, with its sacred peak, Havai'i, the name of the mythical homeland of the Polynesian people who had come to Tahiti from the island of Savai'i in the region of Samoa. When one branch of the Polynesians sailed to the islands of Hawai'i, they altered this name slightly and gave it to the largest Hawai'ian island, an island dominated by the impressive volcanoes of Mauna Loa, Mauna Kea, and the active crater of Kilauea. For the Maori, the Polynesian people who went south to settle New

Ngauruhoe, a volcano revered by the Maori, erupts, displaying the mana or sacred power brought to New Zealand from the earthly paradise of Hawaiki. (Bill Dove)

Zealand, Havai'i became in their language Hawaiki, the supernatural land of the gods, the idyllic place from which their legendary ancestors came. The mountains of this mythical land they saw reflected in the shining shapes of the snow-capped peaks they found deep in the southern reaches of the South Pacific.

Mountains of the Maori

The islands of New Zealand lie along a line that marks the contact between two tectonic plates—the Indian Australian and the Pacific. The pressure of these segments of the earth's crust squeezing together has produced the Taupo Volcanic Zone of the North Island and the Alpine Fault of the South Island. A series of snow-capped volcanoes, many of them venting steam, a few erupting with clouds of ash, release the intense heat accumulating beneath the surface of the North Island, where most of the present-day Maori live. The collision of tectonic plates in the South Island continues to uplift the precipitous ice ranges of the New Zealand Alps. Fed by moist clouds

rising from the Pacific, immense glaciers have carved out deep valleys and fjords, leaving behind alpine peaks that have served as a Himalayan training ground for New Zealand climbers such as Sir Edmund Hillary, who with the Sherpa Tenzing Norgay made the first ascent of Mount Everest in 1953.

When the Maori saw these islands with their immense mountains, so different from anything in their previous experience, their thoughts turned to Hawaiki. There, in the direction of the rising sun, on a sacred mountain named Hikurangi, lay the divine source of life, the earthly paradise where humans were first created and where the gods still dwelled, free from the afflictions of the mortal world. According to Maori mythology, the golden light of dawn, of eternal renewal, hung glowing about the summit of Hikurangi, a place of bliss beyond the reach of sickness and death. With this image in mind, the Maori named a number of mountains in New Zealand after the sacred mountain of their mythic homeland. The most prominent of these Mount Hikurangis stands on the east coast of the North Island. Its summit, 5753 feet above the sea, is the first place in New Zealand to catch the light

of the morning sun, replicating the role of its divine archetype in the distant paradise of Hawaiki.

The Maori identified certain other mountains as the petrified bodies and resting places of legendary ancestors who sailed across the ocean from Hawaiki. One of the sacred canoes in which they came foundered in a storm and turned into a reef off the rugged coast of the South Island. The people who struggled ashore, wet and cold, went off in search of food and firewood. When the sun came up, many of them froze into forms of rock and ice and became the spectacular peaks of the New Zealand Alps. One legend accounts for the height of 12,349-foot Aorangi, or Mount Cook, the highest mountain in New Zealand, by explaining that it was a man who was carrying a boy on his shoulders: the two formed a peak taller than the others. About 150 hills, peaks, and ridges are said to have had their origin in the crew and passengers of the wrecked vessel. Farther down the coast, waves that swept another sacred canoe ashore turned into three mountain ranges whose undulating crests advance like breakers toward the rugged interior of the island.[10]

The Maori associated the fiery energy of the volcanoes on the North Island with the *mana*, or sacred power, of legendary priests who migrated across the ocean from Hawaiki. Ngatoro-i-rangi, the greatest of these ancestral priests, had a particularly close connection with the volcanic peaks of Tongariro, one of the most important sacred mountains in New Zealand. After living for a while by the coast, he took a slave named Ngauruhoe with him and journeyed inland to climb the snow-capped mountain. Although buffeted by wind, sleet, and snow, they managed to reach the summit of its highest peak. The poor slave was freezing to death so Ngatoro called out in a thunderous voice to his sisters in Hawaiki, telling them to bring fire to warm Ngauruhoe. Responding to his call, the women came lighting fires along the way, leaving a string of volcanoes and hot springs behind them. By the time they arrived, however, the slave had perished of cold, and the peak he had climbed became his grave, named Ngauruhoe after him. Ngatoro cast the fire from Hawaiki into the mountain's crater, where it still burns, keeping the volcano hot and lively. The smoke that issues from the summit of Mount Ngauruhoe, which the Maori consider part of Mount Tongariro, bears witness to the power of Ngatoro and the truth of his story.[11]

Another myth tells of a legendary priest named Rakataura who died on the summit of Mount Te Aroha, the Mountain of Longing. He and his wife were traveling through the country, setting up shrines and laying claim to the land for their descendants. She passed away on top of a hill, and he went on to Te Aroha on the eastern boundary of the tribal territory he was establishing. Climbing to the summit, he gazed back in grief to where his wife had perished in the interior of the island and named one peak of the mountain Te Aroha-a-uta, "Longing for Inland Places." Looking the other way, toward the coast where he had left his children, he called another peak Te Aroha-a-tai, "Longing for the Sea Coast." Then, consumed with sorrow, he died, high above the sea and land.[12]

The Maori had particular reverence for mountains like Te Aroha where their ancestors had died or were buried. The people of an area associated themselves with the mountain on which lay the remains of the persons from whom they claimed descent. The *mana* of these ancestors, both legendary and historical, intensified the power of peaks already regarded as *tapu*, or sacred. They made it a practice to place the bones of high-ranking men and women in caves hidden on the mountain identified with their particular tribe. Before going into battle, warriors would often invoke the power of these ancestral caves. The Maori around Mount Edgecumbe, one of New Zealand's most prominent volcanoes, would take the bones of especially important people up to the summit and lower them into a crevice on top of the eastern peak, called Te Tatau o te rangi, the "Door of the Sky."[13]

In atmospheric phenomena taking place around the summit of the mountain associated with their ancestors, the anxious members of a tribe saw signs foretelling what would happen to them. If lightning angled away from the peak in the direction of the territory of a hostile group, that bode well for them and badly for their enemies. If, however, it stabbed straight down on the summit of their mountain, it portended defeat in battle or the death of their chief. A Maori chant laments the fearful sight of this sign of impending doom over Mount Tauwhare:

See the lightning flashing in the sky,
Splitting in two over Tauwhare!
Oh alas, it is the sign of death.[14]

The Maori used the bonds of kinship they felt with sacred mountains to strengthen the ties they established among themselves. On one occasion a chief who wished to make peace with a rival group he had been fighting for a long time announced that he would offer his daughter in marriage to the leader of the hostile tribe. One of his advisors suggested that they

Mount Cook, the highest mountain in New Zealand, seen from the air. Maoris explain the mountain's height by saying that the peak is the petrified body of a legendary ancestor carrying a boy on his shoulders. (Galen Rowell/Mountain Light)

reinforce this tie by making their sacred mountain the wife of the high hill dominating the territory of their enemies. Everyone agreed to the plan, and the marriages took place together—man to woman and peak to peak. The union that the people perceived between the two sacred mountains ensured peace between their tribes for many years thereafter.[15]

Like humans, whose behavior they reflected, mountains also squabbled and fought among themselves. One group of peaks led by Mount Tongariro quarreled over the females in their midst and split up, going their separate ways. Mount Egmont stalked off to the southwest, furrowing out the gorge of the Whanganui River with his massive body. Mount Edgecumbe, another great volcano of the North Island, headed east, toward the Bay of Plenty. He did not get very far, however, because one of his wives stopped to cook food, and when the sun rose, she and her dutiful husband froze in their places, transformed into stationary peaks of stone. This story of the dispersal of mountains in New Zealand bears striking similarities to stories of mountains quarreling and separating in North America—the myths of how Mount Rainier went across Puget Sound to get away from the other wives of Mount Olympus and how Mounts Saint Elias and Fairweather, once married and living together, broke up and moved apart. The explanation of these similarities requires no actual contact between these cultures: family quarrels are a universal phenomenon of human society, and when people identify themselves with mountains, they naturally project onto them salient features of their own behavior.[16]

The phrases that Maori still use to introduce themselves in intertribal gatherings reflect the degree to which they identified themselves and their tribes with the sacred mountains standing over their territories. A person would identify himself by stating the names of his mountain, his river or lake, and his chief, in that order, with the peak first, indicating its greater significance. On that basis his listeners could immediately decide whether he was friend or foe and what to make of what he would say.

When one European missionary who had gained standing among the Maori had to speak at one of these gatherings, an elder gave him this phrase to identify himself in the proper ceremonial manner: "Zion is the mountain; Jordan is the river; Jesus Christ is the man." From the Maori point of view, Mount Zion represented the principal sacred mountain of the Europeans and Jesus the chief priest of their religion. Commenting on the intimate way in which the Maori identified themselves with their tribal mountains, one Western observer remarked, "At times it seems doubtful whether it is the tribe that owns the mountain or river or whether the latter own the tribe."[17]

Above all the Maori regarded mountains as places of awesome power. Only great men endowed with more than mortal abilities could ascend their slopes and stand in the presence of the supernatural beings who controlled their heights. When war parties had to cross the high plateau at the base of Mount Tongariro, they deliberately avoided looking in the direction of its summit for fear their gaze would offend the spirit of the peak and cause it to send blinding snowstorms to punish them for their effrontery. When one group had the audacity to attempt to climb Mount Tarawera, the mountain shrugged them off by growing higher, so legend relates.[18]

The Maori drew on the power of their sacred mountains to resist encroachment on their tribal lands. After Captain James Cook visited New Zealand in 1769, Europeans began to colonize the islands. Although Great Britain signed a treaty in 1840 guaranteeing the territorial rights of the Maori nation, British officials initiated a campaign to get around the restrictions of the agreement by buying up pieces of land from individual Maoris. In 1863 the chiefs of the North Island gathered together to choose a leader with the authority to stop these sales. They met near the foot of Tongariro, the most revered and powerful of all their sacred mountains. There the chief who had called them together showed the Maori leaders a great flagpole set up with ropes hanging from various points

Te Heuheu, a Maori chief buried on Mount Tongariro. Illustration from THE NEW ZEALANDERS ILLUSTRATED *by George French Angas, 1847. (Courtesy of Margaret Orbell)*

along its length. The flagpole, he explained, represented Mount Tongariro and each rope the sacred mountain of each tribe. He instructed them to pull the ropes taut and then stake them to the ground. In so doing the chiefs symbolically brought together the mana of their tribal mountains and concentrated it in the man they selected to be their leader, giving him the power and the prestige he needed to keep the Maori people from selling their lands to the Europeans.[19]

Despite these efforts the chiefs could not stop the British, who resorted to military force to achieve their ends. By the beginning of the twentieth century, the Maori had lost nearly all their tribal lands. Bereft of the very ground of their identity as a people, many of them turned to messianic movements that blended traditional ideas with Christian beliefs. In 1906 a charismatic Maori prophet named Rua Kenna announced to his people that the king of England would give him money to buy back their territory so that he could expel the British from New Zealand and establish a golden age. In anticipation of the impending millennium, Rua founded a utopian settlement on the forested slopes of Maungapohatu, a sacred mountain located on the northern side of the North Island. He and his followers viewed their community as a New Jerusalem and referred to the peak itself as Mount Zion. The white government, however, was not prepared to wait for the Maori millennium. In 1916 it sent police to arrest Rua and to climb the mountain, symbolically breaking the power of the movement. Although he never fully recovered his spiritual authority, his memory lives on among the Maori, many of whom still dream of recovering their sacred lands and cultural integrity. And indeed, in recent years the government of New Zealand has begun to acknowledge their rights and to encourage a revival of their traditional values.[20]

The profound sense of identity with their land and their ancestors that sacred mountains embody for the Maori is beautifully expressed in the words of a contemporary writer and artist by the name of Harry Dansey. Musing on the prospect of dying and being buried with his forefathers, the descendants of those who came from Hawaiki in a legendary canoe called Arawa, he writes:

For we will be in the heart of our own land, in the midst of our own people, which is the only place for the dead to lie. North is Mokoia, grey-blue in the mist; east is Whakapongakau, hill of the longing heart; south is Moerangi where the sky sleeps and Te Tihi-o-tonga, peak of the south; and west is Ngongotaha with the lightning flashing its salute to death on the mountain's gaunt flank. For this is the land of Arawa and we are her people.[21]

Volcanoes of Pele

Like the ancestors of the Maori who sailed south to New Zealand, the Polynesians who went north to Hawai'i found large islands crowned with enormous mountains, far higher and more impressive than any they had known in the places from which they came. The great rounded mass of 13,796-foot Mauna Kea rises some 32,000 feet from the ocean floor, making it the highest mountain in the world—measured from its base to its summit. Near it, sharing the Big Island of Hawai'i, the craters of Mauna Loa and Kilauea boil and surge with red fountains of foaming rock and black waves of molten lava. Elsewhere, on islands such as Maui and Kaua'i, fantastic knife-edged peaks and ridges smoothly draped in gleaming foliage shoot up from the dark Pacific, conjuring up incredible images of alpine ranges encased in green veneers of velvet ice.

The beauty and power of these magical mountains made them a natural abode of deities, an earthly equivalent of Havai'i, the legendary paradise of the Polynesian gods. With thoughts of such a place in mind, the ancestral Hawai'ians who came from the region of Tahiti named the largest island, the one with the highest peaks and active volcanoes, Hawai'i. In the green heights of this and the other Hawai'ian islands they saw the haunts of deities such as Kane, one of the foremost of the ancient Hawai'ian gods. A prayer to him in his various forms opens with the words:

O Kane of the great lightning,
O Kane of the great proclaiming voice,
O Kane of the small proclaiming voice,
Silently listening in the mountains—
In the great mountains,
In the low mountains . . .[22]

Poliahu and her companions, goddesses of cold and snow, dwelled high on the snow-covered summit of Mauna Kea, wrapped in shawls of white. Below them, in the craters of Kilauea, lived an ancient god of fire named Aila'au, the Eater of Trees, who lapped up forests with tongues of burning lava.

Of all the deities specifically associated with mountains, however, the most lively and charismatic was— and continues to be—Pele, the fiery goddess of volcanoes. A late arrival to Hawai'i, having come long after the elder gods, such as Kane, she sailed across the ocean from a mythical homeland called in many sources Kahiki, the Hawai'ian name for Tahiti, although the word seems to refer more to an earthly paradise than a geographical place. A chant that records the myth of her coming tells us:

From Kahiki came the woman, Pele,
From the land of Pola-Pola [Bora Bora],
From the red cloud of Kane,
Cloud blazing in the heavens,
Fiery cloud-pile in Kahiki.[23]

According to certain versions of the myth, Pele had to leave her original home because of disputes with other members of her family. Some attribute her departure to a disconcerting habit she had of setting fire to things that belonged to her relatives. Other accounts say that she had to flee Kahiki to escape the wrath of a jealous sister who suspected Pele of having seduced her husband.

Like the legendary ancestors of the Maori, who also left the region of Tahiti on account of family disputes, Pele sailed across the Pacific in a divine canoe, accompanied by a band of devoted followers composed mostly of friendly relatives. Whereas many of the mythical immigrants to New Zealand froze into mountains, she took up residence inside a volcano and remained alive as the goddess who embodied both its form and power. Pursued by her vengeful older sister, she traveled from island to island, searching for a new home. In each place she dug a great hole with her spade or divining stick, creating the volcanic calderas that pit the Hawai'ian islands from Kaua'i in the northwest to the Big Island of Hawai'i in the southeast. The route that she took coincides with the geological order in which these islands flamed into existence. Every time she attempted to settle down in a new crater, sea water would enter to flood it and put out the fires she had started. Haleakala, an extinct volcano 10,023 feet high, presented a different kind of problem: according to one version of the myth, Pele tried it out but found it too large and drafty to keep comfortably warm. At that point her sister caught up with her and in a savage struggle tore her body to bits, creating a hill on Maui known as the Bones of Pele. But the spirit of the goddess survived this assault and soared into the sky to appear transfigured in fire above the summits of Mauna Kea and Mauna Loa on the Big Island of Hawai'i. There, beside the impressive bulk of Mauna Loa, Pele finally found a permanent home for herself and her followers in the active craters of Kilauea, some 4000 feet above the sea, beyond the reach of its fire-extinguishing waters.[24]

The appearance and temperament of the goddess match the complex and contradictory nature of the volcano she inhabits and animates. Like the graceful fountains of fire and sinuous flows of lava that issue from the craters of Kilauea, Pele can take the form of a beautiful maiden possessed of seductive charms, made even more enticing by the aura of danger that glimmers around her lovely face and tantalizing body.

Appearing in this form, often as a mysterious stranger, she causes mortal men to fall in love with her and burn in the fires of her passionate embrace. Like the unpredictable lava flows that spill out from her crater, suddenly turning to consume houses and fields in their paths, the goddess has a fickle and terrifying nature: she can change in an instant from an alluring young woman to a hideous hag enveloped in flames. A song from the hula has one of her lovers lament,

Alas, there's no stay to the smoke;
I must die mid the quenchless flame—
Deed of the hag who snores in her sleep,
Bedded on lava plate oven-hot.[25]

Although Native Hawai'ians fear her destructive power and call her the Eater of Land, they also regard Pele as a beneficent goddess whose lava flows have created the fertile soil of their island homes and whose divine heat energizes their spirits with the fire of life.

Pele's shifting, multifarious nature blazes forth in a myth that pits her in a cataclysmic struggle against Poliahu, the snow goddess of Mauna Kea. Once Poliahu and her companions came down from her mountain to engage in the ancient Hawai'ian sport of sled racing on a grassy hill. Pele appeared in the form of a beautiful stranger and joined them in their races. When she found herself losing, she became jealous and caused the earth to turn warm with the heat of an impending eruption. Recognizing the mysterious stranger as Pele, Poliahu fled toward the summit of Mauna Kea, pursued by fingers of lava that grabbed at her mantle of snow and started to melt it. The goddess snatched her robe free and cast it over the rest of her mountain. In the great battle that ensued between the goddesses of fire and ice, the island shook with earthquakes, mountains rocked, and great cliffs came crashing down. As Poliahu gradually prevailed, her mantle of snow descended upon the crater of Kilauea and froze Pele's fountains of burning lava to stone, sealing off the passageways through which the molten rock emerged from beneath the earth. Forced to exit elsewhere, it streamed out into the ocean, where it formed the rocky mass of Laupahoehoe and the great arch of Onomea. Defeated by the cool power of her rival from the higher mountain of Mauna Kea, Pele had to settle for control of the southern half of the island of Hawai'i, where she remains today in possession of the active volcanoes of Mauna Loa and Kilauea.[26]

Mortals who challenge Pele do not usually fare as well as the goddess of Mauna Kea. A typical story tells of a Hawai'ian chief who sported with Pele and beat her in a number of games. Growing overconfident and boastful, he presumed to surf on the fiery waves of lava that surged across the crater of Kilauea. When

Pele objected to his desecration of her home, he expressed his contempt for her by riding his surfboard standing on his head. Angered by his impudence, the goddess caused the wave he was poised on to tilt and break. He plunged into the molten lava and perished, punished for his boastful presumption.[27]

At least one person, however, managed to defy the power of Pele and live. In 1824 Kapiolani, a Hawai'ian chief who had converted to Christianity, announced her intention of going to Kilauea to demonstrate the superiority of her new faith over what she regarded as the false worship of Pele. When her subjects tried to discourage her with warnings of disaster, she replied, "If I am destroyed then you may all believe in Pele, but if I am not, you must all turn to the true writings." Proceeding to the rim of the crater, in full view of Pele's seething fires, she ate berries that she should have given as an offering to the goddess. Then, in a gesture of defiance, she threw stones into the boiling lava and prayed to her Christian deity. To the amazement of her subjects, nothing happened. But even so, despite the apparent success of her demonstration, they continued to worship the goddess of the sacred volcano.[28]

To express her defiance of the native gods, Kapiolani mocked one of the most common ritual practices dedicated to Pele—the offering of sacred berries to her in her place of fire in the crater of Kilauea. Priests who sought shamanic powers would come to spend a year on the volcano, making sacrifices and striving to dream a chant, which they would dedicate to the goddess. Hawai'ians also built temples to her on the rim of the crater and beside large streams of molten lava. Those who claimed descent from the embodied spirits of Pele's entourage most commonly worshipped and conducted rituals dedicated to her. When they died, their children would throw their bones into the crater so that their parents might return to live with the goddess as spirits flickering across the flaming surface of the red-hot lava.[29]

The most dramatic and beautiful of the rituals connected with Pele is the performance of the hula. A blend of poetry, music, magic, and religion, these graceful dances evoke and reenact the primordial myths of the gods, especially those associated with the goddess of Kilauea. Specially trained and initiated into their art by teachers versed in sacred knowledge, performers of the hula sing songs traditionally taught long ago to worshippers of Pele by the divinities of her entourage. In performing the most sacred dances, they seek to become one with the deity of the particular dance, usually Laka of the upland forest or Hi'iaka, the sister of Pele. A misstep indicates that the goddess has abandoned the dancer, leaving him or her to stumble in the ordinary world of clumsy mortals.

Many of the songs and dances of the hula reenact or allude to episodes drawn from the well-known myth of Pele and Hi'iaka. Once Pele fell into a sleep and journeyed in spirit to Kaua'i, where she assumed the form of a beautiful woman and fell in love with a young chief named Lohaiu. After becoming his wife, she returned to her sleeping body, promising to send a messenger to bring him back to her home in Kilauea. Hi'iaka agreed to perform this task, but asked Pele to protect her groves of lehua blossoms and her friend, Hopoe, a lovely maiden who had taught her how to dance the hula. Having promised to do so, Pele instructed her to come back in forty days and warned her not to dally with her husband. After a long journey fraught with hardship and danger, Hi'iaka reached Kaua'i to find that Lohaiu had died of grief from being separated from his lover. Using her magic powers as a goddess, she revived him, and the two returned to the island of Hawai'i. However, more than forty days had passed, and enraged at the thought that she had been betrayed, Pele poured out streams of lava to destroy Hopoe and burn the lehua groves of Hi'iaka. Hopoe died dancing the hula, surrounded by flames, and became a rock formation in the shape of a dancer. When Hi'iaka discovered what had happened, she went to the rim of Kilauea and embraced Lohaiu in full view of her sister. Pele flung fire around the two and Lohaiu perished, but Hi'iaka brought him back to life to be reunited with her in the end as her lover.[30]

A hula dedicated to Pele in particular depicts events from this myth and the story of how the goddess came to Hawai'i. Given the importance of the two stories in Hawai'ian religion and mythology, it was performed with great solemnity before the highest chiefs of the islands. Unlike most other hulas, the dancers in this one danced without instrumental accompaniment, evoking the power of the goddess with dignified gestures. The words of a song from her hula reveal the awe in which Hawai'ians held and still hold her, even in her most terrifying manifestations:

Knock-kneed eater of land,
O Pele, god Pele!
O Pele, god Pele!
Burst forth now! Burst forth!
Launch a bolt from the sky!
Let thy lightnings fly![31]

Many of the traditional schools that train students in performing the hula today will go up Kilauea to dance on the rim of the crater and make offerings to Pele, often to obtain her blessings before a contest or major performance.

Of all the traditional deities, Pele is the one who has most successfully survived the conversion of Hawai'i to Christianity. Today many native and non-native Hawai'ians, including Christians, Buddhists, and followers of other religions, believe in the goddess—or else take no chances by doing anything that might offend her if she should turn out to exist. Pele was originally of concern mostly to the people who lived near her fiery abode in the active volcanoes of Kilauea and Mauna Loa, but with the demise of many of the other gods, devotion to her as a major deity spread to other parts of Hawai'i. A belief that she had a special interest in protecting and helping Native Hawai'ians also helped to gain her a larger following.

The popular press reports numerous accounts of people who claim to have encountered her, either in the form of a beautiful young woman or a wizened old lady, often accompanied by a small white dog. A typical story tells of someone who picks her up as a hitchhiker after other vehicles have passed her by. As the person proceeds on his way, he comes on the cars that went before him mysteriously stalled or broken down beside the road. The lady smiles in acknowledgment, sometimes remarking that they have got what they deserved. When the driver turns to speak with her, he finds that she has vanished—a sure sign that his mysterious passenger was Pele. Other stories of a similar nature relate how people who refuse to give food to a woman who comes to their door find their houses destroyed by a lava stream that flows around the home of a neighbor who generously fed her.

Trifling with the goddess can have serious consequences. In 1935, over the objections of local believers in Pele, army aircraft bombed a flow of molten rock

Leatrice Ballesteros, the Lady in Red, makes an offering to Pele at the crater of Kilauea. (J. D. Griggs, courtesy of Hawai'i Volcanoes National Park)

that was advancing on Hilo in an effort to keep it from reaching the city. They succeeded in diverting the lava, but a few weeks later six of the men who had participated in the bombing died in a midair airplane collision. And on their way back to relatives on the mainland, the ashes of one of them mysteriously vanished. The followers of Pele blamed both events on the wrath of the goddess. Even relatively minor infractions can lead to a spate of bad luck attributed to Pele. Many tourists who have innocently picked up pieces of lava on Kilauea as souvenirs have found themselves the victims of uncanny accidents. The National Park Service at the volcano regularly receives packages containing such rocks with instructions to please return them to Pele.

These kinds of accounts, combined with the spectacular nature of her volcanic eruptions and the wealth of traditional beliefs and practices surrounding her, have made Pele a very popular deity. In the 1920s a visitor from San Francisco was so impressed by her activity in Kilauea that he founded a club dedicated to the goddess, called "Hui o Pele Hawaii," which now has tens, perhaps hundreds, of thousands of members. After an earthquake caused by an eruption of Kilauea, a local Lions Club voted to make Pele a member of their chapter, the only woman to have received that honor—dubious as it may be. The tourist industry has also cultivated lore about the goddess as a means of promoting tourism in Hawai'i. The manager of the Volcano House on Kilauea made it a practice to toss bottles of gin into the crater as an offering intended to encourage eruptions for the benefit of spectators.[32]

Practitioners of Native Hawai'ian religion have found much of this commercial interest in Pele crass and sacrilegious. But they have objected most vehemently to plans for geothermal development on the slopes of Kilauea. With the support of powerful interests in the state, power companies have proposed drilling holes in the volcano to inject water and generate electricity from the release of steam. Followers of Pele see this as an extremely offensive and dangerous desecration of the goddess' very form and body. They view the magma, steam, lava, vapor, and heat of Kilauea and Mauna Loa as living manifestations of her power and presence. To use machinery to drill holes in the slopes of the volcano would be equivalent to jabbing spears into the sides of her body, causing irreparable injury to her and her home, as well as the people who depend on her goodwill and beneficence.

Native Hawai'ians foresee terrible consequences from even exploratory drilling. They fear that the unnatural release of steam will drain the goddess of her creative energy so that she will no longer be able to help them in times of need. In fact, the injury inflicted on her body might make her turn on them and destroy their homes and land, something she has often done in the past. In any case, the drilling will make it difficult for them to conduct ritual practices on the sacred mountain and maintain the continuity of their religious and cultural traditions. They also point out that geothermal development would have adverse effects on the environment and nearby rural communities. In addition to releasing poisonous gases and liquids, the project would require the construction of roads and facilities that would threaten virgin forests and endangered species of birds and animals—which devotees of Pele regard as embodiments of deities belonging to her family. Their views of Kilauea as a sacred place focus attention on the interdependent nature of the natural world and the need to respect it for ecological and spiritual reasons.

In a series of hearings beginning in 1983, native Hawai'ians dedicated to preserving the sanctity of Kilauea joined environmental groups such as the Sierra Club and the Audubon Society to contest proposals to do exploratory drilling for geothermal development. Worshippers of Pele testified that the proposed exploitation of the volcano would put an undue burden on their freedom to practice religion, a freedom guaranteed by the First Amendment. A statement by the Pele Defense Fund pointed out the central role of the goddess for practitioners of native Hawai'ian religion: "The Goddess Pele is the heart and life of Hawai'ian religious beliefs and practices today and is indispensable to particular Hawai'ian traditional cultural beliefs and practices." Kilauea itself added weight to their testimony by erupting and covering the area under dispute with lava, highlighting the dangers to anyone trying to drill holes in an active volcano.[33]

Despite additional evidence that the island would have little use for the electricity generated by the project, the Hawai'i Board of Land and Natural Resources in 1986 designated a part of Kilauea a subzone open to geothermal exploration. Native Hawai'ians appealed the decision to the Hawai'i State Supreme Court. In 1987 the judges rejected their arguments, ruling that the geothermal exploration and development would not impose an undue burden on their constitutionally guaranteed freedom to practice religion. In the spring of 1988 the Supreme Court of United States declined to hear an appeal of State Court's decision. Refusing to concede defeat, the Pele Defense Fund initiated a major publicity campaign with announcements in newspapers, both in Hawai'i and on the mainland, urging the public to bring pressure to stop the desecration of Kilauea on religious, cultural, economic, and environmental grounds. Despite these efforts, in 1989 the state allowed the construction of drilling platforms to begin, violating the natural sanctity of the volcano. As of 1990, Native Hawai'ians and local residents have resorted to legal maneuvers and civil disobedience in a desperate attempt to discourage potential investors and limit the size of the project so that it can be dismantled before irreparable damage is done.

Throughout Oceania we see cultures clashing over the heights of sacred mountains. In Australia the Aborigines struggle to preserve the integrity of their Dreamtime sites on Uluru from the degrading effects of tourism. In New Zealand the Maori turn to ancestral mountains as symbols of their identity with the land taken from them by English settlers. In Hawai'i native groups fight to protect the volcanic body of their goddess from the ravages of geothermal drilling and commercial exploitation, powerful expressions of the dominant values of modern civilization.

Although they arise from pain and despair, these conflicts over the status of sacred mountains have beneficial effects. Like the peaks themselves, visible from far away, such conflicts raise into view the beliefs and practices of traditional cultures previously submerged beneath waves of colonialism and modernization. Because their natural beauty and spectacular appearance attract the attention of the outside world, the mountains of Oceania provide opportunities for native peoples to make their concerns seen and heard. Visitors who come to see sacred places such as Uluru and Kilauea come, whether they intend it or not, into contact with other views of reality. These views force modern technological civilization to recognize the existence of indigenous cultures and to begin to make room for their traditional ways of life.

The hand of Pele—a lava flow on Kilauea expresses the fiery power of the volcano goddess of Hawai'i. (©Robert Holmes)

PART II

THE POWER
AND
MYSTERY OF
MOUNTAINS

THE SYMBOLISM
OF SACRED MOUNTAINS

AN EXPLORATION OF CULTURES around the world reveals a bewildering variety of views of sacred mountains. People of different traditions revere hills and peaks as heavens, hells, gods, demons, wombs, tombs, houses, temples, animals, birds, trees, flowers — the list goes on and on. As a well-known scholar of comparative religion, Mircea Eliade, has remarked, "The symbolic and religious significance of mountains is endless."[1]

What do we make of this kaleidoscope of shifting and often contradictory views? Do they exist as a jumble of unrelated visions, each a powerful symbol in its own right? Or do they harmonize in an underlying unity that brings them together and resolves their differences? A number of observers have attempted to reduce this welter of disparate views of sacred mountains to variations on a single, universal archetype — most notably the idea of the mountain as a cosmic axis that stands at the center of the universe, linking together the various levels of existence, from the depths of hell to the heights of heaven. This approach has a powerful appeal, responding as it does to our desire to find unity in the midst of diversity. Many religious traditions, particularly those influenced by Indian conceptions of the cosmos arranged around the axis of Mount Meru, do tend to view sacred peaks as centers of the universe — for example, Hara in Iran, Mount Kailas in Tibet, or Gunung Agung in Bali. Certain mountains in non-Asian cultures — such as Harney Peak among the Sioux in North America — also function as centers of the world, acting as pivots of cosmic order and supernatural power.

The attempt to find a central axis in every sacred mountain, however, runs into a host of exceptions that require ingenious, and even tortured, efforts to force them into a single pattern — efforts that often ignore the ways in which the people who venerate mountains actually view them. What do we do with an important peak such as Mount Sinai, which lies off in the wilderness, far from Jerusalem, the center of the world in the Jewish and Christian traditions? Or with the four sacred mountains that define the outer limits of the Navajo land? In China, the Middle Kingdom where we might expect a central location to be emphasized, the most important of the five imperial peaks is not Sung Shan in the center, but T'ai Shan in the east. Other mountains have nothing to do with cosmic centers: the Vikings of Iceland viewed Helgafell, the "Holy Mountain," as a mead hall of the happy dead. Sometimes a mountain is sacred precisely because it lies on the edge of things, in a realm of inscrutable mystery far from the center of anything.

The attempt to reduce all views of sacred peaks to one underlying theme or archetype, no matter how comprehensive it may seem, actually limits the power of mountains as symbols. The reality revealed by a mountain is what is universal, not any particular view of it. No single image has the range of expression needed to evoke the fullness of that reality and every facet of its significance. Just as one of the blind men in a well-known Eastern parable mistakes the trunk of an elephant for the entire animal, if we try to identify every mountain with a cosmic axis, for example, we run the danger of mistaking what is symbolized for one of its symbols. The very diversity of views that

A Greek Orthodox monastery perches on a pinnacle of rock at Meteora. The snowy heights of the Pindos Range appear in the background. (©Gary Braasch)

seems to confound our efforts to find a common theme points to the existence of a deeper, more meaningful reality and offers us the opportunity to expand and enrich our experience of mountains and everything they symbolize.

MOUNTAIN THEMES

Although no single, universal theme underlies them all, the many disparate views of mountains form more than a random collection of unrelated images. They come together in patterns that help to highlight and clarify the principal roles that sacred peaks play in different cultures. Many of them express particular facets of a few basic themes that we find distributed around the world—the mountain as center, heaven, source of water, place of the dead, and so forth. Just as we get an overall impression of a peak by walking around it and viewing its faces from different angles, we can deepen our understanding of sacred mountains in general by identifying the most important of these views and examining the ways in which they relate to each other.

Even though it lacks complete universality, the view of the mountain as center is one of the most widespread and powerful of all the views we have encountered.[2] It appears in its most comprehensive form as a central axis linking together the three levels of the cosmos—heaven, earth, and hell. Just as the center defines and orients the circle, this axis gives stability and order to the universe around it. As the link between heaven, earth, and hell, it acts as a conduit of power, the place where sacred energies, both divine and demonic, spew into the world of human existence. The most elaborate and influential view of the sacred mountain as cosmic axis appears in Buddhist visions of Mount Sumeru rising eighty thousand miles from the depths of hell to the heights of heaven, surrounded by island continents floating in a vast ocean enclosed within a ring of fire. This vision emerges from visualizations of the universe as a mandala or sacred circle in which everything exists in perfect relationship to everything else.

Less grandiose versions of this theme conceive of the sacred mountain as the center of the world. At the end of his great vision, the Sioux medicine man Black Elk was taken to the top of a rock peak to find everything he had been shown spread around him— people, animals, river, plains, and mountains. He later identified this mountain as Harney Peak, the center of the earth for his people. More than any other place, even the ocean, the view from the summit of an isolated peak offers a concrete sensation of feeling ourselves placed in the very middle of the world,

enclosed within the vast circle of the horizon around us. Some cultures incorporate this kind of experience into a bodily metaphor, referring to a central mountain like Gunung Agung in Bali as the "Navel of the World." The ancient Greeks situated the most famous version of such a navel, or *omphalos,* on the sacred slopes of Mount Parnassus.

Many mountains act as centers of local regions, territories, and countries. The god of Khumbu resides on the summit of Khumbila, a relatively small peak that stands in the middle of the Sherpa homeland in the Himalayas just south of Mount Everest. There, placed where he can survey all of Khumbu, he watches over the Sherpas and protects them from evil. As exalted places of power and stability, peaks regarded as sacred centers make ideal seats for rulers, both human and divine. Many cultures, particularly in Southeast Asia, view the capital of a king or emperor as a mountain situated in the center of the realm. Seated on the summit of such a mountain, the ruler can draw power and authority from the deity who resides there—and whom he often incarnates in human form.

When we look at a mountain, the first thing to impress us is usually not its central location, but its height, which provokes an immediate response of wonder and awe. Poised above the surrounding landscape, set in a fluid realm of drifting clouds and flowing sky, its summit appears to float in another world, higher and more perfect than the one in which we dwell. The names of numerous mountains reflect the impact that the grandeur of their height has made on the people who revere them. The Koyukon Indians who live near Mount McKinley, the highest peak in North America, call it Denali, the "High One." People in many parts of the world venerate mountains as high places where the earth reaches up to touch the sky.

The height that distinguishes mountain peaks from their surroundings tends to single them out as places of divine significance, but not necessarily as centers of the world. Many important peaks whose summits reach into the heavens have little or nothing to do with any conception of a center. Mount Olympus, the heavenly abode of Zeus, god of sky and thunder, rises in northern Greece, far from Athens and Parnassus, the centers of the Greek world. Poets like Homer and Hesiod stress its rugged height and size, but never mention its role as a center or axis of the universe. The Atlas Mountains—viewed as a titan who supports the heavens on his shoulders, a paradigm for the meeting place of heaven and earth—actually stand, according to Hesiod, "at earth's uttermost places."[3]

Many high mountains derive their sanctity not from their role as sacred centers, but from the heavenly

A chorten, or Buddhist shrine, in Tibet symbolizes Mount Sumeru, the mythical mountain at the center of the universe, and the stages of the spiritual path leading to enlightenment. (Edwin Bernbaum)

associations engendered by their height. Reaching up above the clouds, cut off from the world below, their summits become imbued with the celestial attributes of heaven on high. The *Odyssey* describes the top of Mount Olympus as a divine realm of light and bliss, shining with the purity and perfection of the sky itself. Gazing up at the mountain heights cool and serene above the heat and dust of the desert, the ancient Chinese called the range of peaks running along the northern branch of the Silk Route the T'ien Shan, the "Mountains of Heaven." The Kikuyu of Africa worship Mount Kenya as Kere-Nyaga, the "Mountain of Brightness," referring to the brightness of the supreme deity who dwells in the sky.

Some mountains assume the nature of paradises situated on earth rather than in heaven. The monks of Mount Athos regard their sacred mountain as an earthly paradise where they can cultivate their gardens and perfect their souls in an atmosphere of peace and harmony. The Taoists of ancient China viewed the K'un-lun Mountains as the site of a garden with the peaches of immortality. John Muir extolled the Sierra Nevada as an earthly paradise where people weary of

city life might go to refresh body and soul.

As heavens on high, often situated at the center of the world, mountains serve as abodes of gods and goddesses. Shiva, the Hindu deity of destruction and archetypal practitioner of yoga, sits in a state of meditative bliss in a paradise on the summit of Mount Kailas. Zeus, king of the gods, holds court in his heaven perched on the peak of Mount Olympus. Nanda Devi, the goddess of bliss, dwells in a golden pagoda on top of the mountain that bears her name. Sometimes the deity blends with his or her abode, making it difficult to distinguish the two. In North America the inner forms dwelling in the Navajo mountains of the four directions give them life so that the Navajo treat the peaks as living embodiments of the deities themselves.

Mountains often appear in the form of temples housing the deities who reside on or within them. Tibetan texts describe Mount Kailas as the pagoda palace of Demchog, visualized in the center of a mandala formed by the peaks and valleys that encircle the mountain. The Hopis view the San Francisco Peaks as an enormous *kiva* holding their *katsinas* and ancestral spirits. Hindus believe that yogis with supernatural sight can discern a golden temple where the goddess of bliss resides on the shining summit of Nanda Devi.

As centers and high places open to the sky, mountains provide ideal altars for making offerings to gods and spirits. The Chinese emperors chose T'ai Shan as the exalted place to perform sacrifices thanking heaven and earth for the success of their dynasties. The Eka Dasa Rudra sacrifice, one of the greatest and most impressive of all religious ceremonies, occurs once a century on the slopes of Gunung Agung, the sacred mountain of Bali. To test his faith, God commanded Abraham to offer up his son on the summit of Mount Moriah, before having him slay a ram instead. Human sacrifices actually took place on hills and peaks revered by the Aztecs and Incas. By making offerings on the altar of a mountain, close to the deities of heaven and earth, people seek to improve the chances that their gifts will be accepted and their wishes granted.

People also venerate mountains as shrines commemorating the activities of saints and deities. Pilgrims climb Adam's Peak to worship at the footprint left by the Buddha, Shiva, Adam, or Saint Thomas—the particular choice of holy figure depending on the religion to which the worshipper belongs. Catholics revere Croagh Patrick as the place where Saint Patrick began his mission in Ireland by performing the miracle of banishing snakes. Every feature of Ayers Rock in

A Uighur man in the Central Asian oasis of Turfan. The life of his people depends on the waters that flow from the T'ien Shan, the "Mountains of Heaven." (Edwin Bernbaum)

Australia recalls some event performed in the primordial Dreamtime by the human and animal ancestors of the Aborigines.

As powerful and impressive manifestations of the sacred, mountains can be demonic as well as divine. The fear inspired by their eerie atmosphere and terrifying heights has transformed many peaks into hells inhabited by demonic beings. For centuries Europeans regarded the Alps as the sinister haunt of witches, dragons, ogres, ghosts, and other evil spirits. An active volcano like Mount Hekla in Iceland acted as the fiery entrance to the underworld, where the souls of the damned writhed and screamed in endless torment. As cosmic axes, mountains like Sumeru in Buddhist mythology often rise from bottoms in hell to tops in heaven, linking together in one unified image the two extremes of humanity's experience of the sacred.

Whether revered as heavens or feared as hells, mountains have an especially widespread and important role as hallowed places of the dead. The Hopi believe that the spirits of many of their ancestors have gone to dwell inside the San Francisco Peaks. According to the Chinese, the souls of all the dead are gathered at the foot of T'ai Shan, where the god of the sacred mountain passes judgment on them. The Kirghiz of western China looked up to the great ice dome of Muztagh Ata as the tomb of Ali and other important saints of Islam. The Vikings of Scandinavia regarded certain prominent hills as the brightly lit halls of the glorious dead.

The resemblance of mountains to tombs, which often mimic the shape of hills, make them natural places of burial. The largest and most impressive graveyard in Japan lies on top of Mount Koya, where Kobo Daishi, the founder of Shingon Buddhism, is interred, waiting, it is said, to emerge from a meditative trance of suspended animation. The Japanese called the tombs of their ancient emperors "mountains," and even today villagers will refer to a coffin as a "mountain box." Many Jews ask to be buried in Jerusalem, on the summit of Mount Zion, there to await resurrection at the coming of the Messiah. After the bodies of their dead have decomposed beneath the earth, the Chagga of Kilimanjaro dig up the bones and reinter them on the slopes of the sacred mountain with the skulls facing the summit.

In seeing mountains as abodes of the dead, people often regard them as the places from which their ancestors came—or as forms of those ancestors themselves. The Lepchas of Sikkim trace their mythical origins back to a primordial couple born from the glaciers of Kangchenjunga, the great Himalayan peak behind whose veil of ice they go when they die. The Puruhá in the Ecuadorean Andes claimed descent from a sexual union between the masculine mountain of Chimborazo and the feminine peak of Tungurahua. The Maori of New Zealand view many of their mountains, including Mount Cook, as the petrified bodies of ancestral heroes who came to the islands on migrations from the legendary land of Hawaiki.

As points of high ground, often associated with sacred centers, mountains stand out as places of creation and re-creation in numerous flood myths. In Greece the summit of Mount Parnassus, situated just above the *omphalos,* or center of the world, was the first place to emerge from the waters that covered the earth. There Deucalion disembarked from a boat to create a new race of Greeks from stones that he and his wife tossed over their shoulders at Delphi. A similar but better-known myth tells of the repopulation of the world from Noah's ark after it settled on top of Mount Ararat. The people near Nanda Devi in India have identified their sacred mountain as the place where the sage Manu descended after the flood to renew the human race.

People throughout the world look up to mountains as sources of innumerable blessings, often attributed to the ancestral spirits dwelling within them. For many cultures the most important of these blessings is water. The people of the Andes pray to the gods of the high peaks to shower their fields with life-giving rain. Both the Japanese and the Hopi rely on ancestral spirits who reside in mountains to provide them with the water they need to grow their crops. The people of India revere the Himalayas as the divine source of their sacred rivers, the holy Ganges in particular.

Other blessings that flow from sacred mountains include health and peace of mind, as well as treasures of various kinds. Navajo healing rites make use of medicine bundles made with soil gathered from the peaks of the four directions. The singers who perform these ceremonies call on the deities of the mountains to restore the health and serenity of their patients. From North America to Asia people regard sacred peaks as prime places to find medicinal herbs. They also look up to them as sources of wealth and prosperity. The name Kangchenjunga means the "Five Treasuries of Great Snow," referring to material and spiritual treasures said to be stored in its five summits. Kubera, the Hindu god of wealth, dwells on Mount Kailas, the most sacred peak in the Himalayan region.

People look to mountains for blessings because they see them as places of power. This power may come from the mountain itself or from a deity who dwells or descends there. Kakugyo drew directly on the stability of Mount Fuji when he stood immobile in a cave inside the peak in an effort to bring peace and order to a nation rocked by civil wars. Black Elk climbed to the summit of Harney Peak to call on the Grandfathers to hear his plea and help his people. Hawaiians regard the eruptions of Kilauea as fiery manifestations of the wrath of the goddess Pele, who dwells inside the volcano, a source of both blessing and calamity.

As peaks of power, close to heaven, mountains serve as dramatic sites of revelation, vision, and inspiration. In one of the most famous and powerful examples of this type, God descends in a cloud of fire and smoke to reveal the Torah to Moses on Mount Sinai. In a very real sense Islam begins on Mount Hira when Muhammad hears the first words of the Koran and sees the Archangel Gabriel. The Plains Indians of North America climb hills and peaks in quests for visions of power to protect and guide them through their lives. Romantic poets like Wordsworth and Shelley experienced mountains as symbols of the infinite from which they drew poetic inspiration.

The revelation or vision on a mountain often transforms the person who receives it. When Moses comes down from Mount Sinai, his face shines with a divine light. Jesus is transfigured on the mountain where God reveals him to be his chosen son. A Sioux medicine man returns from a successful vision quest with the power to heal and accomplish incredible feats. Japanese practitioners of Shugendo climb sacred peaks to purify themselves and acquire supernatural powers. They visualize themselves dying and being reborn from the womb of the mountain they have ascended. Hermits seek out mountains as places to transform themselves through practices of austerity and meditation. In the process they attempt to transcend the material limitations of their bodies and attain a more spiritual form of existence.

For lay people who do not aspire to the supreme heights of spiritual transcendence, mountains function as popular places of pilgrimage. Millions of Chinese have crept up the winding steps of T'ai Shan, seeking the more worldly attainments of children, prosperity, health, and long life. Buddhist, Hindu, and Muslim pilgrims climb Adam's Peak in Sri Lanka to absorb blessings from the sacred footprint on its summit. Tibetans and Indians make the arduous pilgrimage to Mount Kailas to glimpse the abode of the gods and acquire merit for a good rebirth. Irish Catholics walk barefoot up the painful path of Croagh Patrick as an act of penitence for their sins.

Poets and mystics have visualized the ascent of the sacred mountain as a symbol of the ultimate pilgrimage, leading to the heights of heaven and the final goal of spiritual realization. At the end of the *Mahabharata,* the great epic of Indian literature, the hero, Yudhishthira, decides to leave the world and crosses the Himalayas to ascend Mount Meru. On the slopes of the cosmic mountain, he meets the king of the gods, who puts him through a series of tests before taking him to heaven and the final attainment of a

A line of pilgrims follows a ridge that rises like a pathway to heaven on the pilgrimage to Nanda Devi, a sacred mountain in the Indian Himalayas. (Swami Sundaranand)

celestial body. In climbing a mountain like Omine, Japanese practitioners of Shugendo imagine themselves passing through the stages leading to the ultimate Buddhist goal of enlightenment.

Most other views of sacred mountains relate to the major ones outlined above. Peaks such as the Caucasus serve as places of punishment and martyrdom, commemorating the self-sacrificing activities of heroes and saints like Prometheus. Many cultures venerate mountains for the blessings they provide as sanctuaries and protectors of religion. Such mountains may, like Olympus and Zion, assume the appearance of fortresses and cities. Other peaks, such as those of the Navajo, function as markers defining the boundaries of the sacred land in which a people live in safety. In their role as protectors of religion, mountains also appear as places from which saviors will come to defeat the forces of evil and establish a golden age throughout the world. Sacred peaks often take the form of bodies and parts of bodies, both human and animal. Viewed in this way, they can be either male or female, depending on the body part they resemble, such as a phallus or a breast—for example, Shiv-

ling, a peak venerated by Hindus as the "Phallus of Shiva." People visualize cosmic peaks like Mount Meru in the form of flowers and trees growing at the center of the universe. On a more abstract level, important mountains like Mount Zion can symbolize virtues and ideals such as wisdom and righteousness. Often a single peak will reflect a number of different views, making it shimmer with an ever deeper and brighter aura of sacred power.[4]

MOUNTAIN METAPHORS

When asked what makes a mountain sacred, people will usually respond with one or more of the views discussed above. Most Hindus, for example, will say they revere Mount Kailas because Lord Shiva dwells there and the River Ganges flows from its summit. Buddhists from Tibet will answer that Kailas stands at the center of a mandala and forms the pagoda palace of Demchog. Each of these responses attributes the sanctity of Mount Kailas to a particular view of the peak—as an abode of a deity, a source of a river, a cosmic axis, or a temple. Since mountains can be viewed as almost anything, people offer countless explanations as to what makes them sacred.

Despite their apparent diversity, all these explanations have something in common. They tell us that people experience the sacred nature of mountains in terms of the views they have of them. Whatever the particular view, it serves to imbue the mountain with a special quality that sets it apart. This suggests that the common factor underlying different views of mountains lies not in the views themselves but in the way in which each one awakens a sense of the sacred. To find a unifying principle, we need to look not so much at what is seen as how it is seen—and precisely what effect that has.

Every view considered here brings together two or more images—the mountain and what the mountain is viewed as. When a Tibetan pilgrim gazes on Mount Kailas, the goal of his pilgrimage, he sees both a gleaming peak of snow and the shining temple of a deity. In fact, the two images fuse in his mind so that they become indistinguishable: as far as he is concerned, the mountain *is* the pagoda palace of Demchog. The juxtaposition or fusion of these two images awakens the awareness of something that transcends them both and suffuses Mount Kailas with an aura of sanctity, making it much more than a beautiful and impressive peak. The Tibetan pilgrim feels the power and presence of the deity radiating from the mountain.

Something comparable happens in literary metaphors. Unlike a simile, which merely compares two similar but different things, a metaphor, like a view of a sacred peak, brings them together so that one is seen as or in terms of the other. If we say that a mountain is like a temple, we merely point out an interesting resemblance between the two. But if we say that a mountain *is* a temple, we make a much stronger statement, one that alters our notions of mountains and temples so that we can make sense of what it means to identify one with the other. Recent theories in the fields of philosophy, linguistics, and literature have pointed out that the tension or interaction between such terms, or images, in a metaphor leads to new meanings and observations, even new ways of seeing the world. These theories offer a point of departure for examining the way in which views of mountains awaken a sense of the sacred.[5]

According to scholars such as I. A. Richards and Max Black, the interaction between the terms of a metaphor changes our perceptions of each term, infusing one with the attributes of the other—and vice versa.[6] A similar thing happens in views of sacred mountains, but with a much deeper and more powerful effect. In the Tibetan view of Mount Kailas as the pagoda palace of Demchog, the image of a palace,

or temple, draws out the architectural features of the peak, focusing attention on the symmetrical form of its conical summit and the pillarlike shape of the cliffs that support it. At the same time Kailas invests the visualized dwelling place of the deity with the size and solidity of the mountain itself and highlights the peaklike character of the imagined pagoda roofs. All this has the dual effect of making Mount Kailas more spiritual and the temple of Demchog more physical. Heaven and earth come together in the view of the mountain as the abode of a deity. The Tibetan pilgrim realizes that what he or she seeks in another world lies right here in this one.

But the juxtaposition of images does even more—it can awaken a sense of the sacred itself. At a deeper level the tension created by viewing similar but different things as one and the same opens them up to reveal a vision of something that transcends their differences. A person who regards a mountain as a temple becomes aware of another, deeper dimension of reality hidden beneath the superficial forms of mountains and temples. The action of a stereoscopic viewer provides an illuminating analogy. Such a device brings together two similar but different photographs of a scene so that they fuse to reveal a vision of the third dimension inherent, but not visible, in either picture by itself. The sense of the sacred awakened in a view of a mountain corresponds to the perception of the third dimension disclosed in a stereoscopic viewer. In fact, the actual experience of this sense of the sacred is often characterized by an awareness of luminous depth, as if the peak and the image juxtaposed with it had turned transparent to reveal a limitless vista of incredible significance.[7]

The juxtaposition of images in a view of a mountain also works like the fusion of notes in a chord of music. Hearing the different tones resonate together creates a harmony, a sound with a special quality that no single note can produce by itself. Just as it would be erroneous to claim that one note stands for another, so it would be a mistake to say that a mountain simply represents or symbolizes a temple or center of the universe. Rather, for a religious practitioner the two images work together to awaken an awareness of a sacred reality that they each embody but usually reveal only when they resonate with each other. What the mountain and the temple really symbolize is their own real nature—which people of traditional cultures regard as sacred. And indeed, the perception of this sacred nature of things usually elicits a profound, almost tangible, sense of harmony, a realization of a unity underlying the apparent diversity and discord of the world as we usually know it.

The analogy with music helps to explain why it is so difficult, even impossible, to describe the sacred nature of a mountain—or of anything else, for that matter. To attempt to do so is like trying to produce a chord by playing its notes one at a time: we miss the harmony that makes it what it is. In a similar way any direct, scientific description, by its very nature, can only describe one or the other of the images brought together in a view of a mountain. It cannot represent what the juxtaposition of these images reveals—what issues, for example, from the fusion of peak and temple. That lies beyond the reach of words, unless those words function in another way, as poetic symbols resonating with each other to awaken a sense of the sacred that transcends their literal meanings. This is why the most powerful religious writings, such as the *Bhagavad Gita* and the biblical *Psalms,* usually take the form of poetry.

The analogy with music may lead us to think that the sense of the sacred is only an effect or feeling created by the play of images in a view of a mountain. While views of mountains certainly elicit aesthetic and emotional responses, religious practitioners would argue that they also disclose something real, something that is actually present in the world—not just in the imagination. Just as the sense of depth created in a stereoscopic viewer reflects an actual third dimension in the scene pictured, so, for the person who reveres a mountain, the sense of the sacred awakened in a view of it reveals the true nature of the peak he beholds. Such a person would hold that through this sense of the sacred he experiences a spiritual reality that lies beyond the relatively superficial realm of aesthetic effects and personal feelings. For this experience to take place at a deep and convincing level, the image brought together with the mountain—god, temple, or heaven—must seem at least as real and powerful as the peak itself. Religious traditions help to imbue such images with the divine aura of reality and power necessary for this kind of effect.

Most views of sacred mountains are more complex than the simple case of a peak juxtaposed with a single image. A mountain is usually viewed in a number of different ways. Mount Sinai appears in biblical passages, for example, simultaneously as a place of descent, power, revelation, and ascent—God descends in fire and reveals the Torah to Moses, who climbs up to receive it. The multiplicity of images enhances the effect of the view, helping it to awaken a more profound and powerful sense of the sacred. In a similar way a number of additional notes can enrich a simple chord, making it resonate with an even deeper and more compelling harmony. Often one view of a

mountain will succeed another, like a sequence of chords in a piece of music. A Navajo will see the sacred mountains of the four directions in one context as the beams of a hogan, in another as the abodes of deities, and in yet another as the deities themselves. Sometimes the associations of one view will set off a string of others so that a person will glide naturally from one to the next, experiencing in each a different aspect of the sacred revealing itself in the mountain that appears before him.

Because mountains stand out from their surroundings, attracting attention to their soaring heights, views tend to gather and change around them like clouds about their summits. As the largest features of the natural landscape that we can see and grasp as wholes, they lend themselves to juxtaposition with images of unity and completeness associated with conceptions of the sacred in many different traditions. Isolated mountains that stand out as single unified massifs, such as Mount Kailas or the San Francisco Peaks, are the most apt to be singled out for veneration as places of particular power and sanctity. Since they incorporate aspects of every other feature of the natural landscape—streams, rivers, lakes, forests, and deserts— mountain peaks often function as microcosms of the world, leading cultures to juxtapose them with a wealth of associated images.

Something about a mountain originally attracts attention, providing a reason for first viewing it as sacred. People may notice that a major river flows from its slopes and come to worship it as a holy source of water and fertility. A sage or prophet may have received a commandment on its heights: his followers will venerate the peak as a place of divine revelation. Someone may see a vision of a deity hovering over its summit: the mountain becomes enshrined as the abode of a god. Another person may have an experience of such overwhelming intensity that others come to regard the peak as a place of natural and supernatural power. Whatever originally attracted attention—assuming that we can determine what that original reason was—sets in motion a succession of views that awaken and maintain a sense of the sacred.

Over time the views associated with a mountain develop and change. The original reason for seeing it as sacred may be forgotten or superceded by something else. A peak first worshipped as a source of water may come to be regarded as the center of the universe. A new religion or culture may take over the region and impose its own views on the mountain. Whereas the Kirghiz of the Pamir Mountains revere Muztagh Ata as a tomb of Muslim saints, the people who lived

Lenticular clouds magically transform the sky over the eastern Sierra Nevada of California. (Galen Rowell/Mountain Light)

there before them venerated the snow peak as the *stupa*, or reliquary, of a Buddhist sage.

Often, after many years, the images brought together in a view of a mountain merge so completely that they lose the power to awaken a sense of the sacred. The tension originally involved in seeing a peak as a temple forces us to look beneath their apparent differences and become aware of their deeper nature, in which they are one and the same. When that tension dwindles away, the view of the mountain that had acted as a window revealing a vision of numinous depth and meaning becomes merely a picture. The peak and the temple go flat and opaque, obscuring what they once revealed. A new view must arise to revitalize an awareness of the deeper reality inherent in both. Something similar happens in dead metaphors: when we say "time flies," we no longer think of time as a bird flying swiftly through the sky. It has become a cliché—a hackneyed expression without the power to provoke new insights or meanings.

The shifting views of a mountain, like the shadows of clouds playing over its slopes, keep the sense of the sacred alive in both the peak and the images associated with it—images that often stand at the center of a religious tradition or culture. If we listen too long to a single chord, we may become deadened to the resonance of its harmony. We need a succession of chords to hold our interest and create a piece of music that moves us with its beauty. In a similar way, to maintain a living sense of the sacred, views of a peak need to change, but however much they do so—no matter how far they stray from the original view—the mountain and the viewing of it remain.

The particular views that other people have of mountains may be very different from our own, but the underlying viewing process turns out to be quite similar. The same juxtaposition of images is at work, often with comparable effects. Whether seen as the temple of a god or as a pyramid of light, the mountain evokes a sense of another, more profound reality. The recognition of this fact offers us a means of relating the beliefs and practices of traditional cultures to matters that concern us in the secular world of modern society. The chapters that follow explore the ways in which mountains awaken a sense of the sacred in areas outside the usual province of established religion—in literature, art, and mountaineering, and in our relationship to each other and the world in which we live.

Mountains as a place of meditation. Sitting in a Himalayan cave, Milarepa, Tibet's most famous yogi, uses his spiritual powers to dissuade a hunter from killing a deer. Monastic mural painting. (Edwin Bernbaum)

MOUNTAINS AND THE SACRED
IN LITERATURE AND ART

SOMETIMES DELIBERATELY, often unconsciously, writers and artists draw on traditional views of mountains to awaken a sense of the sacred. Like priests and prophets, they use the powerful symbolism of mountain imagery to evoke a reality more intense and meaningful than that of our ordinary experience. We see people and things in a new and brighter light, one that reveals important aspects of life that we have previously overlooked or ignored. Like the peaks themselves, works of literature and art that use mountains to awaken a sense of the sacred are too numerous to cover in a comprehensive survey. This chapter will, instead, examine a few representative masterpieces by some of the most influential writers and artists of Eastern and Western civilizations.

MOUNTAINS IN LITERATURE

Metaphors embedded in traditional views of sacred mountains play an important role in secular as well as religious literature. They shape the structure of major works of poetry and prose and supply the symbolism that expresses the richness and depth of ideas and emotions. Much of the beauty and power of these works come from the interplay of evocative images in their metaphors. Because of their prevalence and importance in cultures around the world, traditional views of mountains provide a rich source of metaphoric imagery for poets and writers ranging from Kalidasa and Li Po in the East to Dante and Thomas Mann in the West.

Kalidasa, the foremost poet and dramatist of classical Indian literature, drew directly on images of mountains in Hindu mythology to create a landscape of magic peaks that evokes a sense of the sacred in a dazzling variety of ways. The greatest of his long poems, *Kumarasambhava*, or "The Birth of the Young God," composed around the fifth century A.D., opens with an evocative description of the Himalayas, portrayed as both a mountain range and a supernatural person:

There is in the north a supreme king of mountains named Himalaya, possessed of a divine self.
Bathing in the eastern and western oceans, he stretches like a measuring rod across the earth.
. . . .
A source of endless jewels, snow does nothing to diminish his splendor:
Just as a spot on the moon vanishes in a flood of moonlight, so a single blemish disappears beneath a flood of virtues.[1]

The verses that follow amplify this juxtaposition of human, natural, and supernatural images. After describing the gems and minerals found in the range and the magical beings who live beneath its snow-covered summits, the poem tells how the god Himalaya marries a female mountain and has a divine daughter destined to seduce the great god Shiva and give birth to a son who will rid the world of evil. The name of this goddess, Parvati, means literally "Daughter of the Mountain"—a reference to her father, the living embodiment of the Himalayan range. Views of sacred peaks that we have seen in numerous traditions, Western as well as Eastern, flicker across the pages of the poem, like images on the screen of a cinema: mountains as gods, abodes of gods, dwellings of sages, places of treasure, realms of transcendence, heavens on earth.

EARLY SPRING. *Painting by Kuo Hsi. (Collection of the National Palace Museum, Taiwan, Republic of China)*

mountain. His vision of Mount Purgatory stands at the center of the *Divine Comedy*, its ascent forming the subject matter of the middle book, the *Purgatorio*. After descending into the depths of hell in the *Inferno*, Dante climbs this mountain to reach the earthly paradise on its summit and proceed from there to the heights of heaven, described in the *Paradiso*. On the way up the peak, he passes through seven terraces on which he finds souls of the dead purging themselves—in ascending order—of the seven deadly sins that stand between them and God: pride, envy, wrath, slothfulness, greed, gluttony, and lust. The sin represented in the lowest terrace, pride, is the most difficult to overcome: it requires a formidable act of repentance in which the soul must acknowledge its inadequacy and turn away from reliance on itself to dependence on the grace of its creator. The failing symbolized in the highest terrace, lust, is the easiest to purge because it can easily change to divine love that carries one over the final obstacle to salvation—the last feeling of alienation from God. The physical nature of the ascent of Mount Purgatory reflects this spiritual progression of the soul as it sheds the burden of its sins and finds itself being drawn with increasing ease toward the object of its quest. Dante's guide, the Latin poet Virgil, explains to him:

This mountain is such that ever at the beginning below it is toilsome, but the higher one goes the less it wearies. Therefore, when it shall seem to you so pleasant that the going up is as easy for you as going downstream in a boat, then will you be at the end of the path: hope there to rest your weariness; no more I answer, and this I know for true.[9]

In his vision of purgatory as a mountain, Dante brings together images found in many different views of sacred mountains. Mount Purgatory appears as a cosmic axis, positioned between heaven and hell. It rises on the other—or under—side of the earth, exactly opposite Mount Zion in Jerusalem, the spiritual center of the Christian world. By identifying it with purgatory, Dante makes the mountain a place of the dead, another important image commonly associated with sacred mountains. The connotations of these images establish a setting that infuses the principal view of Mount Purgatory with special power and significance. In the *Divine Comedy* the mountain functions primarily as a cosmic place of transformation where the souls of the dead are purged and sanctified, redeemed and made fit to enter the presence of God. The ascent of the peak becomes a striking symbol for the path of repentance that leads to salvation—both in this world and the next. The juxtaposition of all these images in Dante's vision of purgatory contributes not only to the beauty of the *Divine Comedy*,

but to the richness and complexity of its symbolism.

Like Li Po and Basho in Asia, Romantic poets in eighteenth- and nineteenth-century Europe drew on their own experience of mountains to awaken a sense of the sacred in their poetry. Of these Romantics—including Hölderlin, Goethe, Shelley, and Byron—Wordsworth had the most quietly contemplative, almost Eastern, attitude toward nature. He also spent the most time in the mountains, roaming the hills of the Lake District, where he lived. In *The Prelude*, his great autobiographical poem, he describes a night ascent of Snowdon and the view he had from its summit of a vast world suffused with moonlight:

The Moon hung naked in a firmament
Of azure without cloud, and at my feet
Rested a silent sea of hoary mist.
A hundred hills their dusky backs upheaved
All over this still ocean; and beyond,
Far, far beyond, the solid vapours stretched,
In headlands, tongues, and promontory shapes,
Into the main Atlantic, that appeared
To dwindle, and give up his majesty,
Usurped upon far as the sight could reach.[10]

Reflecting on this vision of vastness, filled with the sound of streams roaring in the darkness below, Wordsworth detects the presence of a divine intelligence at work in the realm of nature:

ALLEGORICAL PORTRAIT OF DANTE. *Anonymous, Florentine.*
(National Gallery of Art, Washington, D.C.,
Samuel H. Kress Collection)

There I beheld the emblem of a mind
That feeds upon infinity, that broods
Over the dark abyss, intent to hear
Its voices issuing forth to silent light
In one continuous stream. . . .[11]

In *The Excursion*, projected to be part of his longest and most ambitious work, Wordsworth describes another vision experienced in mountains, but this one has an ecstatic, rather than contemplative quality. The Solitary, one of the characters in the poem, has the following experience as the mist through which he is walking opens to reveal a dramatic view of peaks and clouds:

Through the dull mist, I following—when a step,
A single step, that freed me from the skirts
Of the blind vapour, opened to my view
Glory beyond all glory ever seen
By waking sense or by the dreaming soul!
The appearance, instantaneously disclosed,
Was of a mighty city—boldy say
A wilderness of building, sinking far
And self-withdrawn into a boundless depth,
Far sinking into splendour—without end!
Fabric it seemed of diamond and of gold,
With alabaster domes, and silver spires
. . . .

By earthly nature had the effect been wrought
Upon the dark materials of the storm
Now pacified; on them, and on the coves
And mountain-steeps and summits, whereunto
the vapours had receded, taking there
Their station under a cerulean sky.
. . . .

That which I saw was the revealed abode
Of Spirits in beatitude: my heart
Swelled in my breast—'I have been dead', I cried,
'And now I live! Oh! wherefore do I live?'
And with that pang I prayed to be no more![12]

The images of peaks, clouds, and celestial city fuse to produce a vision of heaven so intense that the Solitary wishes he had died so that he could have lost himself in it forever. In this passage, as well as the one from *The Prelude*, Wordsworth makes explicit use of traditional views of sacred mountains to awaken a powerful sense of a divine presence infusing the natural world of earth and sky.

The writings of the French philosopher and novelist Jean-Jacques Rousseau played a major role in influencing the thought of Romantic poets and inspiring their interest in the divine aspects of nature. His novel *La Nouvelle Héloise*, published in 1761, probably did more than any other single work of literature to awaken

enthusiasm for the Alps and transform European attitudes toward mountains in general. One passage in particular sent waves of people flooding up to the Alps in search of the physical well-being and spiritual renewal Rousseau attributed to mountain environments. In it the hero, Sainte-Preux, describes the beneficial effects of an ascent he has made to a minor summit offering a magnificent view of the range:

It was there that I visibly discerned, in the purity of the air in which I found myself, the true cause of the change in my mood, and the return of the inner peace that I had lost for so long. In effect, it is a general impression experienced by all men, even though they do not all observe it, that on high mountains, where the air is pure and subtle, one feels greater ease in breathing, more lightness in the body, greater serenity in the spirit; pleasures are less ardent there, passions more moderate. Meditations take an inexpressibly grand and sublime character in proportion to the objects that impress us, a tranquil voluptuousness that has nothing to do with anything harsh or sensual. It seems that in rising above the dwellings of men, one leaves behind all low and earthly sentiments, and to the degree that one approaches the ethereal regions, the soul acquires something of their inalterable purity.[13]

Rousseau's hero views the mountains as an earthly paradise where one can be healed and transformed, both physically and spiritually. He even sees them as a sublime site of mystical transcendence in which the soul loses itself in an experience of the divine:

Imagine the variety, the grandeur, the beauty of a thousand astonishing sights. . . . In the end the spectacle has something—I don't know what—of magic, of the supernatural, that ravishes the spirit and the senses; one forgets everything, one forgets oneself, one no longer knows where one is.[14]

After the publication of *La Nouvelle Héloise*, no grand tour of Europe was complete without an obligatory stop to refresh one's jaded spirits with a view of the Alps.

Mountains also play an important role in a major work by another European philosopher, the German Friedrich Nietzsche. They provide a mythic setting and source of imagery for many of the ideas expressed in *Thus Spoke Zarathustra*, his impassioned account of the life and teachings of a prophet he named Zarathustra after the founder of Zoroastrianism, the religion of ancient Iran. Nietzsche begins and ends this lyrical work with references to mountains, enclosing the philosophy presented within its pages in an exalted realm of inspired wisdom set apart from the profane world of ordinary thought. *Thus Spoke Zarathustra* opens with a passage that establishes the context for the story and ideas to follow:

When Zarathustra was thirty years old, he left his home and the lake of his home and went into the mountains. Here he had the enjoyment of his spirit and his solitude and he did not weary of it for ten years. But at last his heart turned—and one morning he rose with the dawn. . . .[15]

Like Moses on Mount Sinai, Zarathustra comes down from the heights with a message for mankind—in this case, the pronouncement of the death of God and his replacement by the superman. The title of a later section, "Of Old and New Law-Tables," makes the comparison with the ascent of the Hebrew prophet even more apparent. In the opening passage Nietzsche draws on traditional views of sacred mountains as places of revelation and transformation, where holy men go for spiritual wisdom and purification. At the end Zarathustra returns to the heights to complete his process of awakening. The book closes with a vision of him coming out of his mountain cave, fully transformed and enlightened, knowing at last what he needs to do:

Thus spoke Zarathustra and left his cave, glowing and strong, like a morning sun emerging from behind dark mountains.[16]

Nietzsche uses the metaphor of mountain climbing to express the idea of self-transcendence basic to his conception of the superman, one who overcomes himself to attain the ultimate heights of superhuman perfection. As he climbs up a ridge, Zarathustra says to himself:

In order to see much one must learn to look away from oneself—every mountain-climber needs this hardness . . . you must climb above yourself—up and beyond, until you have even your stars under you! Yes! To look down upon myself and even upon my stars: that alone would I call my summit, that has remained for me as my ultimate summit![17]

In its notion of self-transcendence, expressed through images of mountain climbing, Nietzsche's conception of the superman has something in common with Buddhist ideas of a Bodhisattva who attains perfection by overcoming himself—or his illusion of self—for the sake of a higher goal. But the exaggerated emphasis on power and the lack of compassion in Nietzsche's image of the ideal human being set some who espoused his philosophy on a very different path from that of Buddhism—one that led to the horrors of Nazism. The extreme view of mountain climbing presented in *Thus Spoke Zarathustra* also inspired German and Austrian mountaineers of the Third Reich to drive themselves so far beyond the limits of their abilities that a number of them perished in attempts

to demonstrate the superiority of the so-called Aryan superman.[18]

Taking a different approach, emphasizing love rather than power, the German author Thomas Mann, one of the most important novelists of the twentieth century, constructed an entire novel around the theme of the sacred mountain as a timeless embodiment of the other world. Conceived as a critique of European society before World War I, *The Magic Mountain* tells the story of a young engineer named Hans Castorp who goes to visit a sick relative at a tuberculosis sanatorium in the Alps. Leaving the bourgeois concerns of life in the flatlands far below, he enters a rarefied realm of sickness, healing, and death, where time ceases to have its usual meaning. Days, weeks, months, seasons—all flow into each other and become indistinguishable, no longer the measure of anything of any real significance. Fascinated by this strange environment, so different from the world to which he is accustomed, Castorp develops a spot of infection on his lungs and winds up staying for seven years. Free from the constraints of time, no longer subject to the obligations of ordinary life, he finds the leisure to explore ideas and discover himself. The mountains become for him a place of awakening.

This process of awakening culminates in a vision high in the Alps that forms the artistic climax of the book. Setting off alone on skis, Castorp climbs up into a gathering snowstorm, drawn to the heights by the kind of fascination and fear that characterize the experience of the sacred:

He pressed on, turning right and left among rocky, snow-clad elevations, and came behind them on an incline, then a level spot, then on the mountains themselves—how alluring and accessible seemed their softly covered gorges and defiles! His blood leaped at the strong allurement of the distance and the height, the ever profounder solitude. At risk of a late return he pressed on, deeper into the wild silence, the monstrous and the menacing, despite that gathering darkness was sinking down over the region like a veil, and heightening his inner apprehension until it presently passed into actual fear.[19]

The storm breaks, and lost in the snow, he slips into a dream in which he beholds an earthly paradise set on sunlit hills beside the Mediterranean—an idyllic landscape from classical mythology, peopled with beautiful men, women, and children radiating the spirit of love. But hidden within this heavenly vision, like a worm in a shining apple, he spies the horror of a demonic sacrifice: he enters a Greek temple and comes upon two hags ripping apart a child and chewing its bloody bones.

Awakening from the dream, he has a revelation

about the significance of what he has seen. Death has great power over man: it appears even in his vision of paradise. But love is stronger: it produces a sweetness of thought that will ultimately prevail. He concludes, with the only sentence italicized in the entire novel,

For the sake of goodness and love, man shall let death have no sovereignty over his thoughts. — And with this — I awake. For I have dreamed it out to the end, I have come to my goal. Long, long have I sought after this word. . . . Deep into the snow mountains my search has led me. Now I have it fast. My dream has given it me, in utter clearness, that I may know it for ever.[20]

From this high point of revelation, Castorp descends to the sanatorium and eventually to the flatlands below. The book ends with a view of him fighting in the hideous battlefields of World War I. The last sentence poses a hopeful question, brought down from the visionary heights of the Magic Mountain:

Out of this universal feast of death, out of this extremity of fever, kindling the rain-washed evening sky to a fiery glow, may it be that Love one day shall mount?[21]

Where Mann glimpsed a hope of love emerging from the ruins of humanity's passion for destruction, the Irish poet William Butler Yeats, one of the greatest poets of the twentieth century, saw something less comforting—a stark awareness of ultimate reality, revealed in a vision of hermits meditating in caves on the icy slopes of a sacred peak. Named for the cosmic mountain of Hindu and Buddhist mythology, his poem "Meru," completed in 1934, uses the image of this peak to express a profound disillusionment with the material accomplishments of civilization, epitomized in the monuments of Egypt, Greece, and Rome:

Civilisation is hooped together, brought
Under a rule, under the semblance of peace
By manifold illusion; but man's life is thought,
And he, despite his terror, cannot cease
Ravening through century after century,
Ravening, raging, and uprooting that he may come
Into the desolation of reality:
Egypt and Greece, good-bye, and good-bye, Rome!
Hermits upon Mount Meru or Everest,
Caverned in night under the drifted snow,
Or where that snow and winter's dreadful blast
Beat down upon their naked bodies, know
That day brings round the night, that before dawn
His glory and his monuments are gone.[22]

A similar expression of disillusionment appears in Yeats's most widely quoted poem, "The Second Coming," composed fifteen years before "Meru" in 1919,

just after the end of the First World War. Both poems open with images of disintegrating circles symbolizing the breakdown of culture and society: the hoop of civilization brought tenuously together by illusion in "Meru" and the expanding spirals of a falcon's flight in "The Second Coming" with its well-known line, "Things fall apart; the centre cannot hold." The nature of what follows this shared symbol of disintegration, however, has changed in the later work. The horrifying image of a sphinxlike beast being born in Bethlehem at the end of "The Second Coming" has become in "Meru" a bleak vision of naked ascetics caverned in snow on a mountain in Tibet.[23]

For all its apparent bleakness, this vision, with its dismissal of man's achievements, has positive implications lacking in the earlier poem. Yeats was well aware that the mountain he chose, Meru, stands in Indian mythology as the supreme symbol of the kind of center whose loss he lamented in "The Second Coming"— a center that gives stability and order to the universe around it. Around the time he composed the later poem, he wrote an introduction to a Hindu swami's account of a pilgrimage to Mount Kailas, the sacred peak in Tibet popularly identified with Meru. In this introduction Yeats discusses the significance of the mountain as a cosmic center and uses the same image of a cave on Meru to describe those who go beyond illusion—"man's glory and his monuments"—to attain the supreme goal of spiritual liberation:

He that moves towards the full moon [taking the path to liberation] may, if wise, go to the Gods (expressed or symbolized in the senses) and share their long lives, or if to Brahma's question — 'Who are you?' he can answer 'Yourself,' pass out of these three penitential circles, that of common men, that of gifted men, that of the Gods, and find some cavern upon Meru, and so pass out of all life.[24]

The desolation of which the poem speaks—and which the mountain embodies—refers to the destruction of "manifold illusion," revealing the underlying reality that is man's ultimate salvation, blissfully free from the painful round of life and death.

Perhaps inspired by his interest in Kailas and Meru, Yeats asked to join a group of England's leading climbers on one of their outings to cliffs in Wales. Geoffrey Winthrop Young, the leader of the party and a poet in his own right, refused to let him come. He had heard that Yeats, who engaged in mystical practices, planned to project himself up the rock in the form of a small green jade elephant. Someone like that, Young felt, would be too dangerous to have as a partner on a climbing rope.[25]

During the Second World War, a few years after Yeats composed "Meru," René Daumal, a French mystic and writer, attempted to construct a cosmic axis for the modern world. Drawing on the symbolism of sacred peaks in Eastern and Western traditions, his allegorical novel *Mount Analogue* posits the existence of a supreme mountain "uniting Earth and Heaven"— a concrete symbol of the way in which people may awake from the slumber of their usual state of mind and ascend to a higher level of consciousness. Writing as a character in his own book, Daumal decides that such a peak must have the following characteristics:

For a mountain to play the role of Mount Analogue, I concluded, its summit must be inaccessible, but its base accessible to human beings as nature has made them. It must be unique, and it must exist geographically. The door to the invisible must be visible.[26]

Having decided that the summits of the highest known mountains lack the requisite inaccessibility and that mythic peaks like Meru lack the necessary geographical reality, Daumal and a group of like-minded characters, led by a professor of mountaineering named Pierre Sogol, determine that Mount Analogue must exist on a huge island in the South Pacific, hidden by a mysterious force field that bends light rays around the peak. They form an expedition and set off to find and climb the mountain.

The nature of the peak and its ascent immediately bring to mind comparisons with Dante's *Purgatorio*. The mountains in both works bear allegorical names that make their symbolism explicit: Mount Analogue and Mount Purgatory. Each rises on an island situated on the opposite side of the earth from places well known to the reader: Paris in *Mount Analogue* and Jerusalem in the *Purgatorio*. When Daumal and his party land at the foot of their mountain, they find a community of people similar to those who reside at the base of Mount Purgatory—procrastinators and others who lack the motivation needed to continue the spiritual quest. Like Mount Purgatory the climb of Mount Analogue requires a profound act of repentance, a purgation of self-willed egotism. Sogol finds the group's first peradam, a nearly invisible crystal needed as payment to ascend the mountain, when he expresses these feelings of contrition and humility:

I have brought you this far, and I have been your leader. Right here I'll take off the cap of authority, which was a crown of thorns for the person I remember myself to be. Far within me, where the memory of what I am is still unclouded, a little child is waking up and making an old man's mask weep. A little child looking for mother and father, looking with you for protection and help—protection from his pleasures and his dreams, and help in order to become what he is without imitating anyone.[27]

Although Dante and Daumal share the basic idea of the mountain as a symbol of the spiritual path, they situate their allegories in the milieux of their times, making for profound differences between the two works. Set in the twentieth century, the French novel tells the story of a mountaineering expedition—inconceivable in the early Renaissance—complete with ropes, crampons, and alpine guides, who represent teachers who have attained higher states of consciousness. Whereas Dante makes the ascent of Mount Purgatory an expression of Christian doctrine regarding the path to salvation in heaven, Daumal uses the climb of Mount Analogue to represent the teaching he considers most relevant for his time—the ideas of the Russian mystic George Gurdjieff concerning the way that people must follow to awaken from the automatism of the human condition. Dante reaches the earthly paradise on top of Mount Purgatory, but we never find out what lies on the summit of Mount Analogue, nor what its heights symbolize. Just at the point that his characters begin the actual ascent of the mountain, Daumal died, leaving the novel to end in midsentence. His failure to complete his work—and the odd fact that nobody has tried after him—may say something about the nature of our times: that the modern world lacks a unified view of the cosmos needed to create a universal allegory comparable to Dante's, with its magnificent vision of a cosmic axis linking the human realm of civilization to a higher order of existence.

MOUNTAINS IN ART

Like literature—but with the more vivid impact of visual imagery—art has the power to transform our views of the world, yielding new visions of reality. By focusing our attention in a particular way, a painting of a familiar subject can reveal the most extraordinary qualities in the most ordinary objects. Depicted in a still life by a master like Cézanne, apples and oranges can assume an awesome grandeur as monumental expressions of rock-hard form revealed through the flickering play of color and light. This transformation can have lasting effects: we may never again see a bowl of fruit in quite the same way. A painting of a strange or fantastic subject, such as a vision of paradise or a surreal landscape, can also change our perspective on the world around us. By sending us soaring off to a realm of the imagination where anything is possible, it liberates us from our usual way of seeing things so that we return to our familiar surroundings with eyes refreshed, open to new possibilities, eager for new perceptions. Whether depictions of ordinary objects or visions of extraordinary scenes, certain works of art have the power to awaken a sense of wonder that reveals a deeper, more meaningful reality—one that underlies and animates the world of everyday life and the universe of fantasy.

Mountains make ideal subjects for works of art imbued with this kind of power. They belong to the material world; yet they evoke the spiritual realm. Their physical height and grandeur inspire a sense of wonder and awe that conjures up images of the sacred enshrined in religious traditions—gods and demons, heavens and hells, visions of revelation, scenes of damnation. When an artist chooses to paint a mountain in an awe-inspiring manner, he automatically calls forth such images from the repository of his own tradition and juxtaposes them with the image of the peak. Acquiring in this way a metaphoric dimension, the work of art takes on a numinous depth that reveals to the viewer a deeper vision of reality, shimmering on the edge of awareness. To create this effect, the artist does not need to make explicit reference to traditional views of the sacred: whether depicted or not, they will spring to mind, evoked by the imposing image of the mountain itself.

The subtle evocation of the sacred through implicit ideas and images is characteristic of what many regard as the supreme genre of mountain art in the world—Chinese landscape paintings. Although they may appear to be secular works devoid of religious imagery, the views they depict of peaks and valleys receding into misty space awaken a sense of mystery as powerful and profound as any explicit portrayal of mountains as temples of the gods—even for someone who knows nothing about the culture from which they come. For a person versed in the traditions of China, these views have an even deeper effect: they call forth ideas and images that enhance the power and significance of the work of art.

The expression used to designate landscape paintings—shan-shui, "mountain-water"—highlights the importance of mountains, or shan, in Chinese thought as one of the two basic constituents of the natural environment. The second element, shui or "water," takes the form of streams and rivers that issue from the heights of peaks to wind about their feet and spread across the plains. To visualize landscapes as compositions created from these two elements recalls ancient views of mountains as sacred sources of water and life. The expression shan-shui also brings to mind the yin-yang theory of complementary opposites basic to Chinese conceptions of reality. The viewer of a landscape sees the male and female principle of this doctrine embodied in the masculine heights of peaks where clouds form and the sun shines and in the feminine depths of valleys where rivers run and shadows lie. Together the two engender the totality of nature and reveal the presence of the Tao, the spiritual essence of all things, flowing through the world like water from the mountains.

For the Chinese who know and love them, landscapes do more than elicit an aesthetic response: a beautiful painting of mountains and rivers serves a higher purpose—to awaken the spirit and disclose the true nature of reality. In the fifth century A.D., Tsung Ping, one of the earliest landscape painters in China, wrote A Preface on Landscape Painting, in which he described his art as a form of spiritual practice:

If truth lies in the satisfaction of both eye and mind, then a picture well executed will also correspond with visual experience and be in accord with the mind. That correspondence will stir the spirit, and when the spirit soars, truth will be achieved. And though one should return again and again to the wilderness, and seek out the lonely cliffs, what more could be added to this?[28]

Landscapes that evoke this correspondence of visual perception and mental imagery "captivate the Tao by their forms." They become expressions of ultimate reality, imbued with the power to transport and transform the person who views them in a sympathetic way.

Tsung Ping's remarks reflect the change of attitude that occurred around the beginning of the fourth century A.D. From awesome abodes of dangerous deities approached only by sages armed with magic powers, mountains became the favorite haunt of poets and painters seeking to cultivate the spirit through contact with the Tao. This shift toward a positive view of the heights as idyllic places of inspiration led to the development of landscape painting as a genre in its own right. Wandering in the mountains, meditating on the views around them, artists acquired the skills and knowledge needed to produce masterpieces of evocative art. During the T'ang Dynasty, between the seventh and tenth centuries, landscapes moved to the foreground to emerge as primary subjects of artistic interest and representation.

One of the most famous and influential artists of the T'ang Dynasty, Chang Tsao, who worked in the eighth century, was a wild-eyed figure who brought the spirit of the mountains directly into the act of painting. A poet who observed him at work described the startling impression he made on his contemporaries:

Right in the middle of the room he sat down with his legs spread out, took a deep breath, and his inspiration began to issue forth. Those present were as startled as if lightning were shooting across the heavens or a whirlwind sweeping up into the sky. Ravaging and pulling, spreading in all directions, the ink seemed to be spitting from his flying brush. He clapped his hands with a cracking sound. Dividing and drawing together, suddenly strange shapes were born. When he had finished, there stood pine trees, scaly and riven, crags steep and precipitous, clear water and turbulent clouds. He threw down his brush, got up, and looked around in every direction. It seemed as if the sky had cleared after a storm, to reveal the true essence of ten thousand things. When we contemplate Master Chang's art, it is not painting, it is the very Tao itself.[29]

More than a painter of landscapes, Chang Tsao transcended his art to become an exemplar of the artist as sorcerer and sage. The dynamic force of his personality and the effortless perfection of his style played an important role in shaping the ideals and aspirations of those who followed him. Unfortunately none of his paintings have come down to us.

Although examples of his work perished, the ideals Chang Tsao espoused endured to develop and reach a peak of perfection in the golden age of landscape painting during the Northern and Southern Sung Dynasties, between 960 and 1279. A number of original works from that period have survived, including one regarded as a classic of Chinese art — *Early Spring*, painted by Kuo Hsi and dated 1072. A description of Kuo Hsi's landscapes from an imperial catalogue of the Northern Sung Dynasty reads:

Winding streams and abrupt banks, craggy cliffs and sheer precipices, rounded heights and sharp peaks rising in abundance; clouds and mists constantly transforming and dissolving, a thousand attitudes and ten thousand forms in the midst of their changing light.[30]

The fluid nature of the landscape depicted in *Early Spring*, with mountains taking the shapes of clouds to drift in and out of the mist, expresses the elusive flow of the Tao, giving rise to all things, yet bound by none of them. Nothing remains fixed, everything is in flux, turning into something else and pointing to a reality beyond form, deeply mysterious and immensely attractive. *Early Spring* invites us to enter the world it depicts and lose ourselves in it, wandering forever through mountains without end.

The author of the most influential treatise on Chinese landscape painting, Kuo Hsi emphasized the need for the artist to bring out the spiritual essence of his subject: "If he fails to get at the essential, he will fail to present the soul of his theme."[31] For Kuo Hsi mountains were not inanimate piles of rock. They were charged with spiritual life and energy, which the aritst had to perceive and infuse in his work. In his essay, compiled and written down by his son, he expressed perhaps most beautifully and simply the rea-

son why the Chinese people have placed so high a value on the art of depicting mountain landscapes:

The din of the dusty world and the locked-in-ness of human habitations are what human nature habitually abhors; while, on the contrary, haze, mist, and the haunting spirits of the mountains are what human nature seeks, and yet can rarely find. . . . Having no access to the landscapes, the lover of forest and stream, the friend of mist and haze, enjoys them only in his dreams. How delightful then to have a landscape painted by a skilled hand! Without leaving the room, at once, he finds himself among the streams and ravines; the cries of the birds and monkeys are faintly audible to his senses; light on the hills and reflection on the water, glittering, dazzle his eyes. Does not such a scene satisfy his mind and captivate his heart? That is why the world values the true significance of the painting of mountains.[32]

Mountains play an equally important role in the art of Japan, where they also elicit associated imagery to evoke a powerful sense of the sacred. Like their counterparts in China, on whose styles they modeled their work, masters such as Sesshu and Gakuo used haunting views of mountainous landscapes to awaken an awareness of a deeper, more spiritual reality hidden in the physical features of the natural world. The depiction of mysterious shapes of irregular crags and peaks found in most of these classical paintings requires a delicacy of brushwork that the Japanese adopted from the Chinese and refined in their own particular way. Among the many mountains available to the artist for awakening a sense of the sacred, however, Japan has one without parallel in China, a particularly evocative peak whose simplicity of form demands a different style of representation — Mount Fuji. Depicted in earlier landscape paintings in largely stylized ways, it became in the nineteenth century a focal point of interest for a Japanese school of art that acquired a special renown in the Western world.

The triangular cone of Fuji with its lack of irregular features made an ideal subject for Ukiyo-e woodblock prints that emphasized smooth geometric shapes and homogeneous masses of color. The name of the school, Ukiyo-e, means "pictures of the floating world" — a reference to the transient world of everyday life portrayed in these prints. This school of art rose in response to the increasing demand of ordinary Japanese for scenes of people and places they knew, such as portraits of geishas and views of Mount Fuji. Where classical landscape paintings depicted idealized visions of reality for the aristocracy, Ukiyo-e prints represented more realistic representations of the world for commoners. The growing popularity of the Fuji devotional cults among the tradespeople of Edo, modern-day Tokyo, further encouraged artists of the new school to represent the peak as an actual mountain set amid scenes of daily life.

One of the greatest masters of Ukiyo-e, Katsushika Hokusai, focused his attention—and much of his devotion—on the inspiring form of Mount Fuji. Between 1831 and 1835, he depicted the mountain in two sets of extraordinary prints—one in color, the other in black and white. The second set, published in book form as the *One Hundred Views of Fuji*, shows that Hokusai was familiar with contemporary religious beliefs and practices concerning the peak. It opens with a portrait of Konohana Sakuya Hime, the goddess of Mount Fuji, followed by two prints depicting a group of onlookers witnessing the miraculous birth of the volcano in a single night and En no Gyoja, the traditional founder of Shugendo, making the legendary first ascent of the mountain. Hokusai himself regarded the peak as sacred, but in his own particular way. He saw Fuji as a symbol of immortality and stability and felt that by repeatedly depicting its perfect form he could attain his cherished goal of living to over the age of a hundred. Referring to an old legend of Taoist origin that must have influenced Hokusai, the author of the preface to the last volume of the *One Hundred Views* wrote:

I have heard that he [Hokusai] has now passed ninety years of age, and yet his sight and hearing are still like that of a youth. Perhaps he was once able to acquire the secret elixir of the Immortals on this miraculous mountain.[33]

Hokusai missed his mark by ten years: he died in 1849 at the age of ninety.

The earlier, more famous set of prints—the *Thirty-Six Views of Mount Fuji*—also brings out the sacred nature of the peak, but without reliance on obvious religious imagery. Unifying them all, the mountain stands in these forty-six color prints—Hokusai added ten to the original thirty-six—as a symbol of stability, set firmly in opposition to scenes of the "floating world" portrayed in the foregrounds of the series. The serene image of Fuji, untouched by the bustle of activity before it, reminds the viewer of both the transience of human life and the permanence of something beyond it. Nowhere in the *Thirty-Six Views* does Hokusai evoke this vision of reality more dramatically than in the *Great Wave off Kanagawa*, probably the single work of Japanese art best known outside of Japan.

In this print, an enormous breaker rears up over three boats of frightened people caught in the swells of its surging power. Writhing shreds of foam reach out like hands toward the small figure of Fuji, viewed through the hollow of the wave. Although the water seems poised to obliterate the mountain, Fuji remains calm and serene, the image of perfect composure in the midst of chaos. All the furious power of the wave succeeds in doing is to captivate the viewer's eye and guide it around to the unshakeable peak that forms the ultimate focus of attention. For all the opposition between them, the two work together, the wild motion of the one accentuating the quiet stillness of the other. White foam on the crest of the breaker, white snow on the summit of the peak, they share the same shade of blue below—mountainous wave and wave-like mountain.

The dramatic scene depicted in the print calls forth ideas basic to the philosophy and practice of Japanese Buddhism. A Buddhist such as Hokusai would see in the curling wave about to overwhelm the terrified people a reference to *samsara*, the turbulent round of life and death, and in the still point of Fuji at the center, an intimation of nirvana, the serene state of freedom from fear and suffering. Just as the mountain appears within the curve of the breaker, so the reality that leads to liberation is to be found in the world of illusion, not somewhere apart from it. The underlying resemblance of wave and peak points to the realization that for all their apparent differences *samsara* and nirvana are one and the same, if we can see them for what they truly are: two ways of experiencing the emptiness of all things, one binding, the other liberating.

Whether a reflection of nirvana or an embodiment of immortality, the serene peak of Fuji represented for Hokusai a center of permanence and stability that he sought in his life and art. His name, one of the many names he chose for himself, means the Northern Studio, a reference to the North Star, the fixed point of the heavens around which the other stars revolve. In a similar way Mount Fuji forms the immovable center around which Hokusai arranged his shifting views of the world of change. The meaning of his name suggests that at some deep level he identified himself with the mountain and the principle of reality it embodied. He strove to realize this reality through the practice of his art:

From the age of six I had a penchant for copying the form of things, and from about fifty, my pictures were frequently published; but until the age of seventy, nothing that I drew was worthy of notice. At seventy-three years, I was somewhat able to fathom the growth of plants and trees, and the structure of birds, animals, insects, and fish. Thus when I reach eighty years, I hope to have made increasing progress, and at ninety to see further into the underlying principles of things, so that at one hundred years I will have achieved a divine state in my art, and at one hundred and ten, every dot and every stroke will be as though alive.[34]

Hokusai's prints of Mount Fuji were among the first works of Eastern art to influence Western artists. The

GREAT WAVE OFF KANAGAWA. *Ukiyo-e print by Katsushika Hokusai. (Art Resource, N.Y.)*

way in which he used simplified colors and shapes to bring out the essence of his subject excited the imagination of Impressionists and Post-Impressionists such as Edouard Manet, Edgar Degas, Paul Gauguin, and Vincent Van Gogh. Speaking of *Fuji in Clear Weather*, the print in the *Thirty-Six Views* that most beautifully captures the sublime serenity of the peak's perfect form, a Western scholar of Japanese art has written,

It is the structure of the mountain itself that the artist wishes to portray, and in this sense such a work is close to the ideals of the Post-Impressionists. They, too, were much concerned with rendering the structure of nature than with any naturalistic likeness, and it is not surprising that painters such as Van Gogh admired and even copied this Japanese artist.[35]

Van Gogh took a particular interest in Hokusai's *One Hundred Views of Fuji* and drew on them to develop his skill as a draftsman. One of his major paintings, *Portrait of Père Tanguy*, shows a Japanese print of Mount Fuji placed in the background.

Landscape painting emerged as a genre in its own right much later in the West than it did in the East.

As in China and Japan, it coincided with a shift toward a positive view of mountains—a view that began to develop in Europe during the Renaissance as poets and painters started to take an active interest in the natural world. Petrarch's account of his ascent of Mount Ventoux in the fourteenth century marks an early harbinger of this change of attitude, which would come to full fruition hundreds of years later in the poetry and art of the Romantic movement. Artists of the Middle Ages, before Petrarch, had shared the negative view of Catholic theologians that nature in general and mountains in particular represented the material world of fallen man corrupted by the demonic influence of Satan.

During the fourteenth and fifteenth centuries, rocky peaks began to appear in the background of paintings depicting scenes from the Bible and the lives of the saints. Derived from Byzantine representations of Mount Sinai, their stylized forms served to enhance the sanctity of the people and events portrayed in the foreground. A dramatic example occurs in *Saint Francis Receiving the Stigmata*, a painting executed by Domenico Veneziano in 1445. Over the jagged moun-

SAINT FRANCIS RECEIVING THE STIGMATA. *Painting by Domenico Veneziano. (National Gallery of Art, Washington, D.C., Samuel H. Kress Collection)*

tains in the background appears a vision of an angel holding the crucified body of Jesus. Rays of light shoot down like laser beams, burning wounds into the hands, feet, and side of the saint. The mountainous setting of the scene brings to mind the love Saint Francis had for nature, where he felt closest to God. Although Veneziano painted their features in an abstract and fanciful way, the peaks themselves possess a size and grandeur that suggest some familiarity with actual views of the Italian Alps.

As Western landscape painting continued to develop, artists drew on their own observations to depict mountains with increasing realism and dramatic effect. At the end of the fifteenth century, Leonardo da Vinci actually climbed a peak in the Alps—the mysterious Monboso, tentatively identified as a spur of Monte Rosa. He drew sketches of the ranges he saw and put mountains in the background of his most famous painting, the Mona Lisa. Albert Dürer also made drawings of the Alps based on his travels and used them in paintings of saints and other religious figures. The *Battle of Alexander* by his contemporary Albrecht Altdorfer presents an extraordinary vision of a mountainous landscape charged with the fiery energy of a supernatural sun, blazing out of a whirlpool of orange and black clouds. For all the increase in the realism and dramatic intensity of their representation, however, mountains still remained in the background, stage props for subjects drawn mostly from Classical and Christian sources. This held true for most of the landscapes painted by artists of the Baroque and Neo-Classical periods, which followed the Renaissance.

With the rise of the Romantic movement at the end of the eighteenth century, mountains moved to the fore as the principal subjects of paintings designed to awaken a sense of the sacred. Inspired by writers such as Rousseau and Wordsworth, artists sought to portray features of the natural landscape as sublime manifestations of the infinite itself. They responded with particular enthusiasm to views of mountains poised above clouds, soaring up toward the limitless heights of heaven. Two tiny figures struggling up toward a cross on the summit of a peak in *Morning in the Reisengebirge* by the German artist Caspar David Friedrich exemplify the Romantic longing to experience the realm of the spirit in the world of nature. Rather than merely provide a setting for their quest, the imposing mountains around them express the essence of the divine reality they seek.[36]

John Ruskin, the influential English critic who most powerfully enunciated the spiritual ideals of Romantic art, was obsessed with the sacred significance of mountains. He regarded them as "great cathedrals of the earth, with their gates of rock, pavements of cloud, choirs of stream and stone, altars of snow, and vaults of purple traversed by the continual stars." The metaphor he used to express this view was for him no fanciful figure of speech: during bouts of depression and religious despair, he would go to the Alps, as to a church, to recover his faith and revive his soul. Mountains, he felt, revealed most clearly the truth that nature is the creation of God — a truth that according to him all great landscape paintings disclose.[37]

Between 1842 and 1860, Ruskin composed *Modern Painters*, a passionate defense of the English artist J. M. W. Turner. This multivolume work, which sets forth Ruskin's philosophy of art, contains entire sections devoted to a study of the nature of mountains and their influence on the human spirit. Like Chinese artists and writers, he felt that paintings of mountainous landscapes should reveal spiritual truths that uplift and refine the moral character of both the individual and society. Although Ruskin knew nothing about Eastern art, a passage in *Modern Painters* comparing the energy of mountains to the muscular action of the body and calling them "the bones of the earth" sounds strangely reminiscent of folk beliefs from ancient China. And his detailed discussions of different kinds of landscape features and how to depict them call to mind similar descriptions from Chinese handbooks such as *The Mustard Seed Garden Manual of Painting*.[38]

The nineteenth-century American artist whose paintings most dramatically embodied the ideals of the Romantic movement was Albert Bierstadt, the leader of the "Rocky Mountain School." In 1859 and 1863, after studying landscape painting in Europe, he made two journeys to the western part of the United States, recording his impressions of the mountains he saw. His companion on the second trip, a Romantic writer named Fitzhugh Ludlow, expressed in words the feelings that Bierstadt rendered in paint:

I confess (I should be ashamed not to confess) that my first view of the Rocky Mountains had no way of expressing itself save in tears. To see what they looked like, and to know what they were, was like a sudden revelation of the truth, that the spiritual is the only real and substantive; that the eternal things of the universe are they which afar off seem dim and distant.[39]

A painting titled *View of the Rocky Mountains* exemplifies the striking way in which Bierstadt used the grandeur of mountain landscapes to awaken a sense of the sacred. Luminous clouds of mottled mist swirling around jagged peaks conjure up images of Zeus casting thunderbolts from Olympus or God descending in fire and smoke on Sinai. Higher up in the blue sky, cut off from the earth below, snow summits appear white and serene, a vision of heaven beyond the reach of mortal effort. In the foreground beneath the mountains, a family of three deer gazes quietly on a grove of gold-lit trees set on a green meadow in front of a still lake. Smooth cliffs hung with shining waterfalls enclose the scene in an atmosphere of primordial perfection that makes one think of the Garden of Eden before the creation of Adam and Eve — or after their expulsion.

Although Bierstadt named his painting *View of the Rocky Mountains*, it has a general, idealized character that fits no actual view of the range. He painted essentially the same vista in a number of other landscapes, including one that he titled *Among the Sierra Nevada Mountains, California*. Indeed, the imposing cliffs and waterfall on the left side of the painting clearly derive their inspiration from views of Yosemite Valley. The spectacular snow peaks in the background, on the other hand, look as though they belong in the Alps, rather than the Rockies or the Sierra Nevada. Only the primeval wilderness of the scene in the foreground has much to do with the Rocky Mountains — and even it seems too neat and clean for the rugged wildness of the range as it actually is. The very lack of specificity and the idealized character of the painting — qualities for which Bierstadt has been often criticized — place the landscape it depicts outside of ordinary time and space, in the eternal realm of the sacred.

VIEW OF THE ROCKY MOUNTAINS. *Painting by Albert Bierstadt. (White House Collection, Washington, D.C.)*

Bierstadt's paintings played a major role in shaping American attitudes toward the West. The sense of the sacred they inspired through evoking images of Eden imbued the Rocky Mountains and Sierra Nevada with the aura of the Promised Land that the Pilgrims had originally sought when they came to colonize the New World in the seventeenth century. For Americans of the 1860s and 1870s, the pristine grandeur of Bierstadt's dramatic landscapes helped to transform the mountains of the West into a contemporary manifestation of the earthly paradise where people might return to their beginnings and renew themselves. The interest his paintings aroused gave an added impetus to the sense of Manifest Destiny that shaped national policies and encouraged the settling of the western part of the United States — often with unfortunate consequences for Native Americans. The Romantic views that Bierstadt encouraged — and from which he personally profited, receiving as much as $25,000 a painting — also contributed to environmental movements dedicated to preserving wilderness areas as shrines of sacred space. It is not surprising that *View of the Rocky Mountains* hangs today in the White House, the symbolic center of the nation.[40]

Around the beginning of the twentieth century, a radically different approach to landscape painting, initiated by the French artist Paul Cézanne, transformed a relatively minor peak in southern France, Mont Sainte-Victoire, into the most celebrated mountain in Western art. At first sight, Cézanne's paintings of this mountain appear to have little to do with our usual conceptions of the sacred. Unlike the evocative landscapes of Bierstadt, their single-minded focus on the representation of color and form, the structure of the peak itself, calls forth no traditional religious associations, such as images of Mount Sinai or the Garden of Eden. And yet the intensity of these paintings, the austere power of the brushwork, gives Mont Sainte-Victoire a monumental presence that speaks of some deeper, more enduring reality hidden like rock beneath the surface of what we see. In a painting such as *Mont Sainte-Victoire from Les Lauves*, nothing intervenes to distract our attention from the mountain whose bold form, abstract in its simplicity, dominates the view, reducing everything else to patches of color that coalesce and take shape in relation to the peak.

MONT SAINT-VICTOIRE FROM LES LAUVES. *Painting by Paul Cézanne, Kunsthaus, Zurich. (Scala/Art Resource, N.Y.)*

As Hokusai had done with Fuji, Cézanne painted numerous views of Sainte-Victoire, depicting it from all sides. Like the Japanese artist, he found in his mountain a center of stability around which he could organize his landscapes—and his life. In the 1880s, after a series of artistic rejections and personal disappointments, he withdrew to his native Aix-en-Provence and devoted himself entirely to art, which he regarded as a "priestly vocation."[41] As he tramped through the countryside, painting and sketching the landscape, Cézanne turned repeatedly to Mont Sainte-Victoire, seeking to extract from its gray limestone something fixed and eternal that would protect him and his work from the vicissitudes of time and emotional turmoil. The mountain became a mysterious obsession that cast its shadow over everything he did—a symbol of what he was trying to attain. As the art historian Kenneth Clark has written, "Of the mountain he made innumerable studies, and we feel that the painting of this motive became for him like a ritual act of worship in which he could achieve perfect self-realisation."[42]

A contemporary who observed Cézanne at work described his painting as a meditation with a paintbrush. Slowly and carefully, with infinite concentration, he immersed himself in his subject, striving to depict in art the essence of what he saw in nature. As he explained to a friend, "The mind of the artist must be like a sensitive plate, a simple receiver at the moment he works, but to prepare the plate and make it sensitive, repeated immersions are needed—long work, meditation, study, sorrow, joy, life." All great art, he felt, came out of a "strong sensation of nature." This sensation, which the artist had to experience with his entire being, functioned for Cézanne as a sense of the sacred, revealing the reality that he dedicated his life to expressing in color and form. The act of depicting a rock or tree he approached as a Tibetan yogi would the visualization of a deity: "If I experience the slightest distraction, the slightest lapse, above all

MOUNT WILLIAMSON, THE SIERRA NEVADA, FROM MANZANAR, CALIFORNIA, 1945. *Photograph by Ansel Adams.*
(Copyright © by the Trustees of The Ansel Adams Publishing Rights Trust. All Rights Reserved)

if I interpret too much, if a theory takes me out of my concentration, if I think while I am painting, if I intervene, then everything collapses and all is lost." The result of his meditation was a kind of mystical experience of unity with the object of his brush: "I feel myself colored by the hues of infinity; I become one with my painting." Out of this kind of experience comes the peculiar power and intensity of Cézanne's art, especially his paintings of Mont Sainte-Victoire.[43]

In the art of the mid-twentieth century, the monumental quality of mountains that makes them natural symbols of eternity has found its most powerful expression in the medium of landscape photography. The sharpness of focus and extraordinary clarity of a photographic image give it a peculiar ability to highlight those aspects of a mountain that imbue it with

an aura of ultimate reality. The photographer whose work most vividly elicits this effect is Ansel Adams. His images of mountains in the Sierra Nevada of California come out of a sensibility shaped by the spiritual values of nature writers such as Henry Thoreau and John Muir. As a young man leading trips for the Sierra Club, Adams wrote:

No matter how sophisticated you may be, a large granite mountain cannot be denied—it speaks in silence to the very core of your being. There are some that care not to listen but the disciples are drawn to the high altar with magnetic certainty, knowing that a great Presence hovers over the ranges. . . .[44]

In *Mount Williamson, Sierra Nevada, from Manzanar,* the presence of which Adams speaks takes the

234

visual form of beams of light pouring in from the upper right corner of the picture. Wreathed in glowing wisps of sunlit clouds, the summit of Mount Williamson appears as a place of revelation, evoking images of God's fiery descent on Mount Sinai. Their chiseled surfaces glittering with reflections of the rays flooding over the mountain, the dark boulders in the foreground stand as altars of stone, drawing the divine presence down from the heights of heaven into the world of human experience. The rocks themselves become materializations of light, their solidity bringing together the realms of spirit and matter in an overwhelming sense of the sacred. Ultimate reality, the picture tells us, is here for us to reach out and touch.

Ansel Adams's images of Yosemite and the Sierra Nevada have assumed the status of religious icons representing the spiritual essence of the American wilderness. The enormous popularity enjoyed by his photographs reflects the secret yearning that Americans have to experience the sense of the sacred missing in so much of modern life. Shortly before his death, Adams intimated in words what he expressed so vividly in his picture of Mount Williamson:

We all move on the fringes of eternity and are sometimes granted vistas through the fabric of illusion.[45]

By creating a realm of the imagination apart from the world of everyday life, works of literature and art make it easier for us to experience other views of reality. No matter how fantastic such a realm may seem, as long as it has a beauty and consistency of its own, we are prepared to entertain it—or rather, to let it entertain us. A mountain can speak as a god in a poem or appear as a paradise in a painting. We suspend disbelief and accept the view of the peak as part of the world created by the poet or artist, just as we do the speech of animals in a fairy tale. In works of literature and art, we are open to possibilities that we would never consider in "real" life. A different set of criteria applies, one that gives us the freedom to explore alternate visions of reality. Through this kind of literary and artistic exploration, we can expand the range of our vision and renew our ability to experience a sense of the sacred in the world around us.[46]

THE SPIRITUAL DIMENSIONS

OF MOUNTAINEERING

WHEN I WAS A BOY, the high peaks of the Andes, where I lived at the time, held a peculiar fascination for me. A snow-capped mountain shining white in the morning sun had the power to move something deep within me, something that urged me to climb up to its summit and enter the pure and magical realm that hung there, gleaming in the sky. I did not know what that something was, nor where it came from, only that it was utterly compelling—and a source of endless daydreams. I began to climb with an Ecuadorean mountaineering club as a way of reaching that high and fantastic world of ice and snow so different from the usual surroundings of my everyday life. Only much later, after I had gone to the Himalayas and learned something of the religious beliefs and practices of the people who lived there, did I realize that what drew me to the heights had much in common with traditional views of sacred mountains and the spiritual goals they symbolize.

Even in the modern world, mountains and mountaineering embody values that many people hold sacred. As the highest point on earth, Mount Everest in particular has become for many a powerful symbol of ultimate goals. The importance the mountain has assumed in the West reflects a modern tendency to attribute ultimate value to the biggest, the best, the finest, the first. Whatever Western society regards as number one tends to take on an aura of ultimacy that makes it seem more real and worthwhile than anything else—in a word, sacred.

Expeditions to climb Mount Everest serve as models of effort and achievement for others to emulate in the world of work and play. The willingness to commit oneself to the highest goal and to do everything in one's power to attain it, even at the risk of sacrificing one's life, reflects values prized in science, business, sports, and other forms of endeavor. In a revealing passage typical of modern views of Everest, a newspaper quoted a member of an expedition to the mountain as saying, "People talk about the corporate world being a pyramid, and so is that mountain. Everest is a fantastic symbol: 'Setting your sights high,' 'going for it,' metaphors like that."[1] As a means of promoting these kinds of values, major corporations have hired Everest climbers to give motivational talks and seminars to their employees.

People commonly refer to ascents of Mount Everest, especially the first ascent by Sir Edmund Hillary and Tenzing Norgay in 1953, as conquests of the peak. In the view expressed in such remarks, climbing the mountain has become a symbol of the value that Western civilization has put on the conquest of nature, a conquest that glorifies the spirit of man and establishes his dominion over the things of this world. Under the influence of modernization with its emphasis on technology and economic development, this value has spread even to the East. When a joint Japanese-Chinese expedition made the first live television transmission from the summit of Everest on a traverse of the peak, a Chinese announcer in Beijing excitedly declared, "Mankind has crossed the highest mountain in a new triumph of the human spirit."[2]

The values of ultimacy, achievement, and conquest revealed in contemporary views of mountains and mountain climbing occupy the place held by religious concerns in traditional cultures. They give many people today the sense of meaning, purpose, and direction that institutional religion used to provide but no

A rock climber clings to a sheer wall beside Yosemite Falls, Sierra Nevada, California. (Galen Rowell/Mountain Light)

longer does. The willingness of business entrepreneurs, for example, to dedicate, even sacrifice, their lives to making a success of their ventures indicates that achieving their objectives has become for them an aim of ultimate concern comparable to the goal of a traditional monk or hermit. The difference lies in the nature of the values directing the lives of each: in the case of the business person, these values are largely material; in the case of the religious seeker, spiritual. One individual pursues his or her aims primarily for self-aggrandizement, the other for self-transcendence. And yet the quest for wealth or fame, whether at the top of the corporate pyramid or on the summit of a mountain, may transform a person's awareness, shifting his or her gaze to other, higher goals.

A spiritual dimension often lies hidden in the materialistic values of ultimacy, achievement, and conquest reflected in modern views of mountaineering. Referring to these values, George Mallory, who later disappeared on Everest, wrote the following words on the experience of reaching the summit of Mont Blanc, the highest peak in the Alps:

Have we vanquished an enemy? None but ourselves. Have we gained success? That word means nothing here. Have we won a kingdom? No . . . and yes. We have achieved an ultimate satisfaction . . . fulfilled a destiny. . . . To struggle and to understand—never this last without the other; such is the law. . . .[3]

Rather than conquer the mountain, the climber vanquishes himself, much as a hermit or yogi overcomes the enemy of his own pride and arrogance on the way to attaining his goal of self-transcendence. Success in a material sense Mallory discounts, and the kingdom he wins in its place carries, for the Christian reader, echoes of the kingdom of heaven described in the New Testament. Finally, in Mallory's view, the struggle, the effort to reach the summit, ends not in a grandiose celebration of victory, but in a quiet understanding that leads to deeper insight and wisdom.

THE CALL OF THE HEIGHTS

Any discussion of Mallory brings up the question: why do climbers do what they do? What makes them risk their lives and endure great discomfort for no immediate material reward? The answer we get depends on the person we ask. Some people climb for fame and glory, to prove themselves, or to conquer a mountain. Others do it for the sense of achievement and satisfaction they experience in using their skills to master something dangerous and difficult. Many go up to the heights simply for pleasure, for the wild delight that comes from being high on a peak, under the open sky. Still others do it to escape from the tedium of everyday life—or the problems that plague them at home. Although few profess to be openly seeking the sacred, a spiritual motivation often lies hidden in their reasons for climbing.

When asked why he wanted to climb Mount Everest, Mallory gave an offhand answer that has become the most famous and commonly cited reason for climbing a mountain: "Because it is there." Why do people quote his reply more often than any other? The answer, perhaps, lies in the intriguing, Zen-like nature of Mallory's response. The enigmatic simplicity of his reply teases the imagination, hinting at some deeper significance hidden beneath its apparent meaning. One feels instinctively that there must be something more to what he said. And if we examine his response from the point of view of what we know of the sacred as experienced in traditional cultures, we can, indeed, find a more profound import in Mallory's words—an import that may well account for the durability of his casual remark.

"Because it is there"—what is the "it" there that provides the motivation for climbing? Is it simply the mountain, or is it something more? If Mallory had been a Tibetan who venerated Everest—or, better, Kailas—as a sacred peak, he might have seen it as an embodiment or expression of some aspect of ultimate reality, whatever form that reality might take, be it a deity, the abode of a god, or something else. Uttered from such a perspective, the "it" in his reply would have referred to what was really there, what for him would have been ultimately real, eternally present, compared to which everything else would have faded into illusion, as the clouds that conceal a summit dissipate into thin air. Although Mallory did not belong to a culture that regarded Everest as sacred in a traditional sense, his words do reverberate with intimations of a deeper perception of the mountain as a manifestation of ultimate reality—or the place on which to encounter it.

Whether Mallory had such a meaning in mind at the moment he spoke—and it seems doubtful that he did, at least consciously—others have certainly heard it in what he said. A literary critic in a review of mountaineering literature wrote:

The distinguishing characteristic of Mallory's words is that they are primarily religious in nature: their exact equivalent in meaning is to be found in the sacred writings of the Hindus, in our own Holy Bible, and no doubt in similar texts. In the Sanskrit it is written, Tat tvam asi, which means, "That Thou art"; and the Lord says to Moses, "I am that I am." All three statements are alike in being ontological—they make the assertion of existence, that it is.

Suffused with the soft red light of alpenglow, the summit of Mount Everest evokes visions of a higher, more perfect reality close to the stars. (Edwin Bernbaum)

Moreover all three statements are identical in meaning, and differ only in the person. In the first person, the statement is, "I am . . . ," in the second, "Thou art . . . ," and in the third, "It is. . . . " Mallory was speaking the language of theology.[4]

As if in support of this interpretation, high on Mount Everest, at over 26,000 feet, British climber Frank Smythe had an intimation of something utterly transcendent, of an awesome reality apart from the world of human existence, a presence that permeated the rock and air, the very substance of the mountain itself:

It was cold. Space, the air we breathed, the yellow rocks were deadly cold. There was something ultimate, passionless, and eternal in this cold. It came to us as a single constant note from the depths of space; we stood on the very boundary of life and death.[5]

Mallory himself vanished into this ultimate reality, the "it" that is there in Mount Everest. In 1924 he and his companion, Andrew Irvine, disappeared into the clouds near the summit, leaving behind them a story and a question that have become an important part of the modern mythology surrounding the world's highest peak: did they reach the top before they died, long before the mountain was officially climbed in 1953? The mysterious circumstances under which they vanished have added immeasurably to the mystique of Everest and its significance for climbers today.

Whatever Mallory intended to mean in his famous reply, in terms of the effort he dedicated to the mountain and the life that he sacrificed there, Everest — or the ascent of it — represented for him something infinitely more real and worthwhile than the things that most people regard as important. Explaining on another occasion why he was going out to climb the peak, he is supposed to have said:

So, if you cannot understand that there is something in man which responds to the challenge of this mountain and goes out to meet it, that the struggle is the struggle of life itself upward and forever upward, then you won't see why we go. What we get from this adventure is just sheer joy. And joy is, after all, the end of life. We do not live to eat and make money. We eat and make money to enjoy life. That is what life means and what life is for.[6]

Standing at the center of the world—the view from 16,000 feet on Denali (Mount McKinley) in Alaska.
(Galen Rowell/Mountain Light)

The joy that he sought on the heights of Mount Everest, in the encounter with the mountain itself, represented for Mallory an ultimate end, his own equivalent to the bliss experienced by mystics in their struggle for union with the objects of their spiritual quests.

F. W. Bourdillon, another British Everest climber, spoke of this kind of joy as the principal reason— usually hidden and unacknowledged—for climbing mountains and linked it explicitly to the sense of the sacred:

One reason is never given openly, rather is disguised and hidden and never even allowed in suggestion, and I venture to think it is because it is really the inmost moving impulse in all true mountain-lovers, a feeling so deep and so pure and so personal as to be almost sacred—too intimate for ordinary mention. That is, the ideal joy that only mountains give—the unreasoned, uncovetous, unworldly love of them we know not why, we care not why, only because they are what they are; because they move us in some way which nothing else does . . . and we feel that a world that can give such rapture must be a good

world, a life capable of such feeling must be worth the living.[7]

The religious motivation concealed in such feelings comes to the surface in the words of Maurice Herzog when, after his ordeal on Annapurna, he explained why he and his companions went out to climb the Himalayas: "The mountains had bestowed on us their beauties, and we adored them with a child's simplicity and revered them with a monk's veneration of the divine."[8]

Some climbers have even attempted to make mountains the source of religion itself. In a speech delivered to the Mountain Club of South Africa, Jan Smuts, who later became prime minister of the Union of South Africa in 1919, declared:

The Mountain is not something externally sublime. It has a great historic and spiritual meaning for us. It stands for us as the ladder of life. Nay, more; it is the ladder of the soul, and in a curious way the source of religion. From it came the Law, from it came the Gospel in the Sermon on the Mount. We may truly say that the highest religion is the Religion of the Mountain.[9]

Few mountaineers would make such an extravagant claim for the objects of their devotion, but Smuts's words do reflect, in extreme form, an attitude that many people bring to the mountains they climb—and contemplate.

SACRED MOUNTAINS AND MOUNTAINEERING

Whether they realize it or not, many mountaineers view the peaks they climb in many of the same ways people of traditional cultures regard the mountains they revere. By examining the roles that the most important of these views play in the practice of mountaineering, we can gain a deeper understanding of what impels climbers to risk their lives on high and dangerous peaks.

In keeping with the individualistic nature of their sport, climbers tend to view mountains as centers in a personal rather than a cosmic sense. Many of them order their lives around a peak that has become the focal point of their attention, subordinating everything to the task of reaching its summit. The mountain becomes the central preoccupation of their waking thoughts and the recurring image of their nightly dreams. This is particularly true of climbers who participate in major expeditions to distant peaks that require vast expenditures of time and effort, both in preparation and actual execution. The members of such an expedition may also unwittingly mimic the circumambulations of pilgrims around a central sacred mountain by circling their peak to look for possible routes up it. In so doing they pay a kind of reverence to the object of their devotion, appreciating the beauties of its faces regardless of whether they offer a way up to the summit.

The view from the top of a peak may, nevertheless, give a climber a concrete experience of the mountain as a cosmic center. As he emerges from the steep confines of a ridge or face and steps on the summit, the great circle of the horizon opens around him, like the rim of a huge mandala. I remember my amazement on reaching the top of a peak near Everest to find myself surrounded by an incredible array of seemingly endless mountains, rolling in wave after wave out to the edge of the sky. I had the distinct impression of standing at the center of a universe far vaster and more mysterious than anything I had ever known or imagined.

The sheer height of a peak constitutes one of its principal sources of attraction for a mountaineer. Most of the importance that climbers attach to Mount Everest comes from its preeminence as the highest point on earth. Other peaks in the Himalayas are more beautiful, difficult, and interesting to climb, but none has the special status and prestige of Everest, which derives from its unsurpassed height. That height represents an ultimate value that endows the mountain with a kind of sacred significance for the modern world, to which mountaineers belong. It also imbues the peak with an aura of otherness, a mysterious reality that transcends the earth and partakes of heaven.

In climbing a mountain, a climber enters another realm of existence, perceived by many as higher spiritually as well as physically. On gazing out a window and seeing the Alps for the first time, Sir Martin Conway, a noted mountaineer and explorer, wrote:

They were not in the least like clouds, nor like anything I had ever beheld or dreamed of. Had they been built of transparent crystal, they could not have been more brilliant. I felt them as no part of this earth or in any way belonging to the world of experience. Here at last was the other world, visible, inaccessible, no doubt, but authentically there; actual yet incredible, veritably solid with an aspect of eternal endurance, yet also ethereal; overwhelmingly magnificent but attractive too.[10]

At first Conway felt no inclination to climb up to this other world: it seemed too lofty and inaccessible for the idea even to enter his mind. But later, in the course of his life, he ventured into its highest reaches on some of the most remote mountain ranges in the world.

The mysterious other world of the mountains may take the form of a heaven in the sky or a paradise on earth. There a mountaineer may see, shining sharply in the sun or glowing softly through the clouds, the things he or she desires but cannot find in this world of gray frustration and imperfection. Many climbers go up to the mountains for a brief escape to paradise, high above the concerns and complications of modern life. Approaching the Himalayas, Gaston Rébuffat, a companion of Herzog on the French expedition that first climbed Annapurna, mused on the mountains in which he found himself:

This place is just like home, only on a larger scale! It is one of those places marked in ochre and white in the atlas, high, sterile and good for nothing; nothing marketable grows there, and higher still nothing can exist at all. It is one of those spots made solely for the happiness of men, in order that in this changing world, grown every day more artificial, they might yet find a few gardens still unspoiled in their silence of forgetfulness, a few gardens full of primal colors that are good for the eyes and for the heart.[11]

Given the monotheistic background of most Western climbers, few of them view mountains as heavenly abodes of various gods, but a number look up to the heights as ideal places to experience the presence of the supreme deity or to commune with the infinite. John Muir regarded the peaks of the Sierra Nevada as cathedrals and temples where he might rejoice in the divine spirit that infused the world with light and life. Reflecting on his ascent of Siniolchu, one of the most beautiful peaks in the Himalayas, the German climber Karl Wien wrote:

So on this day the mountain still stood before us, that we had reached its summit seemed a divine favour which filled us with happiness and gratitude, and all that we had seen and experienced during the hours of our struggle upon the slopes only deepened our reverence for all God's creation.[12]

It is not uncommon for climbers to find themselves overwhelmed with sudden feelings of reverence for the divine creator of the world in which they are privileged to live.

Mountains also have their dark sides, black faces streaked with ice, gouged by rocks and mottled with clouds. Some mountaineers come to see their peaks as hells, horrifying places of suffering and damnation. Many a climber has met a ghastly end, trapped in an avalanche or shattered on rocks. The demons that traditional cultures see haunting the heights climbers often find lurking within themselves—in the passions and fears, envy and rage that emerge under trying conditions to rip expeditions apart. A mountain that seemed divine may suddenly take on the appearance of a demon itself, acting willfully to torment and kill those who try to scale its flanks. To his horror a climber may see its faces, ridges, and clouds come to life, filled with a malice that seems directed at him. At such times it does little good to say to himself that he is only personifying an inanimate object: he feels something there that he cannot dismiss. Struggle as he may, nothing works; an active intelligence seems to be relentlessly impelling him toward his doom.

Mountaineers often see mountains as the hallowed tombs of those who have perished on their heights and lie buried in their snows. When Devi Unsoeld died on Nanda Devi, the mountain after which she had been named, her companions gathered in a circle and, in the words of her father, "laid the body to rest in its icy tomb."[13] A newspaper article about a women's expedition to a peak in western China reported a similar kind of view shared by other climbers: "The group plans to place a memorial plaque on Mt. Kongur to commemorate women mountaineers of all nationalities 'whose souls rest in the mountains of the world.'"[14] When Herzog slipped into a delirium on the way down from his ordeal on Annapurna, he had a vision of his death in which he saw the mountain as his grave:

I looked death straight in the face, besought it with all my strength. Then abruptly I had a vision of the life of men.

Those who are leaving it for ever are never alone. Resting against the mountain, which was watching over me, I discovered horizons I had never seen. There at my feet, on those vast plains, millions of beings were following a destiny they had not chosen.

There is a supernatural power in those close to death. Strange intuitions identify one with the whole world. The mountain spoke with the wind as it whistled over the ridges or ruffled the foliage. All would end well. I should remain there, forever, beneath a few stones and a cross.

They had given me my ice axe. The breeze was gentle and sweetly scented. My friends departed, knowing that I was now safe. I watched them go their way with slow, sad steps. The procession withdrew along the narrow path. They would regain the plains and the wide horizons. For me, silence.[15]

This passage recalls the traditional cry uttered at the beginning of a Japanese funeral procession: "Yama-yuki! We go to the mountain!" It also reflects the sentiments of many climbers who would like, in the end, to die and be buried in the mountains they love.

The possibility of danger and death, always there in the mountains, adds something essential to the experience of climbing. Although climbers seek to minimize the risks they take, to eliminate them altogether would kill what gives life to their sport. The encounter with death has a marvelous power to awaken the spirit and focus the mind. It forces a climber to concentrate on what is important and real, to dispense with the trivial concerns that normally fog his vision and distract his attention. Jolted by the possibility of being no more, he sees what lies around him with a sharper, brighter awareness and feels a sudden burst of appreciation for the life that courses through his body. As a Hindu swami I met in the Himalayas remarked, "Death is the greatest of teachers."

The extreme conditions encountered on mountains—wind, clouds, fatigue, and altitude—predispose climbers to having visions or to seeing unusual phenomena in a visionary light. In 1865, coming down from the first ascent of the Matterhorn, just after four of their companions had fallen to their deaths in the most famous accident in mountaineering history, Edward Whymper and his guides saw this strange sight materialize before them, uncanny in the coincidence of its timing:

When, lo! a mighty arch appeared, rising above the Lyskamm, high into the sky. Pale, colourless, and noiseless, but perfectly sharp and defined, except where it was lost in the clouds, this unearthly apparition seemed like a vision from another world; and, almost appalled, we watched with amazement the gradual development of two vast crosses, one on either side. If the Taugwalders [his local guides] had not been the first to perceive it, I should have *doubted my senses. They thought it had some connection with the accident, and I, after a while, that it might bear some relation to ourselves. But our movements had no effect upon it. The spectral forms remained motionless. It was a fearful and wonderful sight; unique in my experience, and impressive beyond description, coming at such a moment.*[16]

Alpine villagers imbued with Christian beliefs, the Taugwalders interpreted what they saw as a supernatural sign of their companions' deaths, and Whymper himself reacted with feelings of religious awe. Buddhist pilgrims to O-mei Shan in China would have seen this optical phenomenon, created by the projection of light and shadow on mist, as a manifestation of the Buddha's glory.

Other mountaineers have reported experiencing actual visions. High on Mount Everest, his mind wandering from the effects of extreme altitude and fatigue, Frank Smythe had the distinct impression that someone else, a "third man," had joined him on his climbing rope. The feeling was so real that at one point he even turned to offer his mysterious companion something to eat. A Sioux or Cheyenne on a vision quest high on a lonely hill would have interpreted such an experience as a prized encounter with a guardian spirit. Drawing on the mandala symbolism of Tibetan Buddhism, Reinhold Messner, regarded by many as the foremost mountaineer in the world today, gave a vision he experienced at high altitude on Kangchenjunga a decidedly spiritual interpretation:

In the last camp near the summit, I had a very strange vision of all the human parts I am made of. It is very difficult to keep the vision, but I know that I could see a round picture with many pictures inside—not only of my body, but of my whole being. There was a lot of what my life has been, what I did these last years, like seeing my life and my body and my soul and my feelings inside a mandala. But I was not even sure if it was only mine or generally human, yours or anybody's, just a human being's. It was very, very strange.[17]

Whether their experience derives from a vision, inspiration, or just the simple fact of being there, mountaineers regard mountains as a source of blessings, many of them spiritual in nature. From contact with forest and stream, rock and snow, come health, good spirits, and peace of mind, as well as a fresh perspective that can lead to new ideas and ways of seeing things. John Muir, a passionate advocate of mountaineering, urged others to seek such blessings in the mountain heights:

Climb the mountains and get their good tidings. Nature's peace will flow into you as sunshine flows into trees. The winds will blow their own freshness into you, and the

storms their energy, while cares will drop off like autumn leaves.[18]

Some mountaineers have gone even further. Geoffrey Winthrop Young, a noted British climber and poet, regarded mountains as the primary source of inspiration for the intellectual and spiritual development of the human race. In a lecture delivered to the University of Glasgow in 1956, he argued:

In effect, the visible mountain ladder, cloud-compelling and controlling rain and sun, added a third dimension, that of height, to the length and breadth of surface supporting man in the dawn of his intelligence. It gave a new measure to his concrete vision of earth. By so doing, by asserting the existence of a higher world and of a higher order inhabiting it, mountains became the first forces to lift the eyes and thoughts of our branch of animal life above the levels of difficult existence to the perception of a region of spirit, located, as children would locate it, in the sky above.[19]

Although Young was speaking specifically of attitudes he assumed to have originated in Greece, the image

A vision of two crosses materializes before awestruck survivors of the accident on the first ascent of the Matterhorn in 1865. Illustration by Edward Whymper.

of Moses receiving the law on Mount Sinai also lies behind his words.

Inspired in part by images drawn from classical and biblical literature—powerful influences in the Western tradition—mountaineers tend to look up to high peaks as places of personal transformation and purification. When I began to climb as a boy in Ecuador, I carried the hope that in reaching the top of a mountain I would be magically changed and return a new person. An Ecuadorean friend who climbed with me reinforced this idea by expressing the opinion that climbers were better people for climbing. As I discovered, that was not often the case: mountaineers turned out to be just as insecure, confused, and egotistical as any other group of individuals. But the possibility that the experience of climbing a mountain could change a person had a strong appeal and provided a rationale for what we did—as it does for many outdoor courses that seek to develop character and instill self-confidence through exposure to mountains and the practice of mountaineering. Certainly William O. Douglas, a member of the U.S. Supreme Court and a climber himself, had this sort of thing in mind when he wrote:

A people who climb the ridges and sleep under the stars in high mountain meadows, who enter the forest and scale peaks, who explore glaciers and walk ridges buried deep in snow—these people will give their country some of the indomitable spirit of the mountains.[20]

Many climbers are searching for some kind of fulfillment, for something missing in their lives. Like a hermit meditating in a cave or a pilgrim going to a sacred site, they strive to transcend the unsatisfactory world of ordinary life and enter a realm of higher and more perfect existence. The ascent of a peak provides them with a natural symbol of the transcendence they seek. In going up a mountain, a person leaves behind the familiar world of grass and trees to venture into the strange and fantastic realm of the heights. Standing on a crag of rock and ice, gazing down into green valleys far below, he or she has a vivid and concrete sense of having transcended the sphere of ordinary existence. The harsh environment of the heights demands that climbers rise above their physical, mental, and spiritual limitations.

In ascending a mountain, a climber sometimes transcends not only the world, but himself. At moments he may find himself doing things he had never imagined being able to do. While climbing a new route on a Himalayan peak, I started up what appeared to be a very difficult section of a steep face. After a few moments of trepidation, all my fears suddenly dropped away. Poised on a wall of green ice two thousand feet above a glacier, I found myself moving with uncanny precision and certainty—so different from the hesitant way I usually groped my way from hold to hold. It seemed absolutely impossible to fall. I felt connected to the face. If my foot should slip, I knew I could simply reach out a hand and grab an icicle to stay on. I crossed the section in no time at all, and when my companion struggled up and commented on the difficulty of my lead, I felt puzzled: it had seemed so incredibly easy.

In such experiences, when everything becomes clear and simple, a climber momentarily becomes one with himself and the world around him. He tastes the inner freedom and certainty that monks and yogis strive to attain in more traditional and permanent ways. Reflecting on what he and his companions had sought on Annapurna, Herzog wrote, "For us the mountains had been a natural field of activity where, playing on the frontiers of life and death, we had found the freedom for which we were blindly groping and which was as necessary to us as bread."[21] This sense of freedom comes in part from contact with reality, the ultimate reality embodied in a mountain. In the concentration of climbing, becoming one with the rock and snow over which he moves, the climber opens himself to that reality and the wonder that it brings. His senses sharpen to an almost supernatural acuity, and he becomes aware of intense beauty in the smallest things. Having entered this state on an ascent of El Capitan in Yosemite Valley, Yvon Chouinard, a noted American climber, remarked:

Each individual crystal in the granite stood out in bold relief. The varied shapes of the clouds never ceased to attract our attention. For the first time we noticed tiny bugs that were all over the walls, so tiny they were barely noticeable. While belaying, I stared at one for 15 minutes, watching him move and admiring his brilliant red color.[22]

Completely engrossed in the act of climbing, a climber loses awareness of time and enters an almost mystical state of timelessness.

Mountaineers often approach a mountain as a pilgrim does a sacred peak. In its summit they see enshrined the object of their dreams, the goal they long to attain. Climbing becomes a way of getting closer to the mountain they love and experiencing the reality that it embodies. At times, like a pilgrim, they may find it enough to contemplate the peak from its foot—and return from there, satisfied with what they have seen. They may even have the feeling that it would be an act of sacrilege to reach its summit, that in so

Climbing in a realm of fantasy: the ascent of an ice tower on a glacier in the Ecuadorean Andes. (Edwin Bernbaum)

The joy of returning from the heights. A climber exults at the first sight of green grass in the Karakoram Mountains of Pakistan. (Galen Rowell/Mountain Light)

doing they would rob the mountain of the mystery that makes it so attractive. *Everest: The West Ridge*, an account of the first American expedition to reach the summit of Mount Everest, opens with the following quote from the Scottish mountaineer W. H. Murray, recording his thoughts on first seeing the Himalayas he had come to climb:

They were there. An arctic continent of the heavens, far above the earth and its girdling clouds: divorced wholly from this planet. The idea of climbing over such distant and delicate tips, the very desire of it, never entered my heart or head. Had I been born among or in sight of them, I might have been led to worship the infinite beauty they symbolized, but not to set boot on their flanks, or axe on a crest.[23]

Despite these feelings—or perhaps because of them—Murray went on to climb a number of Himalayan peaks, but with a feeling of reverence for the summits he reached.

Many religious traditions see the ascent of a mountain as a powerful symbol of the ultimate pilgrimage—the spiritual path leading to the highest goal. Dante drew deeply on such a view to create his famous allegory of the purgation of the soul as a climb of Mount Purgatory. The *Mahabharata*, the great epic of Indian literature, ends with a king's ascent to heaven via the Himalayas and Mount Meru. This view of mountain climbing as a symbol of the spiritual path brings together and unifies many of the views examined separately above. Seen from a distance, a peak appears as a vision of the goal toward which a person aspires, impossibly high and remote. As the climber approaches the mountain, it takes on texture and definition: a way up it comes into view, a series of gullies and ridges. The actual ascent of the peak requires great effort and concentration, even self-sacrifice. At various points, a route seems to peter out so that the climber must retreat and start again; doubt and uncertainty set in, just as they do for those engaged in following the spiritual path. Finally, the complexities of the way resolve themselves in the simplicity of the summit, where all the ridges come together in a single peak. There at the very top, the climber stands simultaneously at the center of the mountain—conceived as a perfect pyramid—and at its highest point.

The two great themes of the peak as a cosmic axis and as a high place come together in the attainment of the goal, the exalted spot from which a mountaineer views the world as a whole and looks up to the infinite reaches of the open sky.[24]

The symbolism inherent in climbing a mountain influences the thinking and attitudes of many mountaineers. Some are naturally predisposed to view their ascents in this way. Pius XI, an alpinist before he became pope, wrote:

In the laborious efforts to gain the summits where the air is lighter and purer, the climber gains new strength of limb, while in the endeavor to overcome the countless obstacles of the way, the soul trains itself to conquer the difficulties of Duty; and the superb spectacle of the vast horizons, which from the crest of the Alps offers themselves on all sides to our eyes, raises without effort our spirits to the divine Author and Sovereign of Nature.[25]

Guido Rey, an Italian climber who devoted his life to climbing the Alps, spoke of his experience of reaching the summit as the goal and fulfillment of the philosophical or spiritual quest:

. . . and I tasted the fresh, ineffable joy of reaching the highest point—the summit; the spot where the mountain ceases to rise and man's soul to yearn. It is an almost perfect form of spiritual satisfaction, such as is perhaps attained by the philosopher who has at last discovered a truth that contents and rests his mind.[26]

The views that mountaineers share with people of traditional cultures are more than beautiful ideas or poetic analogies: they have the power to awaken a sense of the sacred. The images found in these views can come together and fuse in the mind of the climber, moving him in deep and inexplicable ways. At certain moments his image of heaven, for example, may merge with the mountain on which he stands to stir something deep within him, totally transforming his perception of himself and the world around him. Sir Leslie Stephen described such an experience on the summit of Mont Blanc and the power its memory had to move him years afterward:

Even on the top of Mont Blanc one may be a very long way from heaven. And yet the mere physical elevation of a league above the sea-level seems to raise one by moments into a sphere above the petty interests of everyday life. Why that should be so, and by what strange threads of association the reds and blues of a gorgeous sunset, the fantastic shapes of clouds and shadows at that dizzy height, and the dramatic changes that sweep over the boundless region beneath your feet, should stir you like mysterious

music, or, indeed, why music itself should have such power, I leave to philosophers to explain. This only I know, that even the memory of that summer evening on the top of Mont Blanc has the power to plunge me into strange reveries not to be analyzed by any capacity, and still less capable of expression by the help of a few black marks on white paper.[27]

We climb to hear—whispered in the wind, echoed in the stars—strains of that "mysterious music." Without an intimation of its harmonies reverberating around their heights, mountains become heaps of dust and rock—or glorified pieces of gymnastic equipment. The tendency of modern society to commercialize mountaineering and reduce it to a competitive sport removes its mystique and kills its spirit. Something is lost when hotels in Kathmandu display signs on their marquees welcoming the "such-and-such Mount Everest Expedition" as they would another business convention. Mountaineering itself becomes a one-dimensional activity in which each climber's skill and accomplishments are simply weighed and measured against another's. No longer do evocative images resonate in views of mountains to awaken a sense of deeper mystery. The peaks themselves flatten out and lose their impressive sense of height and grandeur. And a dull emptiness filled with noise replaces the silent music that played about their summits.

Much of the special appeal of mountain climbing comes from the fact that it takes us out of the ordinary world of everyday life to a magical place where we can experience spontaneous feelings of wonder and awe. There, far from the profane realm of routine concerns, we feel free to break out of conventional patterns and discover new and more authentic ways of being. Rather than carry the materialistic, competitive values of modern society into the mountains, converting them into amusement parks and sports stadiums, we need to bring back from the heights a renewed sense of the sacred to transform and revitalize our experience of ourselves and the world back home. Herzog concludes his book on climbing Annapurna with the realization that

Annapurna, to which we had gone emptyhanded, was a treasure on which we should live the rest of our days. With this realization we turn the page: a new life begins. There are other Annapurnas in the lives of men.[28]

What Herzog discovered with great effort and sacrifice on the summit of a distant Himalayan peak, we can find right here, revealed through the fresh, new vision awakened by views of sacred mountains. With this thought we, too, will turn to other Annapurnas in the lives of men and women.

· 15 ·

MOUNTAINS, WILDERNESS,

AND EVERYDAY LIFE

For most of us sacred mountains are remote from our experience of everyday life. They lie far off in space and time, revered by distant cultures, many of which vanished long ago. Even the peaks that we manage to climb and visit rise on the borders of our lives, removed from the cities and plains where most of us live. What is the value, then, of thinking about them? It is simply this: the contemplation of sacred mountains, with their special power to awaken another, deeper way of experiencing reality, opens us to a sense of the sacred in our own homes and communities — a sense that we need to cultivate in order to live in harmony with our environment and with each other. In looking up to the heights and reflecting on the world around them, we discover within ourselves something that enables us to lead deeper and more meaningful lives.

A powerful experience of the sacred high on a mountain can overturn old conceptions and awaken a new awareness of people and things back home. A world we had never noticed or had taken for granted may suddenly appear fresh and bright before our eyes. In describing the long-lasting effects of his experience on Annapurna, Maurice Herzog refers to just such a shift in perspective, one that extended far beyond the mountain to transform his perceptions of everything else, including himself:

In overstepping our limitations, in touching the extreme boundaries of man's world, we have come to know something of its true splendor. In my worst moments of anguish, I seemed to discover the deep significance of existence of which till then I had been unaware. I saw that it was better to be true than to be strong. The marks of the ordeal are apparent on my body. I was saved and I had won my freedom. This freedom, which I shall never lose, has given me the assurance and serenity of a man who has fulfilled himself. It has given me the rare joy of loving that which I used to despise. A new and splendid life has opened out before me.[1]

In a less dramatic manner, exposure to traditional views of sacred mountains may, over time, alter the way we see the world and lead us to a deeper understanding of other people's ways of experiencing reality. Ben Hufford, the lawyer who represented the Hopi and Navajo in their efforts to block development on the San Francisco Peaks in Arizona, told me that many whites attending legal hearings came to appreciate Native American attitudes toward the sacred site after elders and medicine men compared the mountains to a church. That view of the San Francisco Peaks helped them realize that features of the natural landscape could be just as much places of worship as man-made structures like churches, which the general public would never consider desecrating. They could understand more clearly how the Hopi and Navajo could experience the sacred in nature — and were able to gain a deeper appreciation for their beliefs and practices. Unfortunately, United States Supreme Court decisions on the San Francisco Peaks and other Native American religious sites have not shown a comparable respect for this view of reality. One hopes that in time they will.

A man silhouetted against the grandeur of one of the Bhagirathi Peaks in the Indian Himalayas.
(Robert Mackinlay)

MOUNTAINS AND WILDERNESS

When we think of a mountain peak, we usually envision it as a paradigm of wilderness in its wildest and purest form—a spiritually uplifting realm of forests, streams, crags, and snows unspoiled by the works of man. Although villages may cluster around their feet, the summits of the highest peaks lie beyond the reach of human use—and abuse—too steep and high for permanent habitation. Unlike jungles and deserts, two other features of the natural landscape that embody powerful images of wilderness, the heights of mountains cannot be cut down or made to bloom, transformed into cities and farmland. The few huts placed high on their sides for the use of climbers seem to perch there as tiny intruders totally at the mercy of the environment, easily wiped out by a rockfall or avalanche, if the mountain so moves. The forces of untamed nature—wind, cloud, storm, and cold—find their most powerful expression on the tops of mountains, imbuing the heights with an aura of wilderness in its most extreme and inviolable state.

During the Middle Ages, this wild and pristine image of wilderness, which makes mountains so attractive to us today, provoked horror in Europeans. They viewed ranges like the Alps and the Harz Mountains as the haunted realms of witches, dragons, and other demonic beings. There, in the irregular forms of peaks and ridges, lay a world of chaotic forces outside human control, forces opposed to the will of God. With the positive view of wilderness that emerged at the end of the Enlightenment, attitudes toward mountains changed. As the poetry of the Romantic period so clearly demonstrates, the very wildness of nature that the Middle Ages abhorred came to be seen as the manifestation of a divine reality infusing the world and resolving its contradictions. The rugged landscape of the mountains provided the ideal setting for awakening an awareness of this reality, conceived as a kind of infinite mind or universal consciousness.

The rise of science with its interest in nature helped to encourage the development of an appreciation for the spiritual value of wilderness. Whereas the Middle Ages had seen only chaos in the wild landscape of forest and mountains, scientific discoveries revealed the existence of a natural order more subtle and intricate than anything conceived by the human mind. The world of peaks and glaciers, meadows and streams, birds and animals, was seen to form a harmonious system in which each part had a place in the whole. As the writings of naturalists like John Muir so clearly

The Mitten Buttes at sunset reveal the spiritual glory of nature. Navajo National Monument, Arizona.
(Edwin Bernbaum)

attest, the contemplation of this natural order in mountain landscapes awakened a powerful sense of the sacred that influenced people's relationship to the uninhabited parts of their environment. The appearance of the new field of ecology in the twentieth century has reinforced the development of a positive, more spiritual attitude toward wilderness that began in the eighteenth century. Like Chinese sages and poets attempting to become one with the Tao, many people today go to the mountains to put themselves in harmony with what they perceive as a higher, more perfect order infusing the world of nature.

Inspired by the biblical image of Eden, we also tend to view wild places, especially forested peaks, as earthly paradises. Unlike theologians of the Middle Ages, Romantics of the nineteenth century saw the wilderness not as a wasteland, but as a garden cultivated by God—a place of natural abundance, ruled by divine order, rather than chaos. Over this earthly paradise, protecting it from the depredations of humankind, stood the rugged heights of mountains. Seeing the Alps for the first time, John Ruskin wrote:

There was no thought in any of us of their being mere clouds. They were as clear as crystal, sharp on the pure horizon sky, and already tinged with rose by the sinking sun. Infinitely beyond all that we had ever thought or dreamed—the seen walls of lost Eden could not have been more beautiful to us; not more awful, round heaven, the walls of sacred death.[2]

Although we may not retain nineteenth-century conceptions of divinity, many of us have inherited Ruskin's view of mountains as the wild walls of paradise, awful in the sense of inspiring awe.

Whether viewed as the Garden of Eden or as a harsher, more ascetic realm, wilderness functions for many as a sacred space, set apart from the profane realm of everyday life. There, far from the civilized world, lies the mysterious domain of the wholly other, governed by natural forces beyond the reach of human control. By exposing themselves to these forces, wilderness enthusiasts seek to awaken a sense of the sacred that will enable them to transcend their usual preoccupations and know, for a brief time, the taste of a more enduring reality. Like the Garden of Eden, wild places preserve for them the primordial purity of creation, a sacred space that remains undesecrated by humankind. Indeed, the wilderness represents, for many of us, a place of spiritual renewal, where we can go back to the source of our being and recover the freshness of a new beginning. As John Muir put it:

I used to envy the father of our race, dwelling as he did in contact with the new-made fields and plants of Eden; but I do so no more, because I have discovered that I also live in "creation's dawn." The morning stars still sing together, and the world, not yet half made, becomes more beautiful every day.[3]

The idea of wilderness as sacred space has provided the underlying inspiration for numerous conservation movements. John Muir founded the Sierra Club, for example, as a means of preserving the natural sanctity of wild places in the Sierra Nevada of California. Muir's writings protesting the flooding of Hetch Hetchy, a valley comparable in splendor to Yosemite, reveal the essentially spiritual nature of his motivation:

These temple-destroyers, devotees of ravaging commercialism, seem to have a perfect contempt for Nature, and, instead of lifting their eyes to the God of the mountains, lift them to the Almighty Dollar. Dam Hetch Hetchy! As well dam for watertanks the people's cathedrals and churches, for no holier temple has ever been consecrated by the heart of man.[4]

Reflecting a view of mountains as paradigms of wilderness in its most dramatic and pristine state, conservation organizations such as the Sierra Club and the Appalachian Mountain Club have named themselves after ranges that they have held up as models of the natural environment that we need to preserve for spiritual as well as ecological reasons.[5]

A similar attitude has influenced the development of the national park system in the United States. A disproportionate number of parks either bear the names of mountains or lie in mountain ranges, among them Denali, Mount Rainier, Rocky Mountain, Great Smoky Mountains, Grand Teton, Shenandoah, Glacier, Yosemite, Yellowstone, Kings Canyon, and Sequoia. It could be argued that these parks lie in mountainous regions because they were the hardest areas to develop and therefore still had wild parts left when the government initiated the national park system. But this argument only reinforces the point that mountains are not only viewed, but actually act, as paradigms of wilderness in its wildest and most pristine state. In any event, the American public views these national parks as sacred sanctuaries preserved for the spiritual benefit of future generations. People go to them as pilgrims to natural shrines that embody the spirit of the land in its original and pristine condition. An almost religious aura hangs over them: if anyone tries to desecrate a national park, an outcry automatically ensues.

The lure and magic of wilderness, the essence of what makes it so peculiarly attractive, comes from the sense of the sacred that it evokes. There is something fundamentally wild about the sacred itself, the way it eludes all our attempts to control and domesticate it. Like the inaccessible summit of a distant peak, it lies outside our reach, free from the restraints of any artificial order we would try to impose upon it. Its law is its own, not ours. Henry David Thoreau was referring to the sense of the sacred hidden in the wildness of nature when he wrote:

The West of which I speak is but another name for the Wild; and what I have been preparing to say is, that in Wildness is the preservation of world. Every tree sends its fibres forth in search of the Wild. The cities import it at any price. Men plow and sail for it. From the forest and wilderness come the tonics and barks which brace mankind.[6]

In addition to the power of salvation and sustenance, Thoreau attributed to the wild essence of wilderness precisely the qualities of mystery and grandeur that we have seen associated with the sense of the sacred in traditional cultures:

We need the tonic of wildness. . . . At the same time that we are earnest to explore and learn all things, we require that all things be mysterious and unexplorable, that land and sea be infinitely wild, unsurveyed, and unfathomed by us because unfathomable. We can never have enough of nature. We must be refreshed by the sight of inexhaustible vigor, vast and titanic features. . . .[7]

As the last unclimbed mountains are climbed and true wilderness vanishes, replaced with parks and designated "wilderness areas," we will have to turn to the sense of the sacred to find the wildness that Thoreau regarded as essential to the preservation of the world and the maintenance of our own well-being. That wildness, which we associate with unexplored places, actually lies right here, all around us, in the familiar things of our usual surroundings, if we can see them as they truly are, imbued with all the mystery and splendor of the deepest forests and highest peaks.

While doing research in the Himalayas, I went on an expedition to a legendary sacred valley that few, if any, outsiders had ever visited. I wrote of my impressions on reaching this valley after a long and difficult journey:

The freshness of our surroundings brought back childhood fantasies of primeval jungles hidden in the imaginary wilds of my own backyard. This forest had the same remote and mysterious quality, but it also seemed close and oddly familiar, as if I had been here long ago. Although many miles and mountains separated us from the help we would need in case of an accident, I felt at home and secure.[8]

As we grow up, wilderness, the place of mystery, recedes from our everyday lives to distant places, where we feel we must go to encounter it again. But if we know how to look for the essence of it, we can find it where we first experienced it as children—right here in our own backyards, in the wild sense of wonder and awe we felt on seeing everything fresh and new.

Sacred mountains that awaken this sense of wonder and awe rise in the middle of civilization as well as in wilderness. The slopes of T'ai Shan are covered with inscriptions, temples, rest houses, food stalls, tourists, and pilgrims. Western visitors accustomed to regarding mountains as paradigms of wilderness have difficulty understanding the feelings of beauty and reverence that this mountain inspires among Chinese; yet the sight of the peak has as powerful an effect on the latter as any view of an unclimbed spire of rock and ice on the former. One of the most revered mountains in the world, Mount Zion, is actually a city, as densely populated as any place on earth. Some peaks, such as Nanda Devi, are viewed as both wild and civilized: whereas a Western climber feels awe at the sight of a pristine summit of virgin snow, an Indian pilgrim sees with eyes of devotion the golden temple of a Hindu goddess.

These examples of sacred mountains show that we can find in civilization the essence of what we seek in wilderness. If they open us to the mystery of the universe, if they encourage us to address our innermost needs, cities, too, can act as places of spiritual renewal and enrichment. Like a peaceful landscape of meadows and forest, a well-proportioned arrangement of parks and buildings can evoke a feeling of harmony that puts us in tune with the world around us. As a Chinese poet exclaimed,

The wonder and delight of a child. Two boys explore the mysteries of a world hidden in the roots of a giant sequoia, Sierra Nevada, California. (Edwin Bernbaum)

Horsemen approach a remote mountain village in the Himalayas of Ladakh—an example of human habitation in harmony with the environment. (Edwin Bernbaum)

Who need be craving a world beyond this one?
Here, among men, are the Purple Hills![9]

Just as we strive to preserve wilderness areas that embody the sanctity of nature, so we need to cultivate urban environments that nurture the growth of the human spirit.[10]

MOUNTAINS
AND THE ENVIRONMENT

The sense of the sacred awakened by mountains has a crucial role to play in our efforts to respect and protect the environment, both wild and civilized. We usually treat the things we revere with love and respect, seeking to maintain their beauty and integrity. If something has acquired an aura of sanctity in our eyes, we feel little inclination to tamper with it: it seems whole and perfect in its own right. We seek to know it as it is, to enjoy its beauty and perfection— and to feel ourselves transported in its presence. We

admire a flower not to take its petals apart but to appreciate the way they fit together—and to marvel at the delicate intricacy of their construction. If we see the environment in this way, we feel an urge to preserve rather than destroy it. Without such an underlying sense of the sacred to inspire long-term commitment, conservation efforts based only on ecological facts and theories falter in the face of powerful forces determined to use the land and its resources for economic and political purposes. When that commitment does flag, as it will, the mountains provide a place to renew it with a vision of what it is in the world that we really value and need to conserve.

A sense of the sacred by itself, however, does not guarantee preservation of the environment. Although the Japanese revere Mount Fuji as a symbol of their nation, they have polluted its snows and littered its slopes with garbage. Even though the Sherpas regard the Himalayas as the abodes of the gods, they feel no compunction about joining Western climbers in leaving expeditionary refuse strewn about their deities' homes. And despite the sewage that fills the Ganges,

254

Hindus ritually bathe in the river and drink its water, regarding it as pure and holy. Traditional concepts of spiritual purity associated with sacred sites often have little to do with ideas of physical cleanliness that play an important role in modern efforts to clean up and preserve the environment.

Certain conceptions of the sacred can even threaten the environment with destruction. Early Christian missionaries in Europe viewed groves of trees as demonic sites of pagan rituals and cut them down to exorcise the evil spirits lurking within them. Regarding anything—a tree, a rock, or even a person—as divine may also imperil it. The Aztecs dressed their human victims as gods before sacrificing them on mountains. Priests usually sanctify the offerings they destroy in sacrifices. Viewing the environment as sacred may inadvertently set it up as a sacrificial offering for a higher end. And, in fact, we see intimations of such an outcome in pronouncements by the National Academy of Sciences that certain areas of the American Southwest irretrievably ruined by strip-mining may have to be written off as "National Sacrifice Areas," sacrificed for the greater good of society.[11]

All this means that we must be fully aware of the consequences of awakening a sense of the sacred. Because it has to do with matters of ultimate concern—values for which we would sacrifice everything—it has the power to move us for good or ill. It can inspire us to preserve the environment as something that we love and cherish or drive us to destroy it as something that we fear and abhor. It can also lead us to regard parts of nature—for example, trees in the forest—as worthy objects of sacrifice, providentially supplied as resources for improving the quality of life. Only if it encourages us to revere things as valuable in themselves, not as means for higher ends, no matter how noble or exalted, will awakening a sense of the sacred provide a sound basis for efforts to preserve the endangered plants, animals, and features of the world we all inhabit together.

In cutting down our forests, poisoning our rivers, and fouling our cities, we do more than imperil our physical health and livelihood: we impair our ability to experience a deeper reality in our lives. When we kill off wildlife and ravage the landscape, we destroy the beauty and wholeness of nature on which we depend for our spiritual well-being. No longer can we look to trees and streams, meadows and flowers, birds and animals, for images with the power to resonate in our minds and awaken a deep and abiding sense of the sacred. One of the greatest tragedies of desecrating the environment is that we cut ourselves off from the depths of our innermost being—from the source of insight and joy that makes life meaningful and worthwhile.[12]

MOUNTAINS AND LIFE

The sense of the sacred awakened by mountains reveals a reality that has the power to transform our lives. Whatever that reality is, however we may conceive it—as a deity, the ground of being, emptiness, the unconscious, the self, nature, the absolute—our encounter with it frees us from our usual conceptions of ourselves so that we can grow beyond the persons we think we are. Like the view from the summit of a mountain, it opens us to a fresh vision of ourselves and the world around us and, in so doing, gives our life a new meaning and direction. Shaped by the materialistic demands of business and technology, many of the values and goals of modern society have an artificiality that estranges us from our environment, each other, and ourselves. By awakening a sense of the sacred, making us aware of a deeper reality, mountains connect us to the world and make our lives more real.

When we climb to the summit of a mountain, we feel ourselves at the center of the world. We see, spread out around us, the vast circle of the horizon, filled with mountains, rivers, forests, fields, and cities. Features of the landscape that had seemed confused and fragmented seem to fall into place and form, like pieces of a puzzle, interlocking parts of a unified whole. After we descend, the memory of this summit view may remain with us, giving us a sense of where we stand in the greater scheme of things.

The direct experience of the kind of center exemplified in views of sacred mountains can have powerful and far-reaching effects. Becoming aware of such a center within ourselves, we discover a strength and stability that enables us to withstand the pressures of the outside world. We experience a sense of balance and inner serenity that gives us the confidence to be true to ourselves and to do what we feel is right, no matter what others may think or say.

For traditional societies, sacred mountains regarded as cosmic centers commonly function as places of creation. From their summits, like ripples spreading out from a pebble tossed in a pond, a single creative act gives rise to the myriad features of the universe. In standing on top of a mountain and experiencing it as the center of the world around us, we find that place within ourselves from which our actions flow as spontaneous expressions of who we are. In touch with that center, making contact with the reality it reveals, we discover a source of creativity that infuses everything we do, from the most ordinary acts of everyday life to the most extraordinary accomplishments of artistic genius.

Reaching the summit of a mountain, we see the circle of the horizon unfold around us, expanding beyond the limits of sight into the blue haze of the infinite sky. In experiencing ourselves at its center, we become aware of the boundless nature of the universe. Through the experience of its vastness, we discover a sense of space that gives us the freedom to open and grow. No longer do we feel confined within the limits of our narrow views. We know the world and ourselves as greater and more mysterious than we had ever imagined. And in that knowledge, in the awareness of a universe too vast to possess or control, we experience a humility that frees us from the petty demands of egotistic illusions and desires.

This sense of vastness allows us to find the center everywhere, not just on the summit of a mountain. If we remain open to the world around us, aware of its boundless, mysterious nature, then no matter how far we travel, we never come to its edge. The horizon moves away from us so that we always walk in the middle of its circle. The power of the center stays with us, and the fear we have of losing it vanishes. No longer bound to a particular place or occupation, we feel free to go anywhere and do anything, knowing that wherever we go, whatever we do, we will always be in touch with the reality of the world and ourselves.

In discovering a center within ourselves, we become aware of it in all people and things. Looking around us, we see that each person stands in the middle of the world and partakes of the mystery and splendor revealed in his or her particular place. Seeing the center everywhere awakens a sense of the sacred in everyone and everything. We realize that we are all linked together through the mysterious reality of the world we share. Moved by a sense of wonder and awe, we feel a spontaneous love and respect for people and things just as they are. We delight in their joy and share in their sorrow.

Other views of sacred peaks reveal additional ways in which the experience of this reality can transform our lives. As meeting places of heaven and earth, sacred peaks bring together the disparate realms of spirit and matter. To open and grow, we need to deepen our awareness of both and recognize the reality they symbolize within ourselves. Focusing on one without the other leaves us unbalanced. An exclusive emphasis on spiritual matters makes everything we do hazy and insubstantial. A narrow focus on material things hardens our minds and locks us into rigid views of the world and ourselves. To realize our potential as human beings, we need to cultivate both mind and body, thought and emotion—the complementary aspects of our nature that express the underlying totality of who we are.

As places of the ancestral dead, sacred mountains represent the mysterious reality from which we come and to which we return—the source of what has made

us who we are. Contact with that reality, especially on the heights of a mountain, can lead to experiences of death and rebirth. Overwhelmed by the power and mystery of what we encounter, we let go of our habitual selves and find in their place a new person born within us. Momentarily in touch with the source of our being, we see everything fresh and clean, as if for the first time. The Scottish mountaineer W. H. Murray described such an experience on emerging from the Indian plains and finding himself, at last, in the Himalayas:

For a moment dazzled, we suddenly saw spread before us a world made new. All the senses of the soul were not so much refreshed as reborn, as though after death. We were free men once again, for the first time in months really able to live in the present moment.[13]

Viewed as a spiritual path, the ascent of a mountain symbolizes the way that leads to the source of our being. The hardships of the climb represent the efforts and sacrifices required to put ourselves in touch with a higher reality that reveals our deeper nature. In overcoming these difficulties, we overcome our own limitations—the illusions and desires that keep us from realizing who we really are. But we do not do this entirely by ourselves. In climbing a peak as a spiritual path, we open ourselves to the reality we seek on its summit. In so doing, we experience that reality along the way as a mysterious influx of energy and vision that helps to carry us toward our goal. The same spiritual transformation may come about from the act of contemplating a mountain from a distance—or visualizing it in the mind. As a well-known Hindu scripture says of Himachal, or the Himalayas,

He who thinks of Himachal, even if he does not see it, is greater than he who accomplishes all his devotions at Benares. He who thinks of Himachal will be freed from all his sins. . . .[14]

The reality revealed by a mountain can be experienced anywhere. Mountains are no more sacred than any other feature of the landscape. But because of their natural tendency to inspire a sense of wonder and awe, they disclose this reality more easily. Having glimpsed it through a view of a mountain, we can find it elsewhere, through other places, other symbols. When we do, we discover that everyone and everything express the ultimate mystery of life. Although still aware of their importance, we go beyond the distinctions we normally make between the sacred and the profane, the beautiful and the ugly, the good and the bad; we simply experience the world as it is, marvelous in itself.

The curves of a road leading into the Hindu Kush Mountains of Afghanistan invite the viewer to follow the path of the mountain way. (Edwin Bernbaum)

In the end we realize that what we seek on the heights of the highest and most distant peaks we can find right here in the simplest and most ordinary things — in the glow of sunlight in the air, in the aroma of bread in the oven, in the whisper of wind in the trees, in the touch of the person we love. Commenting on the mysterious, inexhaustible nature of the reality concealed in the world and revealed in mountains, the Japanese Zen master Dogen wrote:

As for mountains, there are mountains hidden in jewels; there are mountains hidden in marshes, mountains hidden in the sky; there are mountains hidden in mountains. There is a study of mountains hidden in hiddenness.[15]

Something I experienced in the Himalayas makes this passage come alive for me. In the winter of 1968, when Tibet was still closed to the outside world, I went on a personal pilgrimage to the foot of Mount Everest in Nepal. After watching the sunset fade into a green twilight, evoking a sense of something mysterious and sublime in the mountain itself, I lay outside in my sleeping bag, gazing at the ranges around me. Glimmering with moonlight, the peaks next to Everest stood out black against the stars. A glacier hidden beneath darkness and rubble stretched off to my left, down toward the valleys below. I rolled over and looked at the foot of a slope that rose toward a summit on the border of Tibet.

Something about the smooth sweep of the slope caught me up, and I imagined myself gliding up it to a ridge and on up the ridge to the edge of . . . It suddenly hit me that I was on the edge of Tibet, where there were mountains and valleys that had never been mapped or even explored. *My God*, I thought, *there's still a place left in the world that's really unmapped, unknown, that people haven't labeled and described.*

And suddenly, a feeling that was stronger than a feeling, a feeling with the conviction and certainty of reality, swept through me: the rest of the world is like that, as unmapped and unexplored as Tibet. In some mysterious, inexplicable way, there are valleys and mountains all around us that no one has ever seen or mapped — a world hidden right here, as if in another dimension.

The thought took me back to my childhood, to a memory whose feeling I had long forgotten. When

I was three or four, I would go with my parents to the airport to watch airplanes take off in the early morning sun and fly out over blue-misted mountains. At that age I had never seen a map and anything could have been out there—the world was vast and fresh and endlessly fascinating. And now it felt that way again.

Mountains help us to regain that sense of freshness and wonder possessed by a child. They awaken us to a deeper reality hidden in the world around us, even in cities, far from the sight of the peaks themselves. Moved by the mysterious power of this hidden reality, we recover the vision and delight of childhood, enhanced by the experience and understanding of age. Eyes bright and clear, hearts open and free, we stand once again at the beginning and source of all that is and all that may be.

NOTES

INTRODUCTION

1. Lama Anagarika Govinda, *The Way of the White Clouds* (Berkeley: Shambhala, 1970), 197.

2. Kukai, "Stone Inscription for the Sramana Shodo Who Crossed Mountains and Streams in His Search for Awakening," trans. Allan Grapard, in *The Mountain Spirit*, ed. Michael Charles Tobias and Harold Drasdo (Woodstock, N.Y.: Overlook Press, 1979), 55.

3. Maurice Herzog, *Annapurna*, trans. Nea Morin and Janet Adam Smith (New York: Dutton, 1952), 206–7. Saint Theresa of Avila was a famous sixteenth-century Spanish mystic who wrote extensively about the spiritual path leading to union with God.

4. See Rudolf Otto's classic and influential work *The Idea of the Holy (Das Heilige)*, trans. John W. Harvey (Oxford and New York: Oxford University Press, 1950).

5. Guido Rey, *Peaks and Precipices: Scrambles in the Dolomites and Savoy*, trans. J. E. C. Eaton (New York: Dodd, Mead, 1914), 187–88.

6. From "A Poetic Description of the High Tower," attributed to Sung Yü. For a French translation see Paul Demiéville, "La montagne dans l'art littéraire chinois," *France-Asie/Asia* 20 (1965):13.

7. Gaston Rébuffat, *Mont Blanc to Everest*, trans. Geoffrey Sutton (London: Thames and Hudson, 1956), 83.

8. From Pascal's famous *Mémorial* describing his experience on the night of November 23, 1654. It was found after his death, scribbled on a piece of paper stitched into the lining of his coat.

9. Herbert Tichy heard the story about Smythe from a nurse who treated him. See *Himalaya*, trans. R. Rickett and D. Streatfeild (New York: Putnam's, 1970), 118.

10. Freda Du Faur, *The Conquest of Mount Cook and Other Climbs: An Account of Four Seasons' Mountaineering on the Southern Alps of New Zealand* (London: George Allen & Unwin, 1915), 27.

11. Rudyard Kipling, "The Explorer."

12. Mircea Eliade, *The Sacred and the Profane*, trans. Willard R. Trask (Harcourt, Brace & World, 1959), 12.

13. Ps. 125:1

14. Eccles. 1:14.

15. *Brihadaranyaka Upanishad* 1.3.28, translated in Robert Ernest Hume, trans., *The Thirteen Principal Upanishads*, 2d rev. ed. (Oxford and New York: Oxford University Press, 1931), 80.

16. Herzog, *Annapurna*, 207. On manifestations of the sacred in the profane world, see Eliade, *The Sacred and Profane*, 11–12. For overviews of various approaches to the study of the sacred and sacred places, see Carsten Colpe, "The Sacred and the Profane," trans. Russell M. Stockman, in *The Encyclopedia of Religion*, ed. Mircea Eliade (New York: Macmillan, 1987), 12: 511–26, and Joel P. Brereton, "Sacred Space," in *Ibid*, pp. 526–35.

CHAPTER 1

1. From the *Manasakhanda* of the *Skanda Purana*, adapted from a translation in Edwin T. Atkinson, *Kumaun Hills: Its History, Geography and Anthropology with Reference to Garhwal and Nepal* (Delhi: Cosmo Publication, 1974), 308.

2. *Mahabharata* 3.43.21–25, adapted from a translation in J. A. B. van Buitenen, ed. and trans., *The Mahabharata*, 3 vols. (Chicago: University of Chicago Press, 1975), 2:308.

3. "He who through his powers owns these snowy mountains." *Rig Veda* 10.121.4, a creation hymn dedicated to the Unknown God, the Golden Embryo, translated in Wendy O'Flaherty, trans., *The Rig Veda: An Anthology* (New York: Penguin Books, 1981), 28.

4. Yarlha Shampo is one of a group of four major sacred mountains with protector deities associated with the four directions of the compass: Yarlha Shampo in the east, Kula Kangri in the south, Nöjin Kangsang in the west, and Nyenchen Thanglha in the north.

5. In Nepal, for example, indigenous healing rituals are performed by shamans called *jhankris*. Many of them invoke animistic deities under the name Mahadeo, the Great God, an epithet of Shiva.

6. *Mahabharata* 3.31.39.16–20, translated in van Buitenen, *The Mahabharata*, 2:298–99.

7. For a readable overview of the Himalayas and their religious significance, see Rommel and Sadhana Varma, *The Himalaya: Kailasa-Manasarovar* (Geneva: Lotus Books, 1985).

8. George H. Leigh-Mallory, "The Reconnaissance of the Mountain," in *Mount Everest: The Reconnaissance, 1921*, ed. C. K. Howard-Bury (New York: Longmans, Green, 1922), 192.

9. The usual Tibetan name for Everest is Jomolangma (*jo mo glang ma* in accurate transliteration) rather than Chomolungma. See Alexander W. Macdonald, "The Lama and the General," *Kailash* 1, no. 3 (1973): 227, n. 8. A Tibetan guidebook to a hidden valley near Everest, which I have in my possession, actually refers to the mountain as Miyolangsangma, reinforcing the interpretation of Jomolangma as short for Jomo Miyolangsangma.

10. The five goddesses are called in Tibetan the Tsering Chenga (*tshe ring mched lnga*). On them and Miyolangsangma (*mi g.yo glang bzang ma*), see René de Nebesky-Wojkowitz, *Oracles and Demons of Tibet: The Cult and Iconography of the Tibetan Protective Deities* (1956; repr. Graz, Austria: Akademische Druck-u. Verlagsanstalt, 1975), 177–81. I obtained information on Tashi Tseringma and Chomolhari from villages at the foot of the mountain.

11. When questioned on the subject, Nepalis usually bring up the literal meaning of Sagarmatha, but I suspect that the person or persons who created the name had the secondary associations of the churning stick in mind, especially since they chose a literary, Sanskritized word that would not have been used in ordinary

speech. For the usual Nepali interpretation of the name by the Nepali scholar Baburam Acarya, see Macdonald, "The Lama and the General," 227, n. 8. For a version of the ocean churning myth from the *Mahabharata*, see Wendy O'Flaherty, trans., *Hindu Myths* (Baltimore: Penguin Books, 1975), 274–80. The gods and demons also churned up a dreaded poison that Shiva had to swallow to keep the world from being destroyed; the poison turned his neck blue.

12. From a personal conversation with the Tengboche Rimpoche — the title by which the abbot is known. Additional information on Khumbila and the goddess of Everest appears in a pamphlet prepared by the Tengboche Rimpoche: Ngawang Tenzin Zangbu and Frances Klatzel, *Stories and Customs of the Sherpas* (Kathmandu: Khumbu Cultural Conservation Committee, 1988), 7–8.

13. This myth is the subject of Kalidasa's famous poem *Kumarasambhava*, translated in Kalidasa, *The Origin of the Young God: Kalidasa's Kumarasambhava*, trans. Hank Heifetz (Berkeley and Los Angeles: University of California Press, 1985).

14. Arthur (Sir John Woodroffe) Avalon, trans., *Tantra of the Great Liberation (Mahanirvana Tantra)* (1913; repr. New York: Dover, 1972), 1.

15. *Meghaduta* 64. For another translation see Leonard Nathan, trans., *The Transport of Love: The Meghaduta of Kalidasa* (Berkeley and Los Angeles: University of California Press, 1976), 59.

16. Buddhists and Bonpos (followers of Bon) also see the gully with the line of indentations coming down the south face of Mount Kailas as the vertical segment of a swastika imprinted on the sacred mountain. In these religions swastikas are symbols of auspiciousness and good fortune: the Sanskrit word *swastika* means "being for good." Bonpos refer to Kailas as the "Nine-Storey Swastika Mountain."

17. Adapted from Garma C. C. Chang, trans., *The Hundred Thousand Songs of Milarepa*, 2 vols. (Boston: Shambhala, 1977), 1:262. The story of Milarepa and Naro Bhun Chon appears on 215–24.

18. Demchog is the Tibetan name for the tantric deity whose name in Sanskrit is Chakrasamvara. Most Buddhas and Bodhisattvas in Tibetan Buddhism have both Tibetan and Sanskrit names. Demchog shares a number of characteristics with Shiva. Both wear tiger skins, and just as Shiva has Parvati, Demchog has a female consort, named Dorje Phagmo. For a beautiful description of Kailas, its significance, the pilgrimage around it, and the mandala imposed on the region, see Lama Anagarika Govinda, *The Way of the White Clouds* (Berkeley: Shambhala, 1970), 197–219. More recent descriptions of the circumambulation appear in Stephen Batchelor, *The Tibet Guide* (London: Wisdom Publications, 1987), 357–70; Hugh Swift, *Trekking in Nepal, West Tibet, and Bhutan* (San Francisco: Sierra Club Books, 1989), 221–31; and Russell Johnson and Kerry Moran, *The Sacred Mountain of Tibet: On Pilgrimage to Mount Kailas* (Rochester, Vt.: Park Street Press, 1989) — the last illustrated with beautiful pictures.

19. The Hindu version of Meru rises in the center of our world, called Jambudvipa. Sumeru means "Good Meru." For more on Meru, see the chapters on Central Asia and South and Southeast Asia in this book and I. W. Mabbett, "The Symbolism of Mount Meru," *History of Religions* 23, no. 1 (1983).

20. Govinda, *Way of the White Clouds*, 206. In recent years the Chinese have allowed eight groups of twenty-five Indian pilgrims a year to go to Kailas. More Tibetans and Nepalis visit the mountain on pilgrimage, but no figures are available. Because Tibetan pilgrims can now come in trucks across the plateau, their number has increased in recent years. Pilgrims have also been coming from the West.

21. John Snelling, *The Sacred Mountain* (London and The Hague: East West Publications, 1983), 53. Snelling's book provides one of the most complete accounts in English of Kailas, its religious significance, and its history. Charles Allen in *A Mountain in Tibet: The Search for Mount Kailas and the Sources of the Great Rivers of India* (London: A. Deutsch, 1982) also discusses the history and significance of the mountain, primarily from a Western point of

view.

22. Ibid., 86.

23. Ibid., 120–32, and Herbert Tichy, *Himalaya*, trans. R. Rickett and D. Streatfeild (New York: Putnam's, 1970), 35–45.

24. On the names of the peaks and the seven sages, see H. Adams Carter, "The Goddess Nanda and Place Names of the Nanda Devi Region," *The American Alpine Journal* 21, no. 51 (1977): 24–27.

25. On the Nanda Devi pilgrimage and the story of Rup Kund, see Man Mohan Sharma, *Through the Valley of Gods: Travels in the Central Himalayas*, 2d ed. (New Delhi: Vision Books, 1978), 226–27, and Gurmeet and Elizabeth Thukral, *Garhwal Himalaya* (New Delhi: Frank Bros., 1987), 109–12.

26. On the flood myth, see O'Flaherty, *Hindu Myths*, 181–84; on local variants linking it to Nanda Devi and the Rajis, see Sharma, *Valley of Gods*, 44–45.

27. Eric Shipton, *Nanda Devi*, in *The Six Mountain-Travel Books*, ed. Jim Perrin (1936; repr. Seattle: The Mountaineers, 1985), 79.

28. Personal conversation with H. Adams Carter.

29. Ibid.

30. Louis F. Reichardt and William F. Unsoeld, "Nanda Devi from the North," *The American Alpine Journal* 21, no. 51 (1977): 21–22.

31. From conversations with H. Adams Carter and his article "The Goddess Nanda and Place Names of the Nanda Devi Region," *The American Alpine Journal* 21, no. 51 (1977): 29.

32. For an account of the American women's expedition led by Arlene Blum, see Arlene Blum, *Annapurna: A Woman's Place* (San Francisco: Sierra Club Books, 1980).

33. Diana L. Eck, *Banaras: City of Light* (Princeton: Princeton University Press, 1983), 160–64. Benares is also written Banaras and Varanasi.

34. For a discussion of the local deity of the sanctuary — Pujinim Barahar — and Machapuchare and the effects of tourism on local beliefs and practices, see Stan Stevens, "Sacred and Profane Himalayas," *Natural History* 97, no. 1 (1988): 27–35.

35. Wilfred Noyce, *Climbing the Fish's Tail* (London: Heinemann, 1958).

36. On the treasures and Buddhist deities of Kangchenjunga (*gangs chen mdzod nga* or *gangs chen mched lnga*), see Nebesky-Wojkowitz, *Oracles and Demons*, 216–18, and *Where the Gods Are Mountains: Three Years Among the People of the Himalayas*, trans. Michael Bullock (New York: Reynal, n.d.), 30–31.

37. See Nebesky-Wojkowitz, *Gods*, 29–30, and Halfdan Siiger, "A Cult for the God of Mount Kanchenjunga Among the Lepcha of Northern Sikkim," in *Actes du IVe congrès international des sciences anthropologiques et ethnologiques* (Vienna: Verlag Adolf Holzhausens, 1955), 2:185–89.

38. On the story of Lhatsun Chembo and the masked dances dedicated to the god of Kangchenjunga, see Nebesky-Wojkowitz, *Oracles and Demons*, 217, 402–5, and *Gods*, 31–32, 238–39.

CHAPTER 2

1. By China in this chapter, I refer to the part of the country regarded as the traditional home of the Han Chinese. This excludes the regions of Tibet and Xinjiang and portions of provinces such as Yunnan, Qinghai, and Sichuan occupied by minority peoples with different cultures and traditions. Some of those areas I deal with in other chapters. Because many sources on sacred mountains and religious traditions of China use the older Wade-Giles system of transliteration, I have used that system in this chapter, except for names of well-known places such as Beijing, where I have used the modern pinyin system of the People's Republic.

2. The definition appears in a dictionary written in A.D. 121, which draws on older ideas from the Han Dynasty; see Paul Demiéville, "La montagne dans l'art littéraire chinois," *France-Asie/Asia* 20 (1965): 7, 30, n. 3.

3. *Shu-ching* 2.3, translated in James Legge, *The Chinese Classics,*

7 vols. (London: Trübner, 1865) 3, pt. 1:34–37. The *Shu-ching* refers to T'ai Shan by its other name Tai Tsung, "Tai the Revered."

4. *Analects (Lun Yü)* 6.21.

5. Demiéville, "La montagne," 15.

6. Ko Hung, *P'ao-p'u tzu nei-p'ien* 17.1a. See Demiéville, "La montagne," 14–15, and Paul W. Kroll, "Verses from on High: The Ascent of T'ai Shan," *T'oung Pao* 69, no. 4–5 (1983): 224.

7. For more on the talismans and their role in Taoism and folk religion, consult Edouard Chavannes, *Le T'ai Chan: Essai de monographie d'un culte chinois*, Annales du Musée Guimet 21 (Paris: Ernest Leroux, 1910), 415–24, and William Doub, "Mountains in Early Taoism," in *The Mountain Spirit*, ed. Michael Charles Tobias and Harold Drasdo (Woodstock, N.Y.: Overlook Press, 1979), 131–34. A central peak was added to the four visited by Shun in the *Shu-ching*; see the discussion in the next section of this chapter.

8. Poem by Hsieh Ling-yün (died A.D. 433); See Demiéville, "La montagne," 21–22.

9. *Cold Mountain: 100 Poems by the T'ang Poet Han-shan*, trans. Burton Watson (New York: Columbia University Press, 1970), 87. *Way* here is a reference to the Tao and, probably, to the Buddhist path or *marga*. Han-shan lived sometime between the sixth and ninth centuries A.D., most probably around the end of the eighth.

10. See Demiéville, "La montagne," 10, 25.

11. "My Retreat at Mount Chung-nan," translated in *The Jade Mountain: A Chinese Anthology*, trans. Witter Bynner (1929; repr. Garden City, N.Y.: Anchor Books, 1964), 159.

12. See Demiéville, "La montagne," 29, and *Mao Tsetung Poems* (Peking: Foreign Languages Press, 1976), 18. In pinyin, Mao's name is Mao Zedong.

13. Chinese sources designate these mountains with a special term that sets them apart from other sacred peaks—*yüeh*. Edward Schafer translates this term as "marchmount" to indicate that the mountains stood on the marches or borders of the empire toured by the legendary rule Shun (Edward H. Schafer, *Pacing the Void: Tang Approaches to the Stars* (Berkeley and Los Angeles: University of California Press, 1977), 6. However, it seems to me that each of the four peaks mentioned in the *Shu-ching* stood within each quarter of the realm, marking it as a division of the empire. I therefore refer to them simply as the principal sacred mountains of China, those which represented the nation in the eyes of its people and served as links with heaven—two roles characteristic of many sacred mountains in cultures around the world. For overviews of the five peaks, see William Edgar Geil, *The Sacred 5 of China* (Boston and New York: Houghton Mifflin, 1926), and Anna M. Hotchkis and Mary Augusta Mullikan, *The Nine Sacred Mountains of China: An Illustrated Record of Pilgrimages Made in the Years 1935–1936* (Hong Kong: Vetch and Lee Ltd., 1973). The names of these mountains in pinyin, as they appear on modern maps, are Tai, Heng Bei, Hua, Heng, and Song. *Shan* means "mount" or "mountain."

14. For a beautiful picture essay on Hua Shan with commentary by Wolfram Eberhard, see Hedda Morrison, *Hua Shan: The Taoist Sacred Mountain in West China* (Hong Kong: Vetch and Lee Ltd., 1973). On Hua Shan and embodied spirits of stars, see Schafer, *Pacing the Void*, 132, 136.

15. Tu Fu, "Gazing from Afar at the Sacred Peak." For another translation, see Kroll, "Verses from on High," 228–30.

16. See Chavannes, *T'ai Chan*, 16–26.

17. Michael Loewe, *Ways to Paradise: The Chinese Quest for Immortality* (London and Boston: Allen & Unwin, 1979), 200.

18. On emperors who performed the Feng and Shan sacrifices, see Chavannes, *T'ai Chan*, 20. On K'ang-hsi and Ch'ien-lung, see Dwight Condo Baker, *T'ai Shan: An Account of the Sacred Eastern Peak of China* (Shanghai: Commercial Press, 1925), 183–85, 187–91.

19. Chavannes, *T'ai Chan*, 9, 281–82.

20. On the importance of T'ai Shan in the life of the common people and his role as god of earthly matters, see Henri Maspero, *Taoism and Chinese Religion*, trans. Frank A. Kierman, Jr. (Amherst: University of Massachusetts Press, 1981), 102–4. Chavannes, *T'ai Chan*, 13–16, also discusses T'ai Shan as the place and god of the dead.

21. *Mencius* 7A.24, adapted from the translation in Baker, *T'ai Shan*, 30.

22. According to a personal account by Anna Seidel, a scholar of Chinese religion, who witnessed these happenings on her ascent of T'ai Shan in 1984. On the cult of the Princess of Azure Clouds (Pi-hsia yüan-chün) and her attendants, see Chavannes, *T'ai Chan*, 29–38, and Maspero, *Taoism*, 164–66.

23. Chavannes, *T'ai Chan*, 63, and Baker, *T'ai Shan*, 144–45.

24. Tao-yün, wife of General Wang Ning-chih, translated in Arthur Waley, trans., *A Hundred and Seventy Chinese Poems* (New York: Knopf, 1919), 120.

25. For an overview of the four Buddhist peaks, see Hotchkis and Mullikan, *Nine Sacred Mountains*. The names of these peaks in pinyin are Putuo, Wutai, Emei, and Jiuhua.

26. William Powell has been doing extensive research on Chiu-hua Shan and will be publishing work on it.

27. John Blofeld, *The Wheel of Life: The Autobiography of a Western Buddhist*, 2d ed. (Berkeley: Shambhala, 1972), 121–22.

28. Adapted from Raoul Birnbaum, "The Manifestation of a Monastery: Shen-ying's Experiences on Mount Wu-T'ai in a T'ang Context," *Journal of the American Oriental Society* 106, no. 1 (1986): 122. Birnbaum is the Western scholar who has done the most extensive research on Wu-t'ai Shan, and much of the material in this section is drawn from his work. He is currently writing a book on Wu-t'ai Shan.

29. Ibid. Birnbaum uses "transcendents" for "immortals" as a translation of *hsien*.

30. Hua-yen master Ch'eng-kuan, adapted from a translation in ibid., 119.

31. On Manjushri's role as prince of wisdom, national protector, and lord of the cosmos, see Raoul Birnbaum, *Studies on the Mysteries of Manjusri: A Group of East Asian Mandalas and Their Traditional Symbolism*, Society for the Study of Chinese Religions Monograph Series 2 (Boulder, Colo.: Society for the Study of Chinese Religions, 1983). As Birnbaum points out, Manjushri plays an important role in Buddhist astrology.

32. For descriptions of some of the visions in accordance with which certain monasteries were built, see Birnbaum, "Manifestation of a Monastery" and *Mysteries of Manjusri*.

33. The episode of the red-maned lion appears in the No play *Shakkyo*, described in Carmen Blacker, *The Catalpa Bow: A Study of Shamanistic Practices in Japan* (London: Allen & Unwin, 1975), 316.

34. Adapted from Edwin O. Reischauer, trans., *Ennin's Diary: The Record of a Pilgrimage to China in Search of the Law* (New York: Ronald Press, 1955), 225.

35. Blofeld, *Wheel*, 147–48. Birnbaum told me the story about his experience in the storm on Wu-t'ai Shan.

36. Reported to me by Lewis Lancaster from research at Buddhist monasteries in Korea.

37. The quote appears in Ronald W. Clark, *Men, Myths and Mountains* (New York: Crowell, 1976), 202. The summit of Everest lies on the border between Nepal and Tibet, the latter now part of China.

38. *Another Ascent of the World's Highest Peak—Qomolangma* (Peking: Foreign Languages Press, 1975), 4.

39. Michael Sullivan, *The Birth of Landscape Painting in China* (Berkeley and Los Angeles: University of California Press, 1962), 1–2.

CHAPTER 3

1. On modern maps using the pinyin system of the People's Republic of China, K'un-lun and T'ien Shan are spelled Kunlun and Tian Shan or Tianshan. Taklamakan is also spelled Takla Makan on some maps.

2. On Chang Ch'ien's travels, see Herbert Allen Giles, *A Chinese Biographical Dictionary* (New York: Paragon Book Gallery, 1898), 12–13, and Denis Sinor, *Inner Asia: A Syllabus*, Uralic and Altaic

Series 96 (Bloomington: Indiana University, 1969), 91–92. Sinor's book also provides an overview of the geography, history, and cultures of Central Asia and the Silk Route.

3. Tun-huang is spelled Dunhuang in pinyin.

4. Samuel Beal, trans., *Buddhist Records of the Western World*, 2 vols. in one (1884; repr. New York: Paragon, 1968), 2:325.

5. Genghis Khan is also transliterated, more accurately, as Chingis Khan, but Genghis is the more familiar spelling.

6. Sven Hedin, *My Life as an Explorer*, trans. A. Huebsch (Garden City, N.Y.: Garden City Publishing, 1925), 188.

7. On the Tibetan myth of Shambhala, see Edwin Bernbaum, *The Way to Shambhala* (Garden City, N.Y.: Anchor Press/Doubleday, 1980).

8. *Mahabharata* 3.160.16–23, portion quoted translated in J. A. B. van Buitenen, ed. and trans., *The Mahabharata*, 3 vols. (Chicago: University of Chicago Press, 1975), 2:533. The Grandfather is a reference to Brahma, the supreme deity in the role of creator — i.e., grandfather of all beings. In Buddhist versions of Meru — called Sumeru — Indra, the king of the gods, has his palace on the summit of the mountain; above lie regions of heaven symbolizing more and more refined states of meditation. See Chapters 1 and 5 of this book for more on the mythic mountain.

9. On the symbolism of Meru, or Sumeru, see I. W. Mabbett, "The Symbolism of Mount Meru," *History of Religions* 23, no. 1 (1983): 64–83. For Indian conjectures about the location of the mountain in a lotus blossom configuration, see Shyam Narain Pande, "Identification of the Ancient Land of Uttarakuru," *The Journal of the Ganganatha Jha Research Institute* 26, pts. 1–3 (1970): 727, in particular. For Hindus, Meru rises in the center of our world, known as Jambudvipa; for Buddhists, Sumeru stands at the center of our universe, north of Jambudvipa.

10. Quoted from the *Shan-hai-ching*, translated in Homer C. Dubs, "An Ancient Chinese Mystery Cult," *Harvard Theological Review* 35, no. 4 (1942): 231.

11. Descriptions of the palace of the immortals appear in E. T. C. Werner, *A Dictionary of Chinese Mythology* (Shanghai: Kelly and Walsh, 1932), 163–64, and Henri Maspero, *Taoism and Chinese Religion*, trans. Frank A. Kierman, Jr. (Amherst: University of Massachusetts Press, 1981), 194–96.

12. A full translation of the novel appears in Wu Ch'eng-en, *The Journey to the West*, trans. Anthony C. Yu, 4 vols. (Chicago: University of Chicago Press, 1977–1983). Chapters 4 and 5 deal with the episode of the peach tree.

13. See Dubs, "Chinese Mystery Cult," and Michael Loewe, *Ways to Paradise: The Chinese Quest for Immortality* (London, Boston: George Allen & Unwin, 1979), 86–126.

14. On envoys looking for Hsi wang-mu, see Dubs, "Chinese Mystery Cult," 234, n. 20.

15. James Hilton, *Lost Horizon* (London: Pan Books, 1966), 179.

16. For an account of the history of exploration in the Amnye Machen area and an American ascent of the mountain shortly after the Japanese, see Galen Rowell, *Mountains of the Middle Kingdom* (San Francisco: Sierra Club Books, 1983), 132–68.

17. For a detailed study of the peaks, deities, and people of Amnye Machen, see J. F. Rock, *The Amnye Ma-chhen Range and Adjacent Regions: A Monographic Study*, Serie Orientale Roma 12 (Rome: Istituto Italiano per il Medeo ed Estremo Oriente, 1956). On Machen Pomra, Ganden, and Tsongkhapa, see René de Nebesky-Wojkowitz, *Oracles and Demons of Tibet: The Cult and Iconography of the Tibetan Protective Deities* (1956; repr. Graz, Austria: Akademische Druck-u. Verlagsanstalt, 1975), 209–13.

18. A definitive study of the Gesar Epic and its formulation in the region of Amnye Machen appears in R. A. Stein, *Recherches sur l'épopée et le barde au Tibet* (Paris: Presses Universitaires de France, 1959). Stein also points out the connections between Gesar and Shambhala and the way his name derives from the title of Roman and Byzantine emperors on 524–28, 279–80. On the sword and features of Amnye Machen identified with Gesar, see ibid., 124, and Rock, *Amnye Ma-chhen Range*, 116.

19. Beal, *Buddhist Records*, 2:305. The city, called Wu-sha or U-sha in the text, is in the vicinity of Yarkand at the western end

of the Tarim Basin, according to Stein.

20. M. Aurel Stein, *Ancient Khotan*, 2 vols. (Oxford: Clarendon, 1907), 1:46.

21. Ibid., 1:45, and Sven Hedin, *Through Asia*, 2 vols. (London: Methuen, 1898), 1:218. *Hazrett* (Hedin's spelling) is a term of respect; Muslims often refer to the tomb or shrine of a saint using only his name with *hazrett* or an equivalent word in front of it.

22. On Janaidar and the story of the plums, see Hedin, *Through Asia*, 1:218–19, 221. Because of Hedin's tendencies to exaggerate and romanticize, his reports of Kirghiz beliefs and practices regarding Muztagh Ata may not be very accurate, but they do convey a sense of what the mountain meant to the Kirghiz and to Hedin himself.

23. Ibid., 1:385.

24. M. Aurel Stein, *Sand Buried Ruins of Khotan* (London: T. F. Unwin, 1903), 99.

25. On the climbing history of Muztagh Ata and the ski descent of the mountain, see Rowell, *Mountains of the Middle Kingdom*, 9–11, 24–25.

26. See Jacques Bacot, *Introduction à l'histoire du Tibet* (Paris: Société Asiatique, 1962), 92 n.

CHAPTER 4

1. Dogen, "Treasury of the True Dharma Eye: Book XXIX, The Mountains and Rivers Sutra," trans. Carl Bielefeldt, in *The Mountain Spirit*, ed. Michael Charles Tobias and Harold Drasdo (Woodstock, N.Y.: Overlook Press, 1979), 47.

2. H. Byron Earhart, *A Religious Study of the Mount Haguro Sect of Shugendo: An Example of Japanese Mountain Religion* (Tokyo: Sophia University, 1970), 33.

3. Ichiro Hori, "Mountains and Their Importance for the Idea of the Other World," *History of Religions* 6, no. 1 (1966): 9.

4. On mountain shamanism and Ontake, see Carmen Blacker, *The Catalpa Bow: A Study of Shamanistic Practices in Japan* (London: Allen & Unwin, 1975). Hori discusses the three kind of mountains as well as other categories in his article "Mountains and Their Importance." For more on ancient mountain worship, called *sangaku shinko* by scholars, see Earhart, *Haguro*, 7–16.

5. Dogen, 47.

6. For an overview of Shugendo, its history, doctrines, and practices, see Earhart, *Haguro*, which focuses on the Mount Haguro sect of the religion. Both that work, 110–46, and Blacker, 218–34, describe the ritual ascent of Mount Haguro. An abbreviated description of this ascent appears in Earhart, "Sacred Mountains in Japan: Shugendo as 'Mountain Religion,'" in *The Mountain Spirit*, 114–15. Blacker, 216, describes the ritual of hanging a person over the cliff on Mount Omine.

7. Allan G. Grapard, "Flying Mountains and Walkers of Emptiness: Toward a Definition of Sacred Space in Japanese Religions," *History of Religions* 21, no. 3 (1982): 207.

8. Translated in *The Manyoshu: The Nippon Gakujutsu Shinkokai Translation of One Thousand Poems* (New York: Columbia University Press, 1969), 215.

9. Kentaro Sanari, quoted in Royall Tyler, "A Glimpse of Mt. Fuji in Legend and Cult," *Journal of the Association of Teachers of Japanese* 1, no. 2 (1981): 140. Tyler's article is one of the few sources in a non-Japanese language that goes into any depth on the role of Fuji as a sacred mountain. Much of the following material in this section of the chapter comes from his article.

10. Ibid., 141.

11. Ibid., 143. The original shrine, Fuji Hongu Asama Jinja, is located at Fujinomiya on the south side of Mount Fuji.

12. The poem is by Saigyo, translated in ibid., 151. The romantic nature of the story made it very popular, resulting in many elaborations (see ibid., 150–51). Kaguya Hime seems to have blended with Konohana Sakuya Hime.

13. Ibid., 149–50.

14. Ibid., 145–46. I have adapted Tyler's translation. Dainichi Nyorai is the Japanese translation of Mahavairocana, the original

Sanskrit name for this cosmic Buddha, who also plays an important role in the Buddhism of Tibet.

15. On Fuji's summit as a symbol of the state of perfect concentration (*samadhi*) and a mandala, see Tyler, "Glimpse," 144, 147. Tyler also remarks that Fuji was associated with other paradises, such as the Pure Land of the Buddha Amitabha.

16. On En no Gyoja, also known as En no Ozuno, see ibid., 141–42, and Earhart, *Haguro*, 16–21. *Mt. Fuji* (Tokyo: Japan Times, Ltd., 1970), 101, describes the opening of the Shugendo route on Fuji and the early record of storm and bear deaths on the mountain. Frederick Starr, *Fujiyama: The Sacred Mountain of Japan* (Chicago: Covici-McGee, 1924) mentions the same record of deaths on Fuji and describes the Fuji-ko movement, as well as many other subjects pertaining to the peak. A fascinating early work by an anthropologist, Starr's remains the only full-length book in English on Fuji as a sacred mountain. H. Byron Earhart is working on a book devoted to Mount Fuji and Fuji-ko. For his most recent publication on the subject, see H. Byron Earhart, "Mount Fuji and Shugendo," *Japanese Journal of Religious Studies* 16, no. 2–3 (1989): 205–26.

17. Noritake Tsuda, *Ideals of Japanese Paintings* (Tokyo and Osaka: Sanseido Ltd., 1940), 65.

18. On the history of the Fuji-ko movement, see Tyler, "Glimpse," 153–59. I have also drawn on material from an unpublished article by Tyler on Kakugyo. Kosuke Koyama, *Mount Fuji and Mount Sinai: A Critique of Idols* (Maryknoll, N.Y.: Orbis Books, 1984), 84–87, deals with Jikigyo's theology of Mount Fuji and the role of the mountain in Japanese nationalism. Hanamori Shibata (1809–80) was the Fuji-ko leader responsible for combining Fuji devotion with the cult of the emperor. Starr, *Fujiyama*, 118–20, also deals with the role of Fuji in Japanese nationalism. For a recent article with additional material on Kakugyo, Jikigyo, and the Maitreya cult on Mount Fuji, see Martin Collcutt, "Mt. Fuji as the Realm of Miroku: The Transformation of Maitreya in the Cult of Mt. Fuji in Early Modern Japan," in *Maitreya, the Future Buddha*, ed. Helen Hardacre and Alan Sponberg (Cambridge and New York: Cambridge University Press, 1988), 248–69.

19. Henry D. Smith provided me with a copy of his notes for a lecture he gave on his research on Fuji replicas to the Asiatic Society of Japan on 10 February 1986. With his help I was able to locate and visit several of these replicas in Tokyo.

20. For much of the material on Fuji-ko today I am indebted to Ida Kiyoshige, Fumiko Umezawa, and Minoru Harashida, who discussed it with me and brought articles on the subject to my attention. On the linking of Konohana Sakuya Hime with Fuji and Fuji-ko, see Henry D. Smith II, *Hokusai: One Hundred Views of Mt. Fuji* (New York and Tokyo: George Braziller, 1988), 195.

21. A good account of Kukai's life, along with translations of some of his major works, is Yoshito S. Hakeda, *Kukai: Major Works* (New York: Columbia University Press, 1972), from which I have drawn much of the following material. Grapard, "Flying Mountains," and Hori, "Other World," also discuss Kukai and Koya. Kobo Daishi was a title given to Kukai meaning a "great teacher who spreads widely the Buddhist teachings."

22. From *Indications of the Goals of the Three Teachings*, translated in Hakeda, *Kukai*, 22–23. Kukai wrote this work in his early twenties at the time he was wandering like the ascetic he describes.

23. There are various versions of the legend. I have drawn this account from three sources: Hakeda, *Kukai*, 48; Hori, "Other World," 17; and Yusei Arai, *Odaishi-Sama: A Pictorial History of the Life of Kobo Daishi*, trans. Hiroshi Katayama et al. (Osaka, Japan: Koyasan Shuppansha, 1973), 13.

24. Translated in Hakeda, *Kukai*, 47.

25. Ibid.

26. Ibid., 50.

27. Hakeda, *Kukai*, 7–8.

28. Hakeda's book *Kukai* provides an overview of Shingon doctrine and practice, along with translations of important texts composed by Kobo Daishi. It also examines the projection of mandalas onto Mount Koya.

29. Grapard, "Flying Mountains," 212. In this fascinating article Grapard shows how the process of mandalization led to the divinization of the entire country of Japan.

CHAPTER 5

1. Mount Abu lies in the Aravalli Mountains in the southwest part of Rajasthan, near Gujarat. The statue of Bahubali stands at Shravana Belagola, carved out of rock on the summit of a mountain in Karnakataka.

2. Adapted from a translation in Sri Ramana Maharshi, *Five Hymns to Sri Arunachala*, 3d ed. (Sri Ramanasramam, India: Niranjananda Swamy, 1946), 8.

3. Inscriptions show that the Kailasanatha temple at Kancipuram was earlier called the Rajasimheshvara temple and received its present name at a later date. There is also evidence that the Kailasa temple at Ellora was not originally called by that name either. This makes it even less likely that the actual shape of Mount Kailas influenced the construction of either temple. Robert Del Bontà brought this information to my attention.

4. Two works dealing with the mountain symbolism of Indian temples are George Michell, *The Hindu Temple: An Introduction to Its Meaning and Forms* (New York: Harper & Row, 1977) and Stella Kramrisch, *The Hindu Temple*, 2 vols. (Calcutta: University of Calcutta, 1946). On temples of south India, including Tirupatti and its relation to Meru, see P. V. Jagadisa Ayyar, *South Indian Shrines* (New Delhi: Asian Education Services, 1982).

5. The fourteenth-century Italian monk Giovanni del Marignolli, quoted in Rowland Raven-Hart, *Ceylon: History in Stone* (Colombo: Associated Newspapers of Ceylon, 1964), 102. Ceylon is now known as Sri Lanka.

6. *Mahavamsa* 1.5.77, quoted in Senerat Paranavitana, *The God of Adam's Peak* (Ascona, Switzerland: Artibus Asiae Publishers, 1958), 12.

7. Fa-hsien's account appears in James Legge, *A Record of Buddhist Kingdom* (1886; repr. New York: Paragon Book Reprint and Dover, 1965), 102. On the sapphire, see Raven-Hart, *Ceylon*, 104.

8. On Muslims and Adam's Peak, see Emerson Tennent, *Ceylon: An Account of the Island*, 4th ed. rev., 2 vols. (London: Longman, Green, Longman, and Roberts, 1860), 2:134–36. The manuscript is a Coptic version of a discourse on "Faithful Wisdom" attributed to the Gnostic writer Valentinus.

9. On the Hindu deities of Adam's Peak and the coming of Shiva, see Paranavitana, *The God of Adam's Peak*, 18–21. On Saint Thomas and the naming of Adam's Peak by the Portuguese, see Tennent, *Ceylon*, 132–33.

10. On the Buddhist deity and names of Adam's Peak, see Paranavitana, *The God of Adam's Peak*, 11–12.

11. Ibid., 17.

12. On Vijayabahu and Ibn Batuta, see ibid., 12, 20–21, and Tennent, *Ceylon*, 136.

13. Paranavitana, *The God of Adam's Peak*, 21.

14. On the legend and history of the chains, see ibid., 15.

15. On the installation of the lights, see ibid., 22 and Raven-Heart, *Ceylon*, 103.

16. On the connections between rulers and sacred mountains in Southeast Asia, see Robert Heine-Geldern, *Conceptions of State and Kingship in Southeast Asia*, Data Papers 13 (Ithaca, N.Y.: Southeast Asia Program, Dept. of Far Eastern Studies, Cornell University, 1956); H. G. Quaritch Wales, *The Universe Around Them: Cosmology and Cosmic Renewal in Indianized Southeast Asia* (London: Arthur Probsthain, 1977); and Paul Wheatley, *Nagara and Commandery: Origins of the Southeast Asian Urban Traditions*, Research Papers 207–8 (Chicago: University of Chicago, Department of Geography, 1983). I have used Meru for both the Hindu and Buddhist versions of the cosmic mountain. Buddhists call it Sumeru, which means "Good Meru."

17. For a recent collection of essays on various aspects of Borobudur and its symbolism, see Luis O. Gomez and Hiram W. Woodward, Jr., ed., *Barabudur: History and Significance of a Buddhism Monument*, Berkeley Buddhist Studies Series 2 (Berkeley:

Asian Humanities Press, 1981). Borobudur is also written Barabudur.

18. See A. J. Bernet Kempers, *Ancient Indonesian Art* (Cambridge: Harvard University Press, 1959), 65–66. I am also indebted to Eric Oey for a great deal of information on Java and Bali.

19. On the Balinese myth of the Hindu gods moving from Java, see Miguel Covarrubias, *Island of Bali* (New York: Knopf, 1946), 6.

20. On the significance of Gunung Agung and the northern direction, see ibid., 76.

21. On the meaning of Eka Dasa Rudra, see David J. Stuart Fox, *Once a Century: Pura Besakih and the Eka Dasa Rudra Festival* (Jakarta: Penerbit Citra Indonesia, 1982), 29. I have used this book as my primary source for much of the following material on the successful performance of the ceremony in 1979. Eric Oey, who witnessed the festival, also provided information. For Hindu deities well known outside of Bali, I have used standard pronunciations rather than the particular Balinese—for example, Vishnu, rather than Wisnu. The Balinese see the three principal Hindu deities in the form of Sanghyang Widhi Wasa as manifestations of one of them, Shiva.

22. Windsor P. Booth and Samuel W. Matthews, "Disaster in Paradise," *National Geographic* 124, no. 3 (1963): 453. This article has a detailed account of the eruption of Gunung Agung and its effects on Bali.

23. Imade Budi, a well-known Balinese artist, told me that because of the famine, the governor of Bali—a Communist who had little interest in the ceremony—made the people skimp in such ways as cutting up a coconut that should have been a single offering into a hundred separate offerings.

24. Although the climax of the festival occurred in the great sacrifice on March 28, ceremonies continued until April 23 at Besakih and until May 9 at other places in Bali and Java.

25. These systems of meditation, both Hindu and Buddhist, belong to what is called tantric yoga. Buddhist tantric yoga used to be widely practiced in India and Indonesia. Now in South Asia it is mostly found among people practicing forms of Tibetan Buddhism. Followers of Theravada Buddhism, the kind found in most of Southeast Asia, do not practice tantric yoga. For a classic work on yoga with a discussion of the chakras and the central channel, called the *sushumna*, see Mircea Eliade, *Yoga: Immortality and Freedom*, trans. Willard Trask, 2d ed. (Princeton: Princeton University Press, 1969). Arthur (Sir John Woodroffe) Avalon, trans., *Tantra of the Great Liberation (Mahanirvana Tantra)* (1913; repr. New York: Dover, 1972) is a translation of a Hindu tantric text that goes into the inner symbolism of Meru and Kailas. Many translations of texts dealing with tantric yoga in Tibetan Buddhism are now in print.

CHAPTER 6

1. Drawing on the work of early scholars, Mircea Eliade and Joseph Campbell have popularized the idea that ziggurats symbolized world mountains in which the entire cosmos was seen as a mountain with its peak in heaven and its roots in hell, but more recent research by Near Eastern scholars suggests that ziggurats functioned as cosmic centers linking earth to heaven, centers that may not have been viewed primarily as mountains—see Richard J. Clifford, *The Cosmic Mountain in Canaan and the Old Testament* (Cambridge: Harvard University Press, 1972), 9–25. Clifford also points out that whereas the Egyptians viewed creation as beginning on a primordial hill, probably inspired by the emergence of mounds of land from yearly floods of the Nile, there is no evidence that pyramids were viewed as representations of such a hill or mountain (ibid., 25–29).

2. For a discussion of Zaphon, the Canaanite gods El and Baal, and their influence on biblical conceptions of Sinai and Zion as well as Yahweh or God, see ibid., 34–97, 191–92. Zaphon appears in Ugaritic texts as *spn*.

3. Ps. 125:1–2. Old Testament passages are from the Jewish Publication Society of America translation of the Bible, 1917 edition; New Testament passages are from the King James translation.

4. Deut. 3:25.

5. Ps. 72:3.

6. Ibid., 121:1–2. Some argue that the psalm casts aspersions on mountains by saying that help comes not from them, the reputed abode of Canaanite deities, but from God. However, the view I have given seems more in keeping with the spirit of the psalms and the generally positive view of mountains in the Bible.

7. Ezek. 28:14, referring to the expulsion of Adam and Eve from Eden. This is a later view of the earthly paradise, one not used explicitly in the original description in Genesis.

8. Gen. 22:16–18.

9. Gen. 17:1. El Shaddai appears in most translations of the Bible as "God Almighty," but the *Encyclopedia Judaica*, s.v. "God, Names of," points out that the term is obscure and probably had the original meaning of "El of the Mountain." In an interesting article David Biale argues that *shaddai* comes from a Canaanite word for breasts, suggesting that a fertility goddess of the Canaanite religion was transformed into a masculine god of war in the early Jewish tradition ("The God with Breasts: El Shaddai in the Bible," *History of Religions* 20, no. 3 [1982]): 240–56. This suggests that mountains in the Middle East may themselves have been viewed as female breasts in their role as sources of fertility.

10. Matt. 4:8–10.

11. Matt. 17:1–3.

12. For a detailed study of the role of mountains in the life of Jesus, see Terence L. Donaldson, *Jesus on the Mountain: A Study in Matthean Theology* (Sheffield, England: JSOT Press, 1985).

13. Theodore Studium, translated in Heinz Skrobucha, *Sinai*, trans. Geoffrey Hunt (London and New York: Oxford University Press, 1966), 2.

14. Koran 96:1–5. The Koran is also written Qur'an.

15. Koran 2:93, 7:171; Louis Ginzberg, *the Legends of the Jews*, 7 vols. (Philadelphia: The Jewish Publication Society of America, 1954), 3:92.

16. For accounts of the revelation of the Koran on Mount Hira and the night journey of Muhammad, see Martin Lings, *Muhammad: His Life Based on the Earliest Sources* (London: George Allen & Unwin, 1983), 42–43, 101–3.

17. Koran 5:3.

18. See Richard C. Martin, "Muslim Pilgrimage," in *The Encyclopedia of Religion*, ed. Mircea Eliade (New York: Macmillan, 1986), 11:344.

19. From Al-Ghazzali, *Mishkat Al-Anwar (The Niche for Lights)*, translated in Skrobucha, *Sinai*, 2.

20. James Bryce, *Transcaucasia and Ararat*, 4th ed. rev. (London and New York: Macmillan, 1986), 298. Kaf also appears in written sources as Qaf. For a discussion of the role of Kaf in Sufi mysticism, see Henry Corbin, *Spiritual Body and Celestial Earth: From Mazdean Iran to Shi'ite Iran*, trans. Nancy Pearson, Bollingen Series XCI: 2 (Princeton, N.J.: Princeton University Press, 1977), 73 ff.

21. For an overview of Zoroastrian cosmology and the place of mountains in it, see Mary Boyce, *A History of Zoroastrianism*, Handbuch der Orientalistik, 2 vols. (Leiden, the Netherlands: Brill, 1975), 1:130–46. Kaf appears in the *Bundahishn* 12.17, translated in *Zand-Akasih, Iranian or Greater Bundahishn*, trans. B. T. Ankelsaria (Bombay: 1956), 95. On the identification of Kaf with Elburz, see Corbin, *Spiritual Body*, 74. The date of Zarathustra, also known as Zoroaster, is unclear; Boyce, 190, concludes that Zarathustra lived sometime between 1400 and 1000 B.C.

22. On the crossing of the Chinvat Bridge, see Corbin, *Spiritual Body*, 28; Boyce, *A History*, 237–41; and Mircea Eliade, *A History of Religious Ideas*, trans. Willard Trask, 3 vols. (Chicago: University of Chicago Press, 1978), 1:329–31.

23. On the identification of the modern Elburz Range with Hara Berezaiti, see Boyce, *A History*, 143–44. *Bundahishn* 9.34 mentions Demavend (Damavand) by name and refers to a demon fettered there. Duncan Forbes, *The Heart of Iran* (London: Robert Hale, 1963), 59–61, recounts a recent version of the myth. Older versions refer to Feridun and Zohak as Fredon and Azhdahak.

24. Gen. 8:3–4.

25. References to the land of Ararat occur in 2 Kings 19:37 and Isa. 37:38.

26. For a study of Armenian history and culture with references to Ararat and Urartu, see David Marshall Lang, *Armenia: Cradle of Civilization*, 3d ed. corrected (London: George Allen & Unwin, 1980). On local features and the story of Noah, see Bryce, *Transcaucasia*, 222.

27. Bryce, *Transcaucasia*, 220.

28. Ibid., 309–10.

29. Since the biblical images of Sinai and Zion have played a pivotal role in the development of Western civilization and are the most familiar to people today, I have concentrated on them, rather than on later beliefs and practices concerning these and other sacred mountains in the Middle East.

30. Dr. J. H. Hertz, ed., *The Pentateuch and Haftorahs*, 2d ed. (London: Soncino Press, 1961), 290.

31. Exod. 19:16–20.

32. Ibid., 24:16–18.

33. Ibid., 3:1–13.

34. Deut. 1:2.

35. Emmanuel Anati in *The Mountain of God* (New York: Rizzoli, 1986) argues for the site in the Negev, the southern desert of the present-day state of Israel.

36. From Flavius Josephus, *Antiquitates Iudaeorum*, translated in Skrobucha, *Sinai*, 12.

37. Deut. 4:9–12.

38. Nilus Sinaita, quoted and translated in Skrobucha, *Sinai*, 31.

39. For an excellent history of the Monastery of Saint Catherine and its influence in Europe see Skrobucha, *Sinai*. Skrobucha also describes the monastery and its treasures.

40. Lesley Hazleton, *Where Mountains Roar* (New York and London: Penguin Books, 1981), 37–38.

41. The hill erroneously identified as Mount Zion today was part of the Temple Mount accidentally left out of the present city walls when they were rebuilt in the sixteenth century under Suleiman the Magnificent. Suleiman punished the architect responsible for the error, but the misidentification arising from his mistake remained.

42. Ps. 48:2–3.

43. Ibid., 50:2–3.

44. Isa. 2:2–4.

45. Ps. 2:6.

46. Jon D. Levenson, *Sinai and Zion: An Entry into the Jewish Bible* (Minneapolis, Chicago, New York: Winston Press, Seabury Books, 1985), contrasts the two mountains as Sinai the mountain of the covenant and Zion the mountain of the temple. His book delves deeply into the complex symbolism of the opposition between the two; I have drawn on his work and added ideas of my own. Another book that deals with the symbolism of Sinai and Zion is Robert L. Cohn, *The Shape of Sacred Space: Four Biblical Studies*, AAR Studies in Religion 23 (Chico, Calif.: Scholars Press, 1981).

47. Ps. 43:3.

CHAPTER 7

1. Although not geographically precise, for reasons of convenience and cultural continuity, here Europe includes the British Isles and Iceland.

2. On the female characteristics of mountains and mountain deities in Aegean culture and the orientation of temples toward mountains, see Vincent Scully, *The Earth, the Temple, and the Gods: Greek Sacred Architecture* (New Haven and London: Yale University Press, 1962), 9–19, and Dolores LaChapelle, *Earth Wisdom* (Silverton, Conn.: Finn Hill Arts, 1978), 32–36. Information on the Greek gods and goddesses mentioned in the following section can be found in any number of references on Greek and Roman mythology, such as the *Oxford Classical Dictionary*, the *Larousse World Mythology*, and the *Encyclopaedia of Religion and Ethics*. Scully in his book discusses a number of deities and

their temples in relation to sacred mountains of Greece.

3. *Iliad* 15.189–93, 8.25–26. Edith Hamilton, *Mythology* (1940; repr. New York and Toronto: Mentor Books, New American Library, 1969), 25, points out that since Olympus is common to all three deities, it is not simply heaven.

4. *Iliad* 1.495–99, translated in *The Iliad of Homer*, trans. Richmond Lattimore (Chicago: University of Chicago Press, 1951), 72.

5. For example, *Theogony* 62–63, 793–94.

6. *Iliad* 8.409–12.

7. *Odyssey* 6.44–47, translated in Homer, *The Odyssey*, trans. Robert Fitzgerald (Garden City, N.Y.: Anchor Doubleday, 1963), 100.

8. *Iliad* 14.225–26. On the meaning of Olympus and other mountains bearing the name, see Martin P. Nilsson, *The Mycenaean Origin of Greek Mythology* (Berkeley and Los Angeles: University of California Press, 1972), 234–37.

9. Herodotus, *Histories* 7.141–43.

10. Philip Sherrard, *Athos, The Holy Mountain* (Woodstock, N.Y.: Overlook Press, 1985), 12.

11. Ibid., 14–15. I have used Sherrard's book as the major source for my account of the history and description of Athos in this section of the chapter.

12. Ibid., 22.

13. Konstantine Daponte, *The Garden of Graces*, trans. in ibid., 145.

14. Syméon le Nouveau Théologien, *Hymnes*, ed. Johannes Koder, Sources Chrétiennes 156 (Paris: Les Editions du Cerf, 1969), 1: Hymn 3.

15. Snorri Sturluson, *The Prose Edda of Snorri Sturluson: Tales from Norse Mythology*, trans. Jean Young (Berkeley and Los Angeles: University of California Press, 1954), 37.

16. According to E. O. G. Turville-Petre, *Myth and Religion of the North: The Religion of Ancient Scandinavia* (1964; repr. Westport, Conn.: Greenwood Press, 1975), 64, Hlidskjalf may mean "the hill, rock with an opening in it."

17. *Prose Edda*, 46, 54, 87.

18. For the derivation of *valhalla* see Turville-Petre, *Myth and Religion of the North*, 55.

19. *Eyrbyggja Saga*, trans. Herman Palsson and Paul Edwards (Edinburgh: Southside, 1973), 41.

20. Ibid., 51.

21. On the frost giants, see *Prose Edda*, 51–52.

22. John Lindow, *Swedish Legends and Folktales* (Berkeley and Los Angeles: University of California Press, 1978), 34–35, 98–99.

23. Ibid., 148–51.

24. Translated in Johann Wolfgang von Goethe, *Faust: Part One*, trans. Philip Wayne (Baltimore: Penguin, 1949), 172.

25. Caspar Peucer, quoted in Sigurdur Thorarinsson, *Hekla, A Notorious Volcano*, trans. Jóhann Hannesson (Reykjavik: Almenna Bókafélagid, 1970), 6. Thorarinsson's book gives a history of Hekla's eruptions and beliefs about the mountain.

26. Thorarinsson refers on p. 6 to a travel book written by a Frenchman named De la Martiniere and printed in 1675, which describes the souls on ice.

27. On the Tuatha Dé Danaan and the *sidh*, see J. A. MacCulloch, *The Religion of the Ancient Celts* (Edinburgh: T. & T. Clark, 1911), 63 ff., 372 ff., and Proinsias MacCana, "Sidh," in *The Encyclopedia of Religion*, ed. Mircea Eliade (New York: Macmillan, 1986), 13:314–15.

28. On the cauldron of plenty and its relation to the Grail, see MacCulloch, 382–83, and Roger Sherman Loomis, *The Grail: From Celtic Myth to Christian Symbol* (New York: Columbia University Press, 1963).

29. On the tradition recorded by Geoffrey of Monmouth, see Ann Ross, *Pagan Celtic Britain: Studies in Iconography and Tradition* (New York: Columbia University Press, 1967), 278. On the sacredness of mountains in Snowdonia and legends of Rhita Gawr and Brenin Llwyd, see Amory B. Lovins, *Eryri, the Mountains of Longing* (San Francisco: Friends of the Earth, 1971), 32. A similar belief about a king sleeping within a mountain, waiting to return in time of need, occurs in legends about the holy Roman emperors

Frederick I (Barbarossa) and Frederick II, one or the other said to be waiting within the Kyffhäuser, a mountain in Thoringia, Germany.

30. On the legend of Gwynn (Gwynn ap Nudd) and Saint Collen, see MacCulloch, *Ancient Celts*, 115, and S. Baring-Gould and John Fisher, *The Lives of the British Saints*, 4 vols. (London: Honourable Society of Cymmrodorion, 1908), 2:258-60.

31. On contemporary esoteric interest in Glastonbury Tor and the labyrinth hypothesis, see Geoffrey Ashe, *Avalonian Quest* (Bungay, Suffolk: Fontana, 1984), and Greg Stafford, "The Labyrinth and Tor of Glastonbury," *Shaman's Drum*, no. 9 (1987): 40-43.

32. On Croagh Patrick, see Brian de Breffny, *The Land of Ireland* (New York: Harry N. Abrams, 1979), and *Encyclopaedia Britannica*, 1973 ed., s.v. "Croagh Patrick." Daphne D. C. Pochin Mould, *Irish Pilgrimage* (New York: Devin-Adair, 1957), 133-40, has a detailed description of the annual pilgrimage up the mountain.

33. John Evelyn, *The Diary of John Evelyn*, ed. E. S. de Beer (London: Oxford University Press, 1959), 256.

34. Thomas Burnet, *The Sacred Theory of the Earth*, 6th ed., 2 vols. (London: John Hooke, 1726), 1:194-96.

35. Ibid., 188-89.

36. Joseph Addison, *Remarks on Several Parts of Italy*, in *Works*, ed. George Washington Green (New York, 1854), 2:339-40.

37. On the transformation in European perceptions of mountains, especially English, and their relation to scientific developments in Europe, see Marjorie Hope Nicolson, *Mountain Gloom and Mountain Glory: The Development of the Aesthetics of the Infinite* (New York: Norton, 1963).

38. *The Prelude* 6.624-40.

39. From a letter to Father Denis di Borgo San Sepulcro translated by Henry Reeve and quoted in Francis Gribble, *The Early Mountaineers* (London: T. Fisher Unwin, 1899), 21-23.

40. Ibid.

41. On early German ascents, see Nicolson, *Mountain Gloom*, 49 n. 17.

42. On Roche Melon and its shrine, see *Standard Encyclopedia of the World's Mountains*, ed. Anthony Huxley (New York: Putnam's, 1962), 260, and Gribble, *Early Mountaineers*, 5-13.

43. Letter to Jacob Avienus in Conrad Gesner, *On the Admiration of Mountains*, trans. H. B. D. Soulé (San Francisco: Grabhorn Press, 1937), 5.

44. On the Pilate legend and Mount Pilatus, see Gribble, *Early Mountaineers*, 43-62. For Gesner's own account, see Gesner, *Admiration*, 33-36.

45. On Scheuzer's view of dragons in Grisons, see Yi-fu Tuan, *Landscapes of Fear* (New York: Pantheon Books, 1979), 80, and Gavin Rylands de Beer, *Early Travellers in the Alps* (London: Sidwick & Jackson, 1930), 89-90.

46. Tuan, *Landscapes*, 79-80, 109-10.

47. See Paul Gayet-Tancrède (Samivel), *Hommes, cimes et dieux: les grandes mythologies de l'altitude et la légende dorée des montagnes à traverse le monde* (Paris: Arthaud, 1973), 136-39, for a compilation of these legends and beliefs.

48. Guido Rey, *The Matterhorn*, trans. J. E. C. Eaton (Oxford: Basil Blackwell, 1946), 36-37, and H. Correvon, "Au pied du Cervin," *Bulletin de l'Association pour la protection des plantes*, no. 4 (1896): 19.

49. For an overview of the history and culture of the Alps, along with a discussion of present-day views and problems, see Paul Guichonnet, ed., *Histoire et civilisations des Alpes*, 2 vols. (Toulouse and Lausanne: Privat/Payot, 1980).

50. Gayet-Tancrède (Samivel), *Hommes*, 127-28, quoting a sixteenth-century account by Josias Simmler in W. A. B. Coolidge, *Josias Simler et les origines de l'alpinisme jusqu'en 1600* (Grenoble: Allier Frères, 1904).

51. Letter of 4 November 1799, adapted from English and French translations in Johann Wolfgang von Goethe, *Miscellaneous Travels of J. W. Goethe*, ed. L. Dora Schmitz and trans. A. J. W. Morrison (London: George Bell and Sons, 1884), 30, and *Voyages en Suisse et en Italie*, trans. Jacques Porchat (Paris: Librairie Hacette, 1878),

30-31.

52. Letter of 27 October 1799, in *Miscellaneous*, 25, and *Voyages*, 25.

53. "Mont Blanc: Lines Written in the Vale of Chamouni."

54. Horace Benedict de Saussure, *Voyages dans les Alpes*, 4 vols. (Neuchatel: Louis Fauche-Borel, 1779-1796), translated in R. L. G. Irving, *The Romance of Mountaineering* (London: J. M. Dent and Sons, 1935), 21.

55. Ibid., 39.

56. Gaston Rébuffat, *Mont Blanc to Everest*, trans. Geoffrey Sutton (London: Thames and Hudson, 1956), 13.

57. On the Black Madonna of Czestochowa and her significance, see Leonard W. Moss and Stephen C. Cappannari, "In Quest of the Black Virgin: She Is Black Because She Is Black," in *Mother Worship: Theme and Variations*, ed. James J. Preston (Chapel Hill: University of North Carolina Press, 1982), 57, and James Preston, "Goddess Worship: An Overview," in *The Encyclopedia of Religion*, ed. Mircea Eliade (New York: Macmillan, 1986), 6:42. Preston mentions the fact that members of Solidarity wear badges of the Black Madonna. Newspaper pictures of Lech Walesa show that he always wears her image in public. A fascinating study of the Black Madonna and an account of the pilgrimage to Jasna Góra appear in China Galland, *Invisible Light: Tara and the Black Madonna* (New York: Viking Penguin, 1990). I am indebted to Galland for pointing out the significance of Jasna Góra and helping me with research on the subject.

58. H. E. M. Stutfield, "Mountaineering As a Religion," *The Alpine Journal* 32, no. 218 (1918), and F. T. Wethered, "Correspondence," ibid., no. 219 (1919): 403-4.

59. To be fair to Stephen, the playground he envisioned had a strong aesthetic and spiritual component missing in modern amusement parks like Coney Island.

CHAPTER 8

1. On the Bushmen and their myth of the Tsodilo Hills, see Alf Annenburgh, *The Bushmen* (Cape Town, Johannesburg: C. Struik, 1979), 14-15, 143. For the beliefs of Aborigines concerning Ayers Rock in Australia, see Chapter 11 on Oceania in my book.

2. *Metamorphosis* 4.655 ff., translated in Ovid, *Metamorphoses*, trans. Rolfe Humphries (Bloomington: Indiana University Press, 1955).

3. Henry M. Stanley, *In Darkest Africa*, 2 vols. (New York: Scribner's, 1890), 1: 429.

4. Filippo de Filippi, *Ruwenzori: An Account of the Expedition of H. R. H. Prince Luigi Amedeo of Savoy, Duke of the Abruzzi* (London: Archibald Constable, 1909), 133.

5. See Dian Fossey, *Gorillas in the Mist* (Boston: Houghton Mifflin, 1983), 36-41.

6. On the Masai and Ol Doinyo Lengai, see Peter Matthiessen, *The Tree Where Man Was Born* (New York: Dutton, 1983), 152, and Tepilit Ole Saitoti and Carol Beckwith, *Maasai* (New York: Harry N. Abrams, 1980), 17. For a general overview of Masai religion and folklore, see Hans-Egil Hauge, *Maasai Religion and Folklore* (Nairobi, Kenya: City Printing Works, 1979). Masai is also spelled Maasai.

7. On the Sonjo and their religious beliefs concerning Ol Doinyo Lengai, see Robert F. Gray, *The Sonjo of Tanganyika: An Anthropological Study of an Irrigation-based Society* (London and New York: International African Institute by Oxford University Press, 1963), 24, 97-98, 106-7.

8. The myth was told to me by Willy Makundi, who added that it is still popular among Chagga children today. He remembered that Kibo and Mawanzi were wives, but he was not completely sure that they were married to the god Ruwa, although he thought so. According to other versions that have appeared in print, Kibo and Mawenzi were brothers smoking pipes (Dr. Reusch in *Ice Cap*, no. 1, quoted in Iain Allan, *Guide to Mount Kenya and Kilimanjaro* [Nairobi: Mountain Club of Kenya, 1981], 282) or two men tending fires (Bruno Gutmann, "Chagga Folk-Lore," in *Tanganyika*

Notes and Records, rev. ed., ed. J. A. Hutchinson [Dar Es Salaam: Tanzania Society, 1974], 50). However, as Makundi pointed out, these versions seem to be mistaken because men in Chagga society traditionally have nothing to do with fires—which are only tended by women and children—and they did not smoke pipes in the past. Mawenzi or Kimawenzi means "having a broken top," according to some sources. The Chagga actually pronounce Kibo as Kipo or Kipoo, a word of uncertain meaning that may mean "snow" or "spotted." Kilimanjaro is an artificial word made up by outsiders from words in Swahili and could have a number of meanings. See J. A. Hutchinson, "The Meaning of Kilimanjaro," in *Tanganyika Notes and Records*, 65-67.

9. Willy Makundi. Because of a lack of records, it is unclear when the Chagga arrived on Kilimanjaro. For a history of the Chagga from the nineteenth century on, see Kathleen M. Stahl, *History of the Chagga People of Kilimanjaro* (London and The Hague: Mouton, 1964).

10. John Kesby, "East Africa," in *Mythology: An Illustrated Encyclopedia*, ed. Richard Cavendish (New York: Rizzoli, 1980), 215-16.

11. Charles Dundas, *Kilimanjaro and Its People: A History of the Wachagga, Their Laws, Customs and Legends, Together with Some Account of the Highest Mountain in Africa* (1924; repr. London: Frank Cass & Co., 1968), 33-35. Dundas, a British administrator in the area, provides much information on former Chagga customs and traditions regarding Kilimanjaro.

12. Old customs according to Dundas, *Kilimanjaro*, 38-39, 192. Willy Makundi told me about the altars in Christian churches on Kilimanjaro.

13. Dr. R. Reusch, "Mount Kilimanjaro and Its Ascent," in *Tanganyika Notes and Records*, 131-32.

14. "The Snows of Kilimanjaro," in *The Snows of Kilimanjaro and Other Stories* (New York: Scribner's, 1927), p. 3.

15. Ibid., 27.

16. Jomo Kenyatta, *Facing Mount Kenya: The Traditional Life of the Gikuyu* (London: Secker and Warburg, 1938). Kenyatta tells the origin myth on pp. 3-6. Gikuyu is the spelling closer to the actual pronunciation of the name of the tribe, but it appears in most written sources as Kikuyu, so I have used the latter.

17. Ibid., 203.

18. Ibid., 244-49.

19. Gary Smith, "A Day in the Life of Mount Kenya," *Sports Illustrated* 62, no. 21 (1985): 70.

20. See Kiboi Muriithi and Peter Ndoria, *War in the Forest: The Autobiography of a Mau Mau Leader* (Nairobi: East African Publishing House, 1971), and Smith, "Day in the Life," 68-69.

21. Jacob Kamau, quoted in Smith, "Day in the Life," 82.

22. For two accounts of this episode, see Smith, "Day in the Life," 82, and "Climbers on Mt. Kenya Search for Lost African," *New York Times*, 24 August 1979, 6:3. Iain Allan gave me additional information about the incident. The *New York Times* reported that the man, named Ephraim M'ikiara, was barefoot and that he probably came from the Meru, a tribe living on the east side of Mount Kenya and closely related to the Kikuyu. Whether the man was barefoot or sandal clad, his ascent was a remarkable achievement.

23. Eric Shipton, *Upon That Mountain*, reprinted in *The Six Mountain-Travel Books*, ed. Jim Perrin (Seattle: The Mountaineers, 1985), 360.

24. Felice Benuzzi, *No Picnic on Mount Kenya* (Layton, Utah: Gibbs Smith, Publisher, 1989).

25. Joseph Thomson, *Through Masai Land*, 3d ed. (London: Frank Cass, 1968), 222.

26. For Nyerere's proclamation and a photograph of a Tanzanian climber planting a flag on the summit of Kilimanjaro on the night of independence, see *Tanganyika Notes and Records*, frontispiece. Two years later Tanganyika formed a union with Zanzibar, changing its name to Tanzania.

27. In 1985 Ian Howell had climbed Mount Kenya more times than any other person according to Smith, "Day in the Life," 68.

CHAPTER 9

1. John Muir, *The Yosemite* (1914; repr. San Francisco: Sierra Club Books, 1988), 4.

2. Richard K. Nelson, *Make Prayers to the Raven: A Koyukon View of the Northern Forest* (Chicago: University of Chicago Press, 1983), 45. On the various indigenous names of the mountain, see James Kari, "The Tenada-Denali–Mount McKinley Controversy," *Alaska Native Magazine* (October 1985): 13-14. The Koyukon actually pronounce Denali "deenaalee." Some of the variants on the two basic names are Denaze, Denadhe, Dengadh, and Dghelaay Ce'e.

3. Julius Jetté, "On Ten'a Folklore," *Journal of the Royal Anthropological Institute* 38 (1908): 312-13.

4. Frederica de Laguna, *Under Mount Saint Elias: The History and Culture of the Yakutat Tlingit*, Smithsonian Contributions to Anthropology, vol. 7 (Washington, D.C.: Smithsonian Institution Press, 1972), 1:252.

5. Ibid., 1:237, 323, 440; 3: pls. 144, 216; 3:1303.

6. Ibid., 2:819.

7. Ibid., 2:818-19.

8. *Newsweek*, 25 August 1986, 53.

9. Theodore Winthrop, "Tacoma and the Indian Legend of Hamitchou," in *Mount Rainier: A Record of Exploration*, ed. Edmond S. Meany (New York: Macmillan, 1916), 38-39, quoted from his book *The Canoe and the Saddle*. Ella E. Clark, *Indian Legends of the Northwest* (Berkeley and Los Angeles: University of California Press, 1953), 27-28, has a discussion of the various Indian names of Mount Rainier, ranging in possible meaning from "Breast of the Milk-White Waters" to any "great white mountain." I have used *tacoma* as the most familiar rendition of *takhoma, tacoman*, etc.

10. Sluiskin, "Indian Warning Against Demons," trans. General Stevens, in ibid., 133-34. I have changed the original Takhoma to Tacoma to avoid confusion.

11. Drawn from four myths of the Duwamish, Skikomish, Puyallup, and Nisqually tribes recorded in Clark, *Indian Legends of the Pacific Northwest*, 28-31. The sources Clark used are of varying reliability.

12. Ibid., 31-32. The story suggests Christian influences, either on the Indians themselves or the person who recorded it. The theme of the flood itself is, however, indigenous.

13. Joaquin Miller, *Unwritten History: Life Amongst the Modocs* (Hartford, Conn.: American Publishing, 1874), 264-76.

14. Clark, *Indian Legends of the Pacific Northwest*, 12, and Roland Dixon, "Shasta Myths," *Journal of American Folklore* 23 (1910): 36.

15. W. S. Cervé (pseud. for Harvey Spencer Lewis), *Lemuria: The Lost Continent of the Pacific* (San Jose, Calif.: The Rosicrucian Press, 1931).

16. For a discussion of the Rosicrucians and Lemuria, see Arthur Francis Eichorn, *The Mount Shasta Story*, 2d ed. (Mount Shasta: Mount Shasta Herald, 1971), 62-74.

17. On Ballard and the I AM group, see Eichorn, *Mount Shasta Story*, 100-107.

18. John Muir, *The Mountains of California* (1894; repr. Garden City, N.Y.: The Natural History Library/Anchor Books, 1961), 2-3.

19. Ibid., 2, and John Muir, "Explorations in the Great Tuolumne Cañon," in *Voices for the Earth: A Treasury of the Sierra Club Bulletin*, ed. Ann Gilliam (San Francisco: Sierra Club Books, 1979), 5.

20. On the Sierra Nevada and Mount Sinai, see *John of the Mountains: The Unpublished Journals of John Muir*, ed. Linnie Marsh Wolfe (1938; repr. Madison: University of Wisconsin Press, 1979), 92.

21. John Muir, *My First Summer in the Sierra* (1911; repr. New York: Penguin, 1987), 198. Cathedral Peak was named by the California Geological Survey in 1863. Josiah Dwight Whitney, a geologist with the Survey for whom the highest mountain in the Sierra Nevada is named, wrote of Cathedral Peak (*The Yosemite Book* [New York: J. Bien, 1868], 425), "The majesty of its form and its dimensions are such, that any work of human hands would sink into insignificance if placed beside it."

22. Muir, "Great Tuolumne Cañon," 7-8.

23. Muir, *Mountains of California*, 51–52. For a study of the spiritual dimensions of Muir's life and thought, see Michael P. Cohen, *The Pathless Way: John Muir and American Wilderness* (Madison: University of Wisconsin Press, 1984).

24. For Hoavadunaki's account, see Malcolm Margolin, ed., *The Way We Lived: California Indian Reminiscences, Stories and Songs* (Berkeley: Heyday Books, 1981), 89–91. Hoavadunaki—also known as John Stewart—recounted this story in 1927 or 1928 when he was nearly one hundred years old.

25. On Crazy Horse's vision, see John G. Neihardt, *Black Elk Speaks: Being the Life Story of a Holy Man of the Oglala Sioux* (1932; repr. New York: Pocket Books, 1972), 70–72. For a discussion of the vision quest among the Sioux, see Raymond J. De-Mallie, "Lakota Belief and Ritual in the Nineteenth Century," in *Sioux Indian Religion: Tradition and Innovation*, ed. Raymond J. DeMallie and Douglas R. Parks (Norman and London: University of Oklahoma Press, 1987), 33–42.

26. Peter Nabokov, *Two Leggings: The Making of a Crow Warrior* (New York: Thomas Y. Crowell, 1967), 62–63.

27. Chief John Snow, *These Mountains Are Our Sacred Places: The Story of the Stoney People* (Toronto and Sarasota: Samuel Stevens, 1977), 13.

28. Raymond J. DeMallie, ed., *The Sixth Grandfather: Black Elk's Teachings Given to John G. Neihardt* (Lincoln and London: University of Nebraska Press, 1984), 141. *The Sixth Grandfather* presents the transcripts of the original interviews that Neihardt condensed and rewrote in more poetic form as *Black Elk Speaks*.

29. Ibid., 134.

30. Ibid., 135.

31. Ibid., 296.

32. On the Thunderbird and Harney Peak, see William K. Powers, *Sacred Language: The Nature of Supernatural Discourse in Lakota* (Norman and London: University of Oklahoma Press, 1986), 37.

33. Traditions on Crazy Horse, Custer, and Bear Butte attributed to the Sioux come from Richard B. Williams in a mimeographed paper compiled by him and obtained by David Reigle in Sturgis, South Dakota. This source may or may not be reliable.

34. On the role of Bear Butte, Sweet Medicine, and the sacred arrows in Cheyenne history and contemporary life, including accounts of recent fasts on the mountain, see Father Peter John Powell, *People of the Sacred Mountain*, 2 vols. (San Francisco: Harper & Row, 1981).

35. On the Tewa Pueblos of San Juan, Santa Clara, San Ildefonso, Tsuque and Nambe and their relation to sacred mountains, see Alfonso Ortiz, *The Tewa World: Space, Time, Being, and Becoming in a Pueblo Society* (Chicago and London: University of Chicago Press, 1969) and Alfonso Ortiz, "Look to the Mountaintop," in *Essays on Reflection*, ed. E. Graham Ward (Boston: Houghton Mifflin, 1973). Tesuque and Nambe Pueblos substitute Lake Peak for Truchas Peak.

36. Adapted from Washington Matthews, *Navajo Legends*, Memoirs of the American Folklore Society, vol. 5 (Boston: American Folklore Society, 1897), 78–79.

37. From a conversation with Wilson Aronilth, Jr., at the Navajo Community College at Tsaile, Arizona, on 23 November, 1986.

38. Blessing Way Chant according to Slim Curly, recorded and translated in Leland C. Wyman, *Blessingway* (Tucson: University of Arizona Press, 1970), 150.

39. Philip Hyde and Stephen C. Jett, *Navajo Wildlands* (San Francisco: Sierra Club/Ballantine Books, 1969), 50.

40. George Blue Eyes, quoted in "Native Peoples of the Southwest: The Permanent Collection of the Heard Museum," exhibit in Phoenix, Arizona.

41. Ibid.

42. Wilson Aronilth, Jr., in a conversation we had at Tsaile.

43. John Wood, an anthropologist at the University of Northern Arizona, told me about the Navajo view of the San Francisco Peaks as a woman.

44. For a beautiful and insightful account of Navajo Mountain and its significance for Navajo today, see Karl W. Luckert, *Navajo Mountain and Rainbow Bridge Religion* (Flagstaff: Museum of Northern Arizona, 1977).

45. Edmund Nequatewa, "Chaveyo: The First Kachina," *Plateau* 20, no. 4 (1948): 60–62. For a general discussion of the *katsina* cult in relation to Hopi religion and cosmography, see Armin W. Geertz, "A Reed Pierced the Sky: Hopi Indian Cosmography on Third Mesa, Arizona," *Numen* 31, no. 2 (1984): 216–41. Katsina is often written *kachina*.

46. The Hopi individuals with whom I spoke about the San Francisco Peaks and the *katsinas* requested that I not mention their names.

47. Emory Sekaguaptewa, Jr., quoted in John Dunklee, "Man-Land Relationships on the San Francisco Peaks" (Flagstaff: submitted to the Museum of Northern Arizona Technical Series, n.d.), 75. The Supreme Court seems unable to comprehend the significance of Native American religions and their views of sacred mountains: in 1988 it overturned a ruling by a lower court in favor of preventing construction of the G-O Road through a sacred mountain area in northern California. The Lakota Sioux have also been trying to stop development in the sacred area of the Black Hills in South Dakota.

48. On the Cherokee myth of Mount Mitchell, see James Mooney, *Myths of the Cherokee*, Annual report of the Bureau of American Ethnology to the secretary of the Smithsonian Institution, v. 19, pt. 1 (Washington: Government Printing Office, 1900), 242–44, 431–32.

49. On the Seneca, see Frederick Houghton, "The Traditional Origin and the Naming of the Seneca Nation," *American Anthropologist* 24, no. 1 (1922): 31–33.

50. See Fannie Hardy Eckstorm, "The Katahdin Legends," *Appalachia* 16, no. 1 (1924): 39–52. Many of the legends Eckstorm recorded had suffered a great deal of distortion. According to her understanding, Pamola, the Indian name usually associated with the deity of Katahdin and used for the highest summit of the mountain, applies only to the spirit of the night wind.

51. Henry David Thoreau, *The Maine Woods* (1848; repr. New York: Bramhall House, 1950), 271.

52. Henry James III, ed., *The Letters of William James*, 2 vols. (New York: Kraus Reprint Co., 1969), 2:76–77.

53. David Lujan, director of the Tonantzin Land Institute in Albuquerque, gave me information on the plans to mine at Red Butte and Native American objections to it. On the open-pit uranium mine (Jackpile-Paguate), see John Berger, *Restoring the Earth* (New York: Anchor Doubleday, 1987), 102–5, and *Jackpile-Paguate Uranium Mine Reclamation Project: Environmental Impact Statement*, 2 vols. (U.S. Department of Interior, Bureau of Land Management: Albuquerque, 1986). Paul Robinson, research director at the Southwest Research and Information Center in Albuquerque, gave me information on uranium mining and contamination around Mount Taylor and the state of restoration work at the Jackpile-Paguate mine as of 1989. I also obtained information from conversations with local people and from the New Mexico Museum of Mining in Grants.

CHAPTER 10

1. Latin America comprises the region of Spanish and Portuguese-speaking countries of the Western Hemisphere south of the United States, a region covering Mexico and Central and South America.

2. Robert McCracken Peck of the Academy of Natural Sciences of Philadelphia told me about his research on the beliefs of the Piaroa Indians, who regard the *tepui* Autana in southwestern Venezuela as a tree that once nurtured all creatures of the world.

3. Diego Durán, *Book of the Gods and Rites and the Ancient Calendar*, trans. Fernando Horcasitas and Doris Heyden (Norman: University of Oklahoma Press, 1975), 51.

4. Bernardino de Sahagún, A *History of Ancient Mexico*, trans. Fanny R. Bandelier (Nashville: Fisk University Press, 1932), 45–46.

5. Ibid., 51–52, 72–73.

6. Sahagún, *General History of the Things of New Spain: Florentine Codex, Book 2 — The Ceremonies*, trans. Arthur J. O. Anderson and Charles E. Dibble, 2d ed. rev., Monographs of the School of American Research 14, pt. 3 (Santa Fe; School of American Research, 1981), 44.

7. Durán, *Book of the Gods*, 254, describes the Aztec ascent.

8. The Spanish ascent is described by Henry R. Wagner in "Ascents of Popocatepetl by the Conquistadores," *Sierra Club Bulletin* 26, no. 1 (1940): 88–95.

9. Durán, *Book of the Gods*, 255–57.

10. Ibid., 248–50.

11. Ibid., 250, n. 2, and José Luis Lorenzo, *Las zonas arqueológicas de los volcanes Iztaccíhuatl y Popocatépetl*, Series Publication no. 3 (Mexico City: Instituto Nacional de Antropologia e Historia, 1957), 16–20.

12. George F. Mobley, "Las Sierras, los Volcanes," in *America's Magnificent Mountains* (Washington, D.C.: National Geographic Society, 1980), 149–52.

13. Richard Fraser Townsend, "Pyramid and Sacred Mountain," in *Ethnoastronomy and Archaeoastronomy in the American Tropics*, ed. Anthony F. Aveni and Gary Urton (New York: New York Academy of Sciences, 1982), 37–62, analyzes the relationship between pyramids and sacred mountains in Aztec culture, focusing on the pyramid of Tenochtitlán and the hill of Tetzcotzingo.

14. Durán, *Book of the Gods*, 155.

15. On the archaeological expedition and its findings, see Charles Wicke and Fernando Horcasitas, "Archeological Investigations on Monte Tlaloc, Mexico," *Mesoamerican Notes* 5 (1957): 83–93.

16. Durán, *Book of the Gods*, 156–60.

17. On the reports of sacrifices, see Wicke, "Archeological Investigations," 86–87. The anthropologist was Miguel Barrios of the National School of Anthropology of Mexico (Durán, *Book of the Gods*, 466, n. 1). Winston Crausaz, who helped me with my research on sacred mountains in Mexico, has written a book on various aspects of Orizaba, the highest peak in the country — *Citlaltépetl: A History of Pico de Orizaba Including Exporation, Mountaineering, Natural History, and Geology* (New York: American Alpine Club, 1990).

18. On the relation of Chavín and Tiahuanaco to mountains and mountain deities, see Johan Reinhard, "Chavín and Tiahuanaco: A New Look at Two Andean Ceremonial Centers," *National Geographic Research* 1, no. 3 (1985): 395–422.

19. For the information in this and the next paragraph, see Evelio Echevarría, "The Inca Mountaineers: 1400–1800," in *The Mountain Spirit*, ed. Michael Charles Tobias and Harold Drasdo (Woodstock, N.Y.: Overlook Press, 1979), 121–22. According to Echevarría, Gonzalez and Harseim believed their climb to be the first ascent of the mountain in modern times, but Johan Reinhard informed me that Llullaillaco had already been climbed in 1950 and 1951. According to *The World Almanac and Book of Facts* (New York: World Almanac, 1987), 536, Llullaillaco is the eighth-highest mountain in the Andes; because of different measurements of its altitude, other sources make it the seventh highest.

20. Johan Reinhard, "High Altitude Archeology and Andean Mountain Gods," *The American Alpine Journal* 25, no. 57 (1983): 57 and information provided by Charles Brush, as well as subsequent correspondence from Reinhard.

21. Echevarría, "Inca Mountaineers," 120–21, and Reinhard, "High Altitude Archeology," 59. The sacrifices probably took place in the fifteenth and sixteenth centuries.

22. On Reinhard's research and theories, see his articles "High Altitude Archeology," 54–67, and "Sacred Mountains: An Ethnoarchaeological Study of High Andean Ruins," *Mountain Research and Development* 5, no. 4 (1985): 299–317. *The Highest Altar: The Story of Human Sacrifice* (New York: Viking, 1989), a new book by Patrick Tierney, describes even more recent Andean human sacrifices that have been performed for various other purposes — not just for rain — through 1988.

23. Reinhard has gathered together a survey of Andean mountain beliefs and practices in his two articles, especially the more recent one in *Mountain Research and Development*.

24. Joseph William Bastien, *Mountain of the Condor: Metaphor and Ritual in an Andean Ayllu*, American Ethnological Society Monographs 64 (St. Paul: West, 1978), 72–73. The myth was recorded in the Huarochiri ms. 3169:F. 66.

25. On the legend of the flood in Patagonia and the story of Huascarán and Huandoy, see Evelio Echevarría, "Legends of the High Andes," *The Alpine Journal* 88, no. 332 (1983): 89–90. Huáscar was the name of the Inca ruler whose dispute with his brother Atahualpa split the Inca empire in two and helped the Spanish to conquer it.

26. On Chimborazo, Tungurahua, and the Puruhá, see ibid., 89, and Hermann Trimborn, "South Central America and the Andean Civilizations," in *Pre-Columbian Religions*, ed. W. Krickeberg et al. (New York: Holt, Rinehart and Winston, 1969), 97.

27. See Michael J. Sallnow, *Pilgrims of the Andes: Regional Cults in Cusco* (Washington, D.C.: Smithsonian Institution Press, 1987), 32, 35.

28. The story of the shaman and the preceding material on Ausangate comes from Bernard Mishkin, "Cosmological Ideas Among the Indians of the Southern Andes," *Journal of American Folklore* 53, no. 210 (1940): 237–38.

29. For Qoyllur Rit'i and Randall's analysis, see Robert Randall, "Return of the Pleiades," *Natural History* 96, no. 6 (1987): 42–53. For another description and interpretation of the pilgrimage and the importance of Ausangate in it, see Sallnow, *Pilgrims of the Andes*, 207–42.

30. Bastien, *Mountain of the Condor*, xix.

31. On the Qollahuaya and their relationship to Mount Kaata, see ibid.; for an overview of the same material, see "The Human Mountain," in *Mountain People*, ed. Michael Tobias (Norman and London: University of Oklahoma Press, 1986), 45–57.

32. Johan Reinhard, *The Nazca Lines: A New Perspective on Their Origin and Meaning*, 2d ed. (Lima: Editorial Los Pinos, 1986). For a summary of Reinhard's work and a survey of other theories about the Nazca Lines, see Evan Hadingham, *Lines to the Mountain Gods: Nazca and the Mysteries of Peru* (New York: Random House, 1987). Reinhard's hypothesis helps to explain the lines but, even if proved true, would only provide a partial solution to a complex phenomenon.

33. I have summarized points made by Reinhard in a draft article, "Machu Picchu and Sacred Geography," to be published in Olivier de Montmollin and Michal Saunders, eds., *Contributions to New World Archeology* (New York and London: Oxford University Press, 1990). I am indebted to him for kindly sending it to me. Veronica is the name of a range whose highest summit is 18,865-foot Waqaywillka. Pumasillo, 19,931 feet, is the highest of a series of peaks revered together as a sacred range.

34. On Coropuna and Saint Peter, see Reinhard, "High Altitude Archeology," 61.

35. On the Virgin of Guadalupe, see Ena Campbell, "The Virgin of Guadalupe and the Female Self-Image: A Mexican Case History," in *Mother Worship: Theme and Variations*, ed. James J. Preston (Chapel Hill: University of North Carolina Press, 1982), 6–24, and Eric Wolfe, "The Virgin of Guadalupe: A Mexican National Symbol," *Journal of American Folklore* 71 (1958): 34–39.

36. On crosses on sacred mountains in Chiapas, see Evon Z. Vogt and Catherine C. Vogt, "Lévi-Strauss among the Maya," *Man* 5, no. 3 (1970): 379–92, and Walter F. Morris, Jr., *Living Maya* (New York: Harry N. Abrams, 1987).

CHAPTER 11

1. Some definitions of Oceania limit the area to the three islands groups of Melanesia, Micronesia, and Polynesia, but others also include Australia, as I have done. Melanesia forms a band of islands northeast of Australia, running from New Guinea in the northwest to Fiji in the southeast. Micronesia comprises a parallel band of tiny islands just north of Melanesia. Polynesia forms a rough triangle with corners at New Zealand in the south, Midway Island in the north, and Easter Island in the east.

2. On the John Frum Cult, see Peter Worsley, *The Trumpet Shall Sound: A Study of 'Cargo' Cults in Melanesia* (London: MacGibbon & Kee, 1957), 152–60, and Kal Muller, "Tanna Awaits the Coming of John Frum," *National Geographic* 145, no. 5 (1974): 706–15. Mount Hurun lies in East Sepik Province—see Peter Lawrence, "Cargo Cults," in *The Encyclopedia of Religion*, ed. Mircea Eliade (New York: Macmillan, 1986), 3:78.

3. Robert Layton, *Uluru: An Aboriginal History of Ayers Rock* (Canberra: Australian Institute of Aboriginal Studies, 1986), 5.

4. Ibid., 10, and Charles P. Mountford, *Ayers Rock: Its People, Their Beliefs and Their Art* (Honolulu: East-West Center Press, 1965), 114, 120. Mountford has long, detailed descriptions of Ayers Rock and its myths, along with many illustrative photographs. Both Mountford and Layton have discussions of the significance of the Dreamtime (*tjukurapa* or *tjukurrpa*) and its role in Aboriginal life, particularly as it applies to Ayers Rock.

5. Layton, *Uluru*, 5–7, and Mountford, *Ayers Rock*, 68–114. The demon dingo is called Kurrpanngu (Layton) or Kulphunya (Mountford).

6. Layton, *Uluru*, 7–9, and Mountford, *Ayers Rock*, 31–68.

7. A man of the Kikingkura tribe made the remark about being a "bugger" to Layton (*Uluru*, 15).

8. Layton's book has a detailed discussion of the history of Aboriginal claims to Uluru.

9. Teuira Henry, *Ancient Tahiti*, Bernice P. Bishop Museum Bulletin 48 (Honolulu: Bernice P. Bishop Museum, 1928), 200–201.

10. Margaret Orbell, *Hawaiki: A New Approach to Maori Tradition* (Christchurch: University of Canterbury, 1985), 60–63, and *The Natural World of the Maori* (Auckland: Collins/Bateman, 1985), 111. Both books discuss the important role of Hawaiki in Maori mythology.

11. Orbell, *Hawaiki*, 54–55.

12. Ibid., 53.

13. Orbell, *Natural World*, 85.

14. Ibid., 86.

15. Ibid., 91.

16. Ibid., 129–30.

17. On the importance of mountains and their use in introductory phrases, see Hong-Key Yoon, *Maori Mind, Maori Land: Essays on the Cultural Geography of the Maori People from an Outsider's Perspective* (Berne, New York, Paris: Peter Lang, 1986), 41–59.

18. Ibid., 58, and Orbell, *Natural World*, 84.

19. Orbell, *Natural World*, 91–92.

20. For a fascinating and sympathetic account of Rua and his movement, see Peter Webster, *Rua and the Maori Millennium* (Wellington, New Zealand: Victoria University, 1979).

21. Harry Dansey, "A View of Death," in *Te Ao Hurihuri, the World Moves On: Aspects of Maoritanga*, ed. Michael King (Auckland, New Zealand: Longman Paul, 1981), 141.

22. Martha Warren Beckwith, *Hawaiian Mythology* (New Haven: Yale University Press, 1940), 54–55.

23. Nathaniel Bright Emerson, *Unwritten Literature of Hawaii: The Sacred Songs of the Hula*, Smithsonian Institution, Bureau of American Ethnology, Bulletin 38 (Washington, D.C.: U.S. Government Printing Office, 1909), 188.

24. For a survey and analysis of the various myths pertaining to Pele's travels to Hawai'i, see H. Arlo Nimmo, "Pele's Journey to Hawai'i: An Analysis of the Myths," *Pacific Studies* 11, no. 1 (1987): 1–42.

25. Emerson, *Unwritten Literature*, 195.

26. W. D. Westervelt, *Hawaiian Legends of Volcanoes* (Boston: Ellis Press, 1916), 59–62. Westervelt may have romanticized some of his accounts of Hawai'ian myths.

27. Ibid., 33–34.

28. Ibid., 152–61.

29. Beckwith, *Hawaiian Mythology*, 180, 192.

30. Ibid., 180–81. See also Emerson, *Unwritten Literature*, on the hula.

31. Emerson, *Unwritten Literature*, 201.

32. For numerous stories about encounters with Pele and other matters of related interest, see H. Arlo Nimmo, "Pele, Ancient Goddess of Contemporary Hawaii," *Pacific Studies* 9, no. 2 (1986): 121–79.

33. "Native Hawaiian Position Paper on Geothermal Development," unpublished paper prepared by the Pele Defense Fund. Dr. Emmet Aluli and Palikapu Dedman, who are leading the organization's efforts to preserve Kilauea and Native Hawai'ian culture, gave me additional information on the situation as of 1989. See also Ruth Taswell, "Geothermal Development in Hawaii Threatens Religion and Environment," *Cultural Survival Quarterly* 10, no. 1 (1986): 54–56.

CHAPTER 12

1. Mircea Eliade, *Patterns in Comparative Religion*, trans. Rosemary Sheed (New York: New American Library, 1974), 99. For a relatively recent survey of selected sacred mountains and their significance, see W. Y. Evans-Wentz, *Cuchama and Sacred Mountains*, ed. Frank Waters and Charles L. Adams (Chicago: Swallow Press; Athens, Ohio: Ohio University Press, 1981), 39–83. Evans-Wentz discovered that a mountain he had purchased in southern California, Cuchama or Mount Tecate, had great importance among the Native Americans of the region as a sacred peak (ibid., 5–34). Another, more comprehensive, survey, one that emphasizes the symbolism of mountains as cosmic centers but also looks at other themes, is Paul Gayet-Tancrède (Samivel), *Hommes, cimes et dieux: les grandes mythologies de l'altitude et la légende dorée des montagnes à traverse le monde* (Paris: Arthaud, 1973). Briefer surveys appear in J. A. MacCulloch, "Mountains, Mountain Gods," in *Encyclopedia of Religion and Ethics*, ed. James Hastings (New York: Charles Scribner's Sons, 1915), 8:863–68, and Diana L. Eck, "Mountains," in *The Encyclopedia of Religion*, ed. Mircea Eliade (New York: Macmillan, 1987), 10:130–34.

2. Mircea Eliade, the leading exponent of theories that see all mountains as cosmic axes, has written, "The sacred mountain where heaven and earth meet stands at the center of the world" (*Patterns*, 375). Eliade discusses the theory of mountains as cosmic centers in most of his works, most notably in *The Myth of the Eternal Return or, Cosmos and History*, trans. Willard R. Trask, Bollingen Series 46 (Princeton: Princeton University Press, 1971), 12–17. Other writers influenced by Eliade's ideas on the symbolism of mountains include Joseph Campbell and Paul Gayet-Tancrède (Samivel).

3. *Theogony* 517–18, translated in Richard Lattimore, trans., *Hesiod* (Ann Arbor: University of Michigan Press, 1973), 154.

4. The views surveyed in this section by no means exhaust the list of possibilities. For example, I have not discussed in any detail the important role of mountains as places of caves, usually linked to views of peaks as wombs, sanctuaries, and places of hermitage and meditation.

5. Mark Johnson, ed., *Philosophical Perspectives on Metaphor* (Minneapolis: University of Minnesota Press, 1981), is an anthology that brings together many of these theories. For recent applications of metaphor theory in the field of religious studies, see Sallie McFague, *Metaphorical Theology: Models of God in Religious Language* (Philadelphia: Fortress Press, 1982); Mary Gerhart and Allan Russell, *Metaphoric Process: The Creation of Scientific and Religious Understanding* (Fort Worth: Texas Christian University Press, 1984); and Ian G. Barbour, *Myths, Models and Paradigms: The Nature of Scientific and Religious Language* (London: SCM Press, 1974).

6. For seminal pieces by these two authors, see I. A. Richards, "The Philosophy of Rhetoric," in Johnson, *Philosophical Perspectives*, 48–62, and Max Black, "Metaphor," in ibid., 63–82. Richards was also a mountain climber.

7. Paul Ricoeur has focused on the effects of tension in metaphors—see by him, for example, "The Metaphorical Process as Cognition, Imagination, and Feeling," in Johnson, *Philosophical Perspectives*, 228–47. Whereas Ricoeur says that metaphors create new worlds and meanings, I would say that metaphors reveal the latter—this is truer to the experience of people for whom moun-

tains are sacred. Also the metaphors involved in views of sacred mountains are much more deeply rooted than literary metaphors and are not a matter of conscious choice. They express the most basic assumptions about the nature of reality.

CHAPTER 13

1. Translated from *Kumarasambhava* 1.1, 1.3. For the original Sanskrit text, see Kalidasa, *Kumarasambhava, with the Commentary (the Sanjivini) of Mallinath (1–8 Sargas) and of Siteram (8–17 Sargas)*, ed. Kasinath Pandurang Parab, 3d ed. (Bombay: Nirnayasagara Press, 1893).

2. For a contemporary translation of the entire poem and a discussion of Kalidasa, see Kalidasa, *The Origin of the Young God: Kalidasa's Kumarasambhava*, trans. Hank Heifetz (Berkeley and Los Angeles: University of California Press, 1985).

3. For a French translation of the poem, see Paul Demiéville, "La montagne dans l'art littéraire chinois," *France-Asie/Asia* 20 (1965): 24.

4. Ibid., 30. For a biography of Li Po and a study of his poetry, see Arthur Waley, *The Poetry and Career of Li Po, 701–762 A.D.* (New York: Macmillan, 1950).

5. Matsuo Basho, *The Narrow Road to the North and Other Travel Sketches*, trans. Nobuyuki Yuasa (New York: Penguin Books, 1966), 125.

6. Ibid., 98.

7. For a different translation of the Haiku poem on Fuji, see ibid., 51.

8. *The Manyoshu: The Nippon Gakujutsu Shinkokai Translation of One Thousand Poems* (New York: Columbia University Press, 1969), 215.

9. *Purgatorio* 4.88–96, translated in Dante Alighieri, *The Divine Comedy: Purgatorio*, trans. Charles S. Singleton, 3 vols. (Princeton: Princeton University Press, 1973), 2:41. See Singleton for a commentary on the work.

10. William Wordsworth, *The Prelude* (1850 version) 14.39–49.

11. Ibid., 14.70–74.

12. Wordsworth, *The Excursion* 2:829–77.

13. Translated from Jean-Jacques Rousseau, *Julie ou la Nouvelle Héloise* (Paris: Garnier-Flammarion, 1976), 45.

14. Ibid., 45–46.

15. Friedrich Nietzsche, *Thus Spoke Zarathustra*, trans. R. J. Hollingdale (Baltimore: Penguin Books, 1961), 39.

16. Ibid., 336.

17. Ibid., 174.

18. On the influence of Nazi ideology on some German mountaineers of the Third Reich, see, for example, Elmar Landes, "Mountain Climbers Are Children of Their Age," in *The Big Book of Mountaineering*, ed. Bruno Maravetz, trans. Diana Stone Peters and Frederick G. Peters (Woodbury, N.Y.: Barron's, 1980), 154.

19. Thomas Mann, *The Magic Mountain*, trans. H. T. Lowe-Porter (New York: Alfred A. Knopf, 1927), 606–7.

20. Ibid., 626.

21. Ibid., 900.

22. *The Poems of W. B. Yeats: A New Edition*, ed. Richard J. Finneran (New York: Macmillan, Collier Books, 1989), 289.

23. "The Second Coming," in ibid., 187.

24. William Butler Yeats, "Introduction," in Bhagwan Shri Hamsa, *The Holy Mountain: Manasarovar and the Mount Kailas*, trans. Shri Purohit Swami (1934; repr. Delhi: India Bibliographies Bureau and Balaji Enterprises, 1986), 37.

25. Told to me by David Robertson, who knew the climbers personally and heard the story from one of them, Conor O'Brien. Robertson describes the outings to Wales—but without mention of the Yeats incident—in *George Mallory* (London: Faber and Faber, 1969), his biography of the famous British climber who belonged to the group and disappeared on Mount Everest in 1924.

26. René Daumal, *Mount Analogue: A Novel of Symbolically Authentic Non-Euclidean Adventures in Mountain Climbing*, trans. Roger Shattuck (Baltimore: Penguin Books, 1974), 42.

27. Ibid., 100.

28. From *Hua shan-shui hsü*, translated in Michael Sullivan, *The Birth of Landscape Painting in China* (Berkeley: University of California Press, 1962), 103.

29. The poet Fu Tsai, translated in Michael Sullivan, *Chinese Landscape Painting: The Sui and T'ang Dynasties* (Berkeley and Los Angeles: University of California Press, 1980), 2:66.

30. Catalogue of the collection of the emperor Hui-tsung, translated in John Hay, *Kernels of Energy, Bones of Earth: The Rock in Chinese Art* (New York: China Institute in America, 1985), 53.

31. Kuo Hsi, *An Essay on Landscape Painting*, trans. Shio Sakanishi (London: John Murray, 1935), 33.

32. Ibid., 30–31.

33. Shipposanka Rojin Shoryu, translated in Henry D. Smith II, *Hokusai: One Hundred Views of Mt. Fuji* (New York and Tokyo: George Braziller, 1988), 212. In the introduction to his book, Smith proposes his thesis that Hokusai revered and depicted Fuji as a means of attaining longevity.

34. Ibid., 7. Smith discusses the significance of Hokusai's name and its bearing on his relationship to Fuji.

35. Hugo Munsterberg, *The Landscape Painting of China and Japan* (Rutland, Vt., and Tokyo: Charles Tuttle, 1955), 123.

36. On the development of landscape painting in the West, see Kenneth Clark, *Landscape into Art*, new ed. (New York: Harper & Row, 1976). For a history of Western art focused on mountains, see Helmuth Zebhauser, "Man Has Always Painted Mountains," in *The Big Book of Mountaineering*, 99–104. On Friedrich's mountain paintings specifically, see Linda Seigel, "The Riesengebirge as a Transcendental Image in Friedrich's Art," in *The Mountain Spirit*, ed. Michael Charles Tobias and Harold Drasdo (Woodstock, N.Y.: Overlook Press, 1979), 189–93. Friedrich painted *Morning in the Reisengebirge* in 1810.

37. John Ruskin, *Modern Painters*, 2d ed., 5 vols. (New York: Wiley, 1883), 4:350. On Ruskin's relation to the Romantic movement and the religious and spiritual dimensions of his thought, see David Anthony Downes, *Ruskin's Landscape of Beatitude* (New York and Berne: Peter Lang, 1984).

38. Ruskin, *Modern Painters*, 1:271–72. Michael Sullivan, *Symbols of Eternity: The Art of Landscape Painting in China* (Stanford, Calif: Stanford University Press, 1979), 4–5, points out the values that Ruskin shared with Chinese artists.

39. Quoted in Ralph A. Britsch, *Bierstadt and Ludlow: Painter and Writer in the West* (Provo, Utah: Brigham Young University, 1980), 31.

40. On Bierstadt and his influence, see William H. Goetzmann and William N. Goetzmann, *The West of the Imagination* (New York: Norton, 1986).

41. On art as a priestly vocation, see Cézanne's letter to Ambrose Vollard, 9 January 1903, in Paul Cézanne, *Correspondance*, ed. John Rewald (Paris: Editions Bernard Grasset, 1937), 252.

42. Clark, *Landscape*, 226.

43. Quotes translated from Joachim Gasquet, *Cézanne* (Paris: Les Editions Bernheim-Jeune, 1926), 131, 134, 136. Cézanne scholars such as John Rewald maintain that Gasquet, a poet, romanticized the artist and put his own words and ideas into the latter's mouth. Cézanne did, however, feel unusually close to Gasquet and may well have confided feelings and insights that he shared with no one else. The contemporary who observed Cézanne at work was Emile Bernard; see Emile Bernard, *Souvenirs sur Paul Cézanne et lettres* (Paris: A La Rénovation Esthétique, 1946), 27. Cézanne spoke of his "strong sensation of nature" in a letter to Lous Aurenche, 25 January 1904, in *Correspondance*, 257.

44. Ansel Adams and Mary Street Alinder, *Ansel Adams: An Autobiography* (New York and Boston: New York Graphic Society Books, Little, Brown, 1985), 143.

45. Ibid., 382. For a study of the spiritual dimensions of twentieth-century art, see Roger Lipsey, *An Art of our Own: The Spiritual in Twentieth-Century Art* (Boston and Shaftesbury: Shambhala, 1988); pp. 390–94 deal specifically with the work of Ansel Adams.

46. This chapter has focused on literature and painting, where the role of mountains is most apparent. Mountains also awaken a sense of the sacred in other forms of art, such as architecture, music, drama, and cinema. Skyscrapers and cathedrals, for example, call to mind images of mountain peaks soaring up toward the sky with their heavenly connotations. Certain forms of religious architecture, such as Buddhist *stupas* and Hindu temples in Asia, make associations with mythical mountains, most notably Mount Meru, explicit. Works of music such as Modest Mussorgsky's *Night on Bald Mountain* and Alan Hovhaness's *Mysterious Mountain* attempt to evoke the spirit of sacred mountains, both demonic and divine. The Austrian Alps and Devil's Tower in Wyoming play important roles in *The Sound of Music* and *Close Encounters of a Third Kind*, symbolizing a realm of spiritual freedom in the first film and the other world of aliens from outer space, with their promise of salvation, in the second. The examples given in this chapter show how the approach I have taken can be applied to other forms of literature and art.

CHAPTER 14

1. Ellen Lapham quoted by Maitland Zane in "Bay Woman Taking Aim at Everest," *San Francisco Chronicle*, 13 August 1986.

2. *San Francisco Chronicle*, 6 May 1988. The television transmission took place on 5 May 1988.

3. George Mallory, "Mont Blanc from the Col du Géant by the Eastern Buttress of Mont Maudit," *The Alpine Journal* 32, no. 218 (1918): 162.

4. Isaac Rosenfeld, "Speaking of Books," *The New York Times Book Review*, 23 January 1955.

5. F. S. Smythe, *Camp Six: An Account of the 1933 Mount Everest Expedition*, 2d ed. (London: Hodder and Stoughton, 1938), 243.

6. George Leigh Mallory quoted in Thomas F. Hornbein, *Everest: The West Ridge* (San Francisco and New York: Sierra Club/Ballantine Books, 1968), 24. I was unable to track down the source for this quote. David Robertson, who wrote the authoritative biography of Mallory, told me that he did not recognize the quote. He suspected, however, that it might have come from a newspaper reporter's account of a lecture Mallory gave at the Broadhurst Theater in New York on 4 February 1923. There is the possibility that Mallory did not actually say these words, but they do reflect his sentiments—and those of other climbers.

7. F. W. Bourdillon, "Another Way of Mountain Love," *The Alpine Journal* 24, no. 180 (1908): 160.

8. Maurice Herzog, *Annapurna*, trans. Nea Morin and Janet Adam Smith (New York: Dutton, 1952), 311.

9. Jan Smuts, "Speech on Table Mountain," quoted in R. L. G. Irving, ed., *The Mountain Way: An Anthology in Prose and Verse* (New York: Dutton, 1938), 123–24.

10. Sir William Martin Conway, *Mountain Memories: Pilgrimage of Romance* (New York: Funk & Wagnalls, 1920), 21–22.

11. Gaston Rébuffat, *Mont Blanc to Everest*, trans. Geoffrey Sutton (London: Thames and Hudson, 1956), 11–12.

12. Quoted from Wien's diary by Paul Bauer in *Himalayan Quest: The German Expedition to Siniolchum and Nanga Parbat*, trans. E. G. Hall (London: Nicholson & Watson, 1938), 71.

13. Louis F. Reichardt and William F. Unsoeld, "Nanda Devi from the North," *The American Alpine Journal* 21, no. 51 (1977): 22.

14. Albany Woman Heads for the Himalayas," *The Berkeley Voice*, 14 May 1986.

15. Herzog, *Annapurna*, 276.

16. Edward Whymper, *Scrambles Amongst the Alps in the Years 1860-69*, 5th ed. (London: John Murray, 1900), 387–88.

17. "Reinhold Messner: Why Does Man Climb?" in *Nepal*, ed. John Gottberg Anderson, 7th ed. (Singapore: APA Publications, 1989), 287.

18. John Muir, *Our National Parks* (Boston and New York: Houghton Mifflin, 1901), 56.

19. Geoffrey Winthrop Young, *The Influence of Mountains upon the Development of Human Intelligence*, Glasgow University Publications. The W. P. Ker Memorial Lectures 17 (Glasgow: Jackson, Son & Co., 1957), 14.

20. William O. Douglas, *Of Men and Mountains* (London: Victor Gollancz, 1951), 328.

21. Herzog, *Annapurna*, 311.

22. Yvon Chouinard quoted in Doug Robinson, "The Climber as Visionary," in *Voices for the Earth: A Treasury of the Sierra Club Bulletin*, ed. Ann Gilliam (San Francisco: Sierra Club Books, 1979), 290. Robinson's article examines the causes and effects of this visionary state in climbing.

23. W. H. Murray, *The Scottish Himalayan Expedition* (London: J. M. Dent & Sons, 1951), 23.

24. On the spiritual symbolism of mountain climbing, see Marco Pallis, *The Way and the Mountain* (London: P. Owen, 1960), 13–35. Other works on the subject are Roger Godel, *Essais sur l'expérience libératrice* (Sisteron: Editions Présence, 1976), 191–204; Charles Meade, *High Mountains* (London: Harvill, 1954); and various essays in Michael Charles Tobias and Harold Drasdo, eds., *The Mountain Spirit* (Woodstock, N.Y.: Overlook Press, 1979). A collection of writings on mountaineering, many of them spiritual in nature, appears in Irving, *The Mountain Way*.

25. From a letter by Pope Pius XI (Abate Achille Ratti) to the Bishop of Annécy on the occasion of the celebration of the millenary of St. Bernard of Menthon, translated in "Alpine Notes," *The Alpine Journal* 35, no. 227 (1923): 296. I am in debt to Patricia Fletcher for tracking down the source of this quote, which appears without attribution in Hornbein, *Everest: The West Ridge*.

26. Guido Rey, *The Matterhorn*, trans. J. E. C. Eaton (Oxford: Basil Blackwell, 1946), 126–27.

27. Sir Leslie Stephen, *The Playground of Europe* (Oxford: Basil Blackwell, 1936), 194–95.

28. Herzog, *Annapurna*, 311.

CHAPTER 15

1. Maurice Herzog, *Annapurna*, trans. Nea Morin and Janet Adam Smith (New York: Dutton, 1952), 12.

2. John Ruskin, *Praeterita* (Boston: Dana Estes, 1885), 97.

3. John Muir, "Explorations in the Great Tuolumne Cañon," in *Voices for the Earth: A Treasury of the Sierra Club Bulletin*, ed. Ann Gilliam (San Francisco: Sierra Club Books, 1979), 6.

4. John Muir, *The Yosemite* (1914; repr. San Francisco: Sierra Club Books, 1988), 196–97.

5. See Linda H. Graber, *Wilderness as Sacred Space*, Association of American Geographers Monograph Series 8 (Washington: Association of American Geographers, 1976). For a classic study of American attitudes toward wilderness and their influence on environmental movements, see Roderick Nash, *Wilderness and the American Mind*, 3d ed. (New Haven and London: Yale University Press, 1982).

6. Henry David Thoreau, "Walking," in *The Natural History Essays*, ed. Robert Sattelmeyer (Salt Lake City: Peregrine Smith Books, 1984), 112.

7. Henry D. Thoreau, *Walden*, ed. J. Lyndon Shanley (Princeton: Princeton University Press, 1971), 317–18.

8. Edwin Bernbaum, *The Way to Shambhala* (Garden City, N.Y.: Anchor Press/Doubleday, 1980), 60.

9. Han Hung translated in Witter Bynner, trans., *The Jade Mountain: A Chinese Anthology* (Garden City, N.Y.: Anchor Books, 1964), 18.

10. The American poet and environmentalist Gary Snyder has written on the relationship of wilderness, civilization, and the sacred in *Good Wild Sacred* (Madley, Hereford: Five Seasons Press, 1984) and "The Etiquette of Freedom," *Sierra* 74, no. 5 (1989): 74 ff.

11. On the strip-mined areas of the Southwest as national sacrifice areas, see *The Four Corners: A National Sacrifice Area?*, a film produced by Christopher McLeod, Glenn Switkes, and Randy Hayes (Oley, Penn.: Bullfrog Films, 1983).

12. A number of contemporary environmental conservation

groups—for example, the deep ecology movement, inspired by the writings of Norwegian philosopher Arne Naess—have expanded their concerns to include the spiritual and cultural dimensions of our relationship to the environment. For works on deep ecology, see Bill Devall and George Sessions, eds., *Deep Ecology* (Salt Lake City: G. M. Smith, 1985); Dolores LaChapelle, *Earth Wisdom* (Silverton, Colo.: Finn Hill Arts, 1978) and *Sacred Land, Sacred Sex, Rapture of the Deep: Concerning Deep Ecology and Celebrating Life* (Silverton, Colo.: Finn Hill Arts, 1988); and Michael Tobias, ed., *Deep Ecology* (San Diego, Calif.: Avant Books, 1984)—the last work has only a few articles actually by deep ecologists

such as Naess.

13. W. H. Murray, *The Scottish Himalayan Expedition* (London: J. M. Dent & Sons, 1951), 34.

14. From the *Manasakhanda* of the *Skanda Purana*, adapted from a translation in Edwin T. Atkinson, *Kumaun Hills: Its History, Geography and Anthropology with Reference to Garhwal and Nepal* (Delhi: Cosmo Publication, 1974), 307–8.

15. Dogen, "Treasury of the True Dharma Eye: Book XXIX, The Mountains and Rivers Sutra," trans. Carl Bielefeldt, in *The Mountain Spirit*, ed. Michael Charles Tobias and Harold Drasdo (Woodstock, N.Y.: Overlook Press, 1979), 48–49.

SELECTED BIBLIOGRAPHY

ALLEN, CHARLES. *A Mountain in Tibet: The Search for Mount Kailas and the Sources of the Great Rivers of India.* London & Sydney: Futura, Macdonald, 1983.

ANATI, EMMANUEL. *The Mountain of God.* New York: Rizzoli, 1986.

ANKELSARIA, B. T., trans. *Zand-Akasih: Iranian or Greater Bundahishn.* Bombay: Rahnumae Mazdáyasnan Sabha, 1956.

ARAI, YUSEI. *Odaishi-Sama: A Pictorial History of the Life of Kobo Daishi.* Translated by Hiroshi Katayama, Karl Kinoshita, and Alberta Freidus. Osaka, Japan: Koyasan Shuppansha, 1973.

ARTHUR, CLAUDEEN, et al. *Between Sacred Mountains: Navajo Stories and Lessons from the Land.* Tucson: Sun Tracks and the University of Arizona Press, 1984.

ASHE, GEOFFREY. *Avalonian Quest.* Bungay, Suffolk, England: Fontana, 1984.

ATKINSON, EDWIN T. *Kumaun Hills: Its History, Geography and Anthropology with Reference to Garwhal and Nepal.* Delhi: Cosmo, 1974.

AVALON, ARTHUR (SIR JOHN WOODROFFE), trans. *Tantra of the Great Liberation (Mahanirvana Tantra).* 1913. Reprint. New York: Dover, 1972.

BAKER, DWIGHT CONDO. *T'ai Shan: An Account of the Sacred Eastern Peak of China.* Shanghai: Commercial Press, 1925.

BASTIEN, JOSEPH W. "The Human Mountain." In *Mountain People,* edited by Michael Tobias, 45–57. Norman and London: University of Oklahoma Press, 1986.

———. *Mountain of the Condor: Metaphor and Ritual in an Andean Ayllu.* American Ethnological Society Monographs, no. 64. St. Paul: West, 1978.

BATCHELOR, STEPHEN. *The Tibet Guide.* London: Wisdom, 1987.

BEAL, SAMUEL, trans. *Buddhist Records of the Western World.* 1884. Reprint (2 vols. in 1). New York: Paragon, 1968.

BECHKY, ALLEN. *Adventuring in East Africa: The Sierra Club Travel Guide to the Great Safaris of Kenya, Tanzania, Eastern Zaire, Rwanda, and Uganda.* San Francisco: Sierra Club Books, 1989.

BECKEY, FRED. *Mountains of North America.* San Francisco: Sierra Club Books and American Alpine Club, 1982.

BECKWITH, MARTHA WARREN. *Hawaiian Mythology.* New Haven: Yale University Press, 1940.

BEGAY, HARRISON. *The Sacred Mountains of the Navajo in Four Paintings by Harrison Begay.* Explanatory text by Leland C. Wyman. Flagstaff: Museum of Northern Arizona, 1967.

BERNBAUM, EDWIN. *The Way to Shambhala.* Garden City, N.Y.: Anchor Press/Doubleday, 1980.

BIRNBAUM, RAOUL. "The Manifestation of a Monastery: Shen-ying's Experiences on Mount Wu-T'ai in a T'ang Context." *Journal of the American Oriental Society* 106, no. 1 (1986): 119–37.

———. *Studies on the Mysteries of Manjusri: A Group of East Asian Mandalas and Their Traditional Symbolism.* Society for the Study of Chinese Religions Monograph Series, no. 2. Boulder, Colo.: Society for the Study of Chinese Religions, 1983.

BLACKER, CARMEN. *The Catalpa Bow: A Study of Shamanistic Practices in Japan.* London: Allen & Unwin, 1975.

BLOFELD, JOHN EATON CALTHORPE. *The Wheel of Life: The Autobiography of a Western Buddhist.* 2d ed. Berkeley: Shambhala, 1972.

BOYCE, MARY. *A History of Zoroastrianism.* 2 vols. Handbuch der Orientalistik. Leiden, The Netherlands: Brill, 1975–1982.

BRYCE, JAMES. *Transcaucasia and Ararat.* 4th ed. rev. London and New York: Macmillan, 1896.

CARTER, H. ADAMS. "The Goddess Nanda and Place Names of the Nanda Devi Region." *The American Alpine Journal* 21, no. 51 (1977): 24–29.

CHANG, GARMA C. C., trans. *The Hundred Thousand Songs of Milarepa.* 2 vols. Boston: Shambhala, 1977.

CHAVANNES, EDOUARD. *Le T'ai Chan: Essai de monographie d'un culte chinois.* Annales du Musée Guimet, no. 21. Paris: Ernest Leroux, 1910.

CLARK, ELLA E. *Indian Legends of the Northwest.* Berkeley and Los Angeles: University of California Press, 1953.

CLARK, KENNETH. *Landscape into Art.* New ed. New York: Harper & Row, 1976.

CLARK, RONALD, W. *Men, Myths and Mountains.* New York: Crowell, 1976.

CLIFFORD, RICHARD J. *The Cosmic Mountain in Canaan and the Old Testament.* Cambridge: Harvard University Press, 1972.

COHEN, MICHAEL P. *The Pathless Way: John Muir and American Wilderness.* Madison: University of Wisconsin Press, 1984.

COHN, ROBERT L. *The Shape of Sacred Space: Four Biblical Studies.* AAR Studies in Religion, no. 23. Chico, Calif.: Scholars Press, 1981.

COLLCUTT, MARTIN. "Mt. Fuji as the Realm of Miroku: The Transformation of Maitreya in the Cult of Mt. Fuji in Early Modern Japan." In *Maitreya: The Future Buddha,* edited by Helen Hardacre and Alan Sponberg, 248–69. Cambridge and New York: Cambridge University Press, 1988.

CORBIN, HENRY. *Spiritual Body and Celestial Earth: From Mazdean Iran to Shi'ite Iran.* Translated by Nancy Pearson. Princeton, N.J.: Princeton University Press, 1977.

COVARRUBIAS, MIGUEL. *Island of Bali.* New York: Knopf, 1946.

CRAUSAZ, WINSTON. *Citlaltépetl: A History of Pico de Orizaba Including Exploration, Mountaineering, Natural History, and Geology.* New York: American Alpine Club, 1990.

DAUMAL, RENÉ. *Mount Analogue: A Novel of Symbolically Authentic Non-Euclidean Adventures in Mountain Climbing.* Translated by Roger Shattuck. Baltimore: Penguin Books, 1974.

DE BEER, G. R. *Early Travellers in the Alps.* London: Disgwick and

Jackson, 1930.

DeMallie, Raymond J. "Lakota Belief and Ritual in the Nineteenth Century." In *Sioux Indian Religion: Tradition and Innovation*, edited by Raymond J. DeMallie and Douglas R. Parks, 25–43. Norman and London: University of Oklahoma Press, 1987.

_____, ed. *The Sixth Grandfather: Black Elk's Teachings Given to John G. Neihardt*. Lincoln and London: University of Nebraska Press, 1984.

Demiéville, Paul. "La montagne dans l'art littéraire chinois." *France-Asie/Asia* 20 (1965): 7–32.

Dixon, Roland. "Shasta Myths." *Journal of American Folklore* 23, no. 87 (1910): 8–37.

Dogen. "Treasury of the True Dharma Eye: Book XXIX, The Mountains and Rivers Sutra." Translated by Carl Bielefeldt. In *The Mountain Spirit*, edited by Michael Charles Tobias and Harold Drasdo, 41–49. Woodstock, N.Y.: Overlook Press, 1979.

Doub, William. "Mountains in Early Taoism." In *The Mountain Spirit*, edited by Michael Charles Tobias and Harold Drasdo, 129–35. Woodstock, N.Y.: Overlook Press, 1979.

Downes, David Anthony. *Ruskin's Landscape of Beatitude*. New York and Berne: Peter Lang, 1984.

Dubs, Homer C. "An Ancient Chinese Mystery Cult." *Harvard Theological Review* 35, no. 4 (1942): 221–40.

Dundas, Charles. *Kilimanjaro and Its People: A History of the Wachagga, Their Laws, Customs and Legends, Together with Some Account of the Highest Mountain in Africa*. 1924. Reprint. London: Frank Cass, 1968.

Dunklee, John. *Man-Land Relationships on the San Francisco Peaks*. Flagstaff: Submitted to the Museum of Northern Arizona Technical Series, n.d.

Duran, Diego. *Book of the Gods and Rites and the Ancient Calendar*. Translated by Fernando Horcasitas and Doris Heyden. Norman: University of Oklahoma Press, 1975.

Earhart, H. Byron. "Mt. Fuji and Shugendo." *Japanese Journal of Religious Studies* 16, nos. 2–3 (1989): 205–26.

_____. *A Religious Study of the Mount Haguro Sect of Shugendo: An Example of Japanese Mountain Religion*. Tokyo: Sophia University, 1970.

Echevarría, Evelio. "The Inca Mountaineers: 1400–1800." In *The Mountain Spirit*, edited by Michael Charles Tobias and Harold Drasdo, 117–24. Woodstock, N.Y.: Overlook Press, 1979.

_____. "Legends of the High Andes." *The Alpine Journal* 88, no. 332 (1983): 85–91.

Eck, Diana L. *Banaras: City of Light*. Princeton: Princeton University Press, 1983.

_____. "Mountains." In *The Encyclopedia of Religion*, edited by Mircea Eliade, 10:130–34. New York: Macmillan, 1987.

Eckstorm, Fannie Hardy. "The Katahdin Legends." *Appalachia* 16, no. 1 (1924): 39–52.

Eichorn, Arthur Francis. *The Mount Shasta Story*. 2d ed. Mount Shasta, Calif.: Mount Shasta Herald, 1971.

Eliade, Mircea. *The Myth of the Eternal Return or, Cosmos and History*. Translated by Willard R. Trask. Princeton, N.J.: Princeton University Press, 1971.

_____. *Patterns in Comparative Religion*. Translated by Rosemary Sheed. New York: New American Library, 1974.

_____. *Yoga: Immortality and Freedom*. 2d ed. Translated by Willard Trask. Princeton, N.J.: Princeton University Press, 1969.

Emerson, Nathaniel Bright. *Unwritten Literature of Hawaii: The Sacred Songs of the Hula*. Smithsonian Institution, Bureau of American Ethnology Bulletin no. 38. Washington: Government Printing Office, 1909.

Evans-Wentz, W. Y. *Cuchama and Sacred Mountains*. Edited by Frank Waters and Charles L. Adams. Chicago: Swallow Press; Athens: Ohio University Press, 1981.

Farquhar, Francis Peloubet, and Phoutrides, Aristides E. *Mount Olympus*. San Francisco: Johnck & Seeger, 1929.

Fox, David J. Stuart. *Once a Century: Pura Besakih and the Eka Dasa Rudra Festival*. Jakarta: Penerbit Citra Indonesia, 1982.

Galland, China. *Invisible Light: Tara and the Black Madonna*. New York: Viking Penguin, 1990.

Gayet-Tancrède (Samivel), Paul. *Hommes, cimes et dieux: les grandes mythologies de l'altitude et la légende dorée des montagnes à traverse le monde*. Paris: Arthaud, 1973.

Geertz, Armin W. "A Reed Pierced the Sky: Hopi Indian Cosmography on Third Mesa, Arizona." *Numen* 31, no. 2 (1984): 216–41.

Geil, William Edgar. *The Sacred 5 of China*. Boston and New York: Houghton Mifflin, 1926.

Gesner, Conrad. *On the Admiration of Mountains*. Translated by H. B. D. Soulé. San Francisco: Grabhorn Press, 1937.

Godel, Roger. *Essais sur l'expérience libératrice*. Sisteron, France: Editions Présence, 1976.

Goetzmann, William H., and Goetzmann, William N. *The West of the Imagination*. New York: Norton, 1986.

Govinda, Lama Anagarika. *The Way of the White Clouds*. Berkeley: Shambhala, 1970.

Graber, Linda H. *Wilderness as Sacred Space*. Association of American Geographers Monograph Series, no. 8. Washington: Association of American Geographers, 1976.

Grapard, Allan G. "Flying Mountains and Walkers of Emptiness: Toward a Definition of Sacred Space in Japanese Religions." *History of Religions* 21, no. 3 (1982): 195–221.

Gribble, Francis. *The Early Mountaineers*. London: T. Fisher Unwin, 1899.

Guichonnet, Paul, ed. 2 vols. *Histoire et civilisations des Alpes*. Toulouse and Lausanne: Privat/Payot, 1980.

Gutmann, Bruno. "Chagga Folk-Lore." In *Tanganyika Notes and Records*, rev. ed., edited by J. A. Hutchinson, 50–55. Dar Es Salaam: Tanzania Society, 1974.

Hakeda, Yoshito S. *Kukai: Major Works*. New York: Columbia University Press, 1972.

Hamsa, Bhagwan Shri. *The Holy Mountain: Manasarovar and the Mount Kailas*. Translated by Shri Purohit Swami. 1934. Reprint. Delhi: India Bibliographies Bureau and Balaji Enterprises, 1986.

Hay, John. *Kernels of Energy, Bones of Earth: The Rock in Chinese Art*. New York: China Institute in America, 1985.

Hedin, Sven. *Through Asia*. London: Methuen, 1898.

Heine-Geldern, Robert. *Conceptions of State and Kingship in Southeast Asia*. Data Papers, no. 13. Ithaca, N.Y.: Southeast Asia Program, Dept. of Far Eastern Studies, Cornell University, 1956.

Herzog, Maurice. *Annapurna*. Translated by Nea Morin and Janet Adam Smith. New York: Dutton, 1952.

Hori, Ichiro. "Mountains and Their Importance for the Idea of the Other World." *History of Religions* 6, no. 1. (1966) 1–23.

Hornbein, Thomas F. *Everest: The West Ridge*. San Francisco and New York: Sierra Club/Ballantine Books, 1968.

Hotchkis, Anna M., and Mullikan, Mary Augusta. *The Nine Sacred Mountains of China: An Illustrated Record of Pilgrimages Made in the Years 1935–1936*. Hong Kong: Vetch and Lee Ltd., 1973.

Huxley, Anthony, ed. *Standard Encyclopedia of the World's Mountains*. New York: Putnam's, 1962.

Hyde, Philip, and Jett, Stephen C. *Navajo Wildlands*. San Francisco: Sierra Club/Ballantine Books, 1969.

Irving, R. L. G., ed. *The Mountain Way: An Anthology in Prose and Verse*. New York: Dutton, 1938.

Jetté, Julius. "On Ten'a Folklore." *Journal of the Royal Anthropological Institute* 38 (1908): 298–367.

Johnson, Mark, ed. *Philosophical Perspectives on Metaphor*. Minneapolis: University of Minnesota Press, 1981.

Johnson, Russell, and Moran, Kerry. *The Sacred Mountain of Tibet: On Pilgrimage to Mount Kailas*. Rochester, Vt.: Park Street Press, 1989.

Koyama, Kosuke. *Mount Fuji and Mount Sinai: A Critique of Idols*. Maryknoll, N.Y.: Orbis Books, 1984.

Kroll, Paul W. "Verses from on High: The Ascent of T'ai Shan." *T'oung Pao* 69, nos. 4–5 (1983): 223–60.

Kuo Hsi. *An Essay on Landscape Painting*. Translated by Shio Sakanishi. London: John Murray, 1935.

LaChapelle, Dolores. *Earth Wisdom*. Silverton, Colo.: Finn Hill Arts, 1978.

LAGUNA, FREDERICA DE. *Under Mount Saint Elias: The History and Culture of the Yakutat Tlingit.* Smithsonian Contributions to Anthropology, vol. 7, pts. 1–3. Washington, D.C.: Smithsonian Institution Press, 1972.

LAYTON, ROBERT. *Uluru: An Aboriginal History of Ayers Rock.* Canberra: Australian Institute of Aboriginal Studies, 1986.

LEHRMAN, FREDERIC. *The Sacred Landscape.* Berkeley: Celestial Arts, 1988.

LEVENSON, JON D. *Sinai and Zion: An Entry into the Jewish Bible.* Minneapolis, Chicago, New York: Winston Press, Seabury Books, 1985.

LINDOW, JOHN. *Swedish Legends and Folktales.* Berkeley and Los Angeles: University of California Press, 1978.

LIPSEY, ROGER. *An Art of Our Own: The Spiritual in Twentieth-Century Art.* Boston and Shaftesbury: Shambhala, 1988.

LOEWE, MICHAEL. *Ways to Paradise: The Chinese Quest for Immortality.* London, Boston: George Allen & Unwin, 1979.

LOOMIS, ROGER SHERMAN. *The Grail: From Celtic Myth to Christian Symbol.* New York: Columbia University Press, 1963.

LUCKERT, KARL W. *Navajo Mountain and Rainbow Bridge Religion.* Flagstaff: Museum of Northern Arizona, 1977.

MABBETT, I. W. "The Symbolism of Mount Meru." *History of Religions* 23, no. 1 (1983): 64–83.

MacCULLOCH, J. A. *The Religion of the Ancient Celts.* Edinburgh: T. & T. Clark, 1911.

_____. "Mountains, Mountain Gods." In *Encyclopedia of Religion and Ethics,* edited by James Hastings, 8:863–68. New York: Charles Scribner's Sons, 1915.

MACDONALD, ALEXANDER W. "The Lama and the General." *Kailash* 1, no. 3 (1973): 225–33.

MARGOLIN, MALCOLM. *The Way We Lived: California Indian Reminiscences, Stories and Songs.* Berkeley: Heyday Books, 1981.

MASPERO, HENRI. *Taoism and Chinese Religion.* Translated by Frank A. Kierman, Jr. Amherst: University of Massachusetts Press, 1981.

MATTHEWS, WASHINGTON. *Navajo Legends.* Memoirs of the American Folklore Society, vol. 5. Boston: American Folklore Society, 1897.

MEADE, CHARLES. *High Mountains.* London: Harvill, 1954.

MEANY, EDMOND S., ed. *Mount Rainier: A Record of Exploration.* New York: Macmillan, 1916.

MILLER, JOAQUIN. *Unwritten History: Life Amongst the Modocs.* Hartford, Conn.: American Publishing, 1874.

MISHKIN, BERNARD. "Cosmological Ideas Among the Indians of the Southern Andes." *Journal of American Folklore* 53, no. 210 (1940): 225–41.

MORAVETZ, BRUNO, ed. *The Big Book of Mountaineering.* Translated by Diana Stone Peters and Frederick G. Peters. Woodbury, N.Y.: Barron's, 1980.

MORRISON, HEDDA. *Hua Shan: The Taoist Sacred Mountain in West China.* Hong Kong: Vetch and Lee Ltd., 1973.

MOULD, DAPHNE D. C. POCHIN. *Irish Pilgrimage.* New York: Devin-Adair, 1957.

MOUNTFORD, CHARLES P. *Ayers Rock: Its People, Their Beliefs and Their Art.* Honolulu: East-West Center Press, 1965.

MUIR, JOHN. "Explorations in the Great Tuolumne Cañon." In *Voices for the Earth: A Treasury of the Sierra Club Bulletin,* edited by Ann Gilliam, 4–8. San Francisco: Sierra Club Books, 1979.

_____. *The Mountains of California.* 1894. Reprint. San Francisco: Sierra Club Books, 1988.

_____. *The Yosemite.* 1914. Reprint. San Francisco: Sierra Club Books, 1988.

MURIITHI, KIBOI, and NDORIA, PERTER. *War in the Forest: The Autobiography of a Mau Mau Leader.* Nairobi: East African Publishing House, 1971.

NABOKOV, PETER. *Two Leggings: The Making of a Crow Warrior.* New York: Crowell, 1967.

NASH, RODERICK. *Wilderness and the American Mind.* 3d ed. New Haven and London: Yale University Press, 1982.

NEBESKY-WOJKOWITZ, RENÉ DE. *Oracles and Demons of Tibet: The Cult and Iconography of the Tibetan Protective Deities.* 1956. Reprint. Graz, Austria: Akademische Druck-u. Verlagsanstalt, 1975.

_____. *Where the Gods Are Mountains: Three Years Among the People of the Himalayas.* Translated by Michael Bullock. New York: Reynal, n.d.

NEIHARDT, JOHN G. *Black Elk Speaks.* 1932. Reprint. New York: Pocket Books, 1972.

NELSON, RICHARD K. *Make Prayers to the Raven: A Koyukon View of the Northern Forest.* Chicago: University of Chicago Press, 1983.

NEQUATEWA, EDMUND. "Chaveyo: The First Kachina." *Plateau* 20, no. 4 (1948): 60–62.

NICOLSON, MARJORIE HOPE. *Mountain Gloom and Mountain Glory: The Development of the Aesthetics of the Infinite.* New York: Norton, 1963.

NIMMO, H. ARLO. "Pele, Ancient Goddess of Contemporary Hawaii." *Pacific Studies* 9, no. 2 (1986): 121–79.

_____. "Pele's Journey to Hawai'i: An Analysis of the Myths." *Pacific Studies* 11, no. 1 (1987): 1–42.

ORBELL, MARGARET. *Hawaiki: A New Approach to Maori Tradition.* Christchurch: University of Canterbury, 1985.

_____. *The Natural World of the Maori.* Auckland: Collins/Bateman, 1985.

ORTIZ, ALFONSO. *The Tewa World: Space, Time, Being, and Becoming in a Pueblo Society.* Chicago and London: University of Chicago Press, 1969.

OTTO, RUDOLF. *The Idea of the Holy.* Translated by John W. Harvey. Oxford and New York: Oxford University Press, 1950.

PALLIS, MARCO. *The Way and the Mountain.* London: P. Owen, 1960.

PARANAVITANA, SENERAT. *The God of Adam's Peak.* Ascona, Switzerland: Artibus Asiae Publishers, 1958.

POWELL, FATHER PETER JOHN. *People of the Sacred Mountain.* 2 vols. San Francisco: Harper & Row, 1981.

PRICE, LARRY W. *Mountains & Man: A Study of Process and Environment.* Berkeley and Los Angeles: University of California Press, 1981.

RANDALL, ROBERT. "Return of the Pleiades." *Natural History* 96, no. 6 (1987): 42–53.

REICHARDT, LOUIS F., and UNSOELD, WILLIAM F. "Nanda Devi from the North." *The American Alpine Journal* 21, no. 51 (1977): 1–23.

REINHARD, JOHAN. "Chavin and Tiahuanaco: A New Look at Two Andean Ceremonial Centers." *National Geographic Research* 1, no. 3 (1985): 395–422.

_____. "High Altitude Archeology and Andean Mountain Gods." *The American Alpine Journal* 25, no. 57 (1983): 54–67.

_____. "Sacred Mountains: An Ethno-archaeological Study of High Andean Ruins." *Mountain Research and Development* 5, no. 4 (1985): 299–317.

REUSCH, DR. R. "Mount Kilimanjaro and Its Ascent." In *Tanganyika Notes and Records,* rev. ed., edited by J. A. Hutchinson, 131–34. Dar Es Salaam: Tanzania Society, 1974.

REY, GUIDO. *The Matterhorn.* Translated by J. E. C. Eaton. Oxford: Basil Blackwell, 1946.

ROBERTSON, DAVID. *George Mallory.* London: Faber and Faber, 1969.

ROBINSON, DOUG. "The Climber as Visionary." In *Voices for the Earth: A Treasury of the Sierra Club Bulletin,* edited by Ann Gilliam, 289–93. San Francisco: Sierra Club Books, 1979.

ROCK, J. F. *The Amnye Ma-chhen Range and Adjacent Regions: A Monographic Study.* Serie Orientale Roma, no. 12. Rome: Istituto Italiano per il Medeo ed Estremo Oriente, 1956.

ROWELL, GALEN. *Mountains of the Middle Kingdom.* San Francisco: Sierra Club Books, 1983.

RUSKIN, JOHN. *Modern Painters.* 2d ed. 5 vols. New York: Wiley, 1883.

_____. *Praeterita.* Boston: Estes, 1885.

SAHAGUN, BERNARDINO DE. *General History of the Things of New Spain: Florentine Codex, Book 2 — The Ceremonies.* 2d ed. rev. Translated by Arthur J. O. Anderson and Charles E. Dibble. Santa Fe, N.M.: School of American Research and the University of Utah, 1981.

_____. *A History of Ancient Mexico*. Translated by Fanny R. Bandelier. Nashville: Fisk University Press, 1932.

Sallnow, Michael J. *Pilgrims of the Andes: Regional Cults in Cusco*. Washington, D.C.: Smithsonian Institution Press, 1987.

Sculley, Vincent. *The Earth, the Temple, and the Gods: Greek Sacred Architecture*. New Haven and London: Yale University Press, 1962.

Seigel, Linda. "The Riesengebirge as a Transcendental Image in Friedrich's Art." In *The Mountain Spirit*, edited by Michael Charles Tobias and Harold Drasdo, 189–93. Woodstock, N.Y.: Overlook Press, 1979.

Sharma, Man Mohan. *Through the Valley of Gods: Travels in the Central Himalayas*. 2d ed. New Delhi: Vision Books, 1978.

Sherrard, Philip. *Athos: The Holy Mountain*. Woodstock, N.Y.: Overlook Press, 1985.

Shipton, Eric. *Nanda Devi*. In *The Six Mountain-Travel Books*, edited by Jim Perrin, 15–151. 1936. Reprint. Seattle: The Mountaineers, 1985.

Siiger, Halfdan. "A Cult for the God of Mount Kanchenjunga Among the Lepcha of Northern Sikkim." In *Actes du IVe congrès international des sciences anthropologiques et ethnologiques*, vol. 2, 185–89. Vienna: Verlag Adolf Holzhausens, 1955.

Skrobucha, Heinz. *Sinai*. Translated by Geoffrey Hunt. London: Oxford University Press, 1966.

Smith, Gary. "A Day in the Life of Mount Kenya." *Sports Illustrated* 62, no. 21 (1985): 66–82.

Smith, Henry D. II. *Hokusai: One Hundred Views of Mt. Fuji*. New York and Tokyo: George Braziller, 1988.

Snelling, John. *The Sacred Mountain*. London and the Hague: East West Publications, 1983.

Snow, Chief John. *These Mountains Are Our Sacred Places: The Story of the Stoney People*. Toronto and Sarasota: Samuel Stevens, 1977.

Starr, Frederick. *Fujiyama: The Sacred Mountain of Japan*. Chicago: Covici-McGee, 1924.

Statler, Oliver. *Japanese Pilgrimage*. New York: William Morrow, 1983.

Stein, M. Aurel. *Ancient Khotan*. 2 vols. Oxford: Clarendon, 1907.

Stein, R. A. *Recherches sur l'épopée et le barde au Tibet*. Paris: Presses Universitaires de France, 1959.

Stephen, Sir Leslie. *The Playground of Europe*. Edited by H. E. G. Tyndale. Oxford: Basil Blackwell, 1936.

Stevens, Stan. "Sacred and Profane Himalayas." *Natural History* 97, no. 1 (1988): 26–35.

Sturluson, Snorri. *The Prose Edda of Snorri Sturluson: Tales from Norse Mythology*. Translated by Jean Young. Berkeley and Los Angeles: University of California Press, 1954.

Stutfield, H. E. M. "Mountaineering as a Religion." *The Alpine Journal* 32, no. 218 (1918): 241–47.

Sullivan, Michael. *Chinese Landscape Painting: The Sui and T'ang Dynasties*. Berkeley and Los Angeles: University of California Press, 1980.

_____. *Symbols of Eternity: The Art of Landscape Painting in China*. Stanford, Calif.: Stanford University Press, 1979.

Swift, Hugh. *Trekking in Pakistan and India*. San Francisco: Sierra Club Books. 1990.

_____. *Trekking in Nepal, West Tibet, and Bhutan*. San Francisco: Sierra Club Books, 1989.

Tennent, Emerson. *Ceylon: An Account of the Island*. 4th ed. rev. London: Longman, Green, Longman, and Roberts, 1860.

Thorarinsson, Sigurdur. *Hekla: A Notorious Volcano*. Translated by Jóhann Hannesson. Reykjavik: Almenna Bókafélagid, 1970.

Thoreau, Henry David. *The Maine Woods*. 1848. Reprint. New York: Bramhall House, 1950.

_____. *Walden*. Edited by J. Lyndon Shanley. Princeton, N.J.: Princeton University Press, 1971.

Tichy, Herbert. *Himalaya*. Translated by Richard Rickett and David Streatfeild. New York: Putnam's, 1970.

Tobias, Michael, ed. *Mountain People*. Norman and London: University of Oklahoma Press, 1986.

Tobias, Michael Charles, and Drasdo, Harold, eds. *The Mountain Spirit*. Woodstock, N.Y.: Overlook Press, 1979.

Townsend, Richard Fraser. "Pyramid and Sacred Mountain." In *Ethnoastronomy and Archaeoastronomy in the American Tropics*, edited by Anthony F. Aveni and Gary Urton, 37–62. New York: New York Academy of Sciences, 1982.

Tuan, Yi-Fu. *Landscapes of Fear*. New York: Pantheon Books, 1979.

Turville-Petre, E. O. G. *Myth and Religion of the North: The Religion of Ancient Scandinavia*. 1964. Reprint. Westport, Conn.: Greenwood Press, 1975.

Tyler, Royall. "A Glimpse of Mt. Fuji in Legend and Cult." *Journal of the Association of Teachers of Japanese* 1, no. 2 (1981): 140–65.

Van Buitenen, J. A. B., ed. and trans. *The Mahabharata*. 3 vols. Chicago: University of Chicago Press, 1973–1978.

Varma, Rommel, and Varma, Sadhana. *The Himalaya: Kailasa-Manasarovar*. Geneva: Lotus Books, 1985.

Wales, H. G. *The Universe Around Them: Cosmology and Cosmic Renewal in Indianized Southeast Asia*. London: Arthur Probsthain, 1977.

_____. *The Mountain of God: A Study in Early Religion and Kingship*. London: B. Quaritch, 1953.

Webster, Peter. *Rua and the Maori Millennium*. Wellington, New Zealand: Victoria University, 1979.

Wenkam, Robert. *The Edge of Fire: Volcano and Earthquake Country in Western North America and Hawaii*. San Francisco: Sierra Club Books, 1987.

Westervelt, W. D. *Hawaiian Legends of Volcanoes*. Boston: Ellis Press, 1916.

Wicke, Charles, and Horcasitas, Fernando. "Archeological Investigations on Monte Tlaloc, Mexico." *Mesoamerican Notes* 5 (1957): 83–96.

Wu, Nelson I. *Chinese and Indian Architecture: The City of Man, the Mountain, and the Realm of the Immortals*. New York: Braziller, 1963.

Wyman, Leland C. *Blessingway*. Tucson: University of Arizona Press, 1970.

Yoon, Hong-Key. *Maori Mind, Maori Land: Essays on the Cultural Geography of the Maori People from an Outsider's Perspective*. Berne, Francfort-s. Main, New York, Paris: Peter Lang, 1986.

Young, Geoffrey Winthrop. *The Influence of Mountains upon the Development of Human Intelligence*. Glasgow University Publications. The W. P. Ker Memorial Lectures, no. 17. Glasgow: Jackson, Son & Co., 1957.

Zebhauser, Helmuth. "Man Has Always Painted Mountains." In *The Big Book of Mountaineering*, edited by Bruno Moravetz, 99–104. Translated by Diana Stone Peters and Frederick G. Peters. Woodbury, N.Y.: Barron's, 1980.

INDEX